Logjam

The Earthscan Forest Library

Jeffrey A. Sayer, Series Editor

Logjam:
Deforestation and the Crisis of Global Governance
David Humphreys

Forest Quality:
Assessing Forests at a Landscape Scale
Nigel Dudley, Rodolphe Schlaepfer, Jean-Paul Jeanrenaud, William Jackson and Sue Stolton

Forests in Landscapes:
Ecosystem Approaches to Sustainability
Jeffrey A. Sayer and Stewart Maginnis (eds)

The Politics of Decentralization:
Forests, Power and People
Carol J. Pierce Colfer and Doris Capistrano

Plantations, Privatization, Poverty and Power:
Changing Ownership and Management of State Forests
Mike Garforth and James Mayers (eds)

The Sustainable Forestry Handbook 2nd edition
Sophie Higman, James Mayers, Stephen Bass, Neil Judd and Ruth Nussbaum

The Forest Certification Handbook 2nd edition
Ruth Nussbaum and Markku Simula

Logjam

Deforestation and the Crisis of Global Governance

David Humphreys

David Humphreys is Senior Lecturer in Environmental Policy at the Open University. He was an adviser to the World Commission on Forests and Sustainable Development, has been a member of United Kingdom delegations to the United Nations Forum on Forests and is currently serving on the Scientific Advisory Board of the European Forest Institute. He is the author of *Forest Politics: The Evolution of International Cooperation* (Earthscan, 1996) and has written several journal articles and papers on global environmental politics.

First published by Earthscan in the UK and USA in 2006
Paperback edition first published in 2009

For a full list of publications please contact:

Earthscan
2 Park Square, Milton Park, Abingdon, Oxon OX14 4RN
711 Third Avenue, New York, NY 10017

Earthscan is an imprint of the Taylor & Francis Group, an informa business

ISBN: 978-1-84407-301-6 hardback
978-1-84407-611-6 paperback

Typesetting by JS Typesetting Ltd, Porthcawl, Mid Glamorgan
Cover design by Susanne Harris

A catalogue record for this book is available from the British Library

Library of Congress Cataloging-in-Publication Data

Humphreys, David, 1957–
Logjam : deforestration and the crisis of global governance / David Humphreys.
p. cm.
Includes bibliographical reference to (p.).
ISBN-13: 978-1-84407-301-6 (hardback)
ISBN-10: 1-84407-301-7 (hardback)
1. Deforestation. 2. Forest policy. 3. Forest management. I. Title.
SD418.H86 2006
333.75'137–dc22

2006013947

WITH LOVE AND THANKS TO
COLETTE
WHO PROVIDES A SUSTAINABLE HABITAT

We carry in our heart the true country
And that cannot be stolen
We follow in the steps of our ancestry
And that cannot be broken
Mining companies
Pastoral companies
Uranium companies
Collected companies
Got more rights than people
Got more say than people

(Midnight Oil, 'The Dead Heart,'
from the album *Diesel and Dust*, 1988)

As we in the academy begin to use business-speak fluently, we become accustomed to thinking in commercialized terms about education. We talk no longer as public intellectuals, but as entrepreneurs. And we thus encourage instead of fight the disturbing trend that makes education a consumer good rather than a public good.

(Michele Tolela Myers, 'A Student is Not on Input,' *New York Times*,
26 March 2001)

...if we fail to act we will become the willing executioners of the millions whose lives are lost every year as a result of the degradation of nature. We will willingly sentence billions of people – whose lives are most dependent on nature – the poorest of the poor – to lifelong poverty, hunger, and disease.

(Don Melnick, Statement to the United Nations Forum on Forests
on behalf of the United Nations Millennium Project Task Force on
Environmental Sustainability, 25 May 2005)

I have often asked myself, 'What did the Easter Islander who cut down the last palm tree say while he was doing it?' Like modern loggers, did he shout: 'Jobs, not trees!'? Or: 'Technology will solve our problems, never fear, we'll find a substitute for wood'? Or: 'We don't have proof that there aren't palms somewhere else on Easter, we need more research, your proposed ban on logging is premature and driven by fear-mongering?'

(Jared Diamond, *Collapse: How Societies Choose to Fail or to Survive*,
2005, p.114)

Contents

List of Tables and Boxes

Tables

Boxes

Foreword

Forest conservation has captured the attention of the public, at least in developed countries, in recent decades. The media in these countries publicize the latest estimates of forest destruction. They assail us with accounts of illegal logging, forest fires and species extinctions. They are fond of making extravagant judgements about the success or failure of initiatives to conserve forests. They hail the signature on yet another agreement to tackle forest problems as a major step forward. They lambaste yet another UN talk-shop that involves the expense of taking hundreds of well paid people to spend weeks in Geneva or New York, and then fails to agree on anything.

All of us who are concerned with forest conservation have our own views on what might be done to promote the sustainable management and halt the inexorable decline of the world's forests. Some of us believe that the only useful way forward is through practical action on the ground. Others believe in the stroke of pen solutions that might emerge from intergovernmental negotiations. We all have differing bases for making these judgements depending on our personal backgrounds and experience. Throughout the 1980s and 1990s there was a groundswell of support for the idea that international political action was the key to success. Now, in the early years of the 21st century, people are tiring of endless meetings and negotiations that produce few, if any, visible outcomes. The groundswell is moving towards local action. Community management, devolution and decentralized governance are seen as the new beacons of hope.

In the late-1980s and early-1990s all the conservation non-governmental organizations (NGOs) descended in force on the biannual meetings of the International Tropical Timber Organization. Today the NGOs are almost totally absent from its meetings. In the mid-1990s the campaigning groups divided their attention between the United Nations forest institutions and the Convention on Biological Diversity. Each of these initiatives had its band of supporters. Those who backed the Convention on Biological Diversity heaped derision on the UN forest institutions and vice versa. Now, most activist groups are abandoning all international gatherings. Many continue to work through the media; surprisingly few find it useful to get involved on the ground.

The missing ingredient in all of this is any objective way of measuring the impact of all this activity. How do we know whether a Euro spent on an international meeting in New York achieves more than a Euro spent promoting certification of a forest in Cameroon. What is lacking is any objective evidence to help us to decide how progress might be made and how we should invest our time and resources. Few people have the opportunity or capacity to step back from the fray and evaluate the relative costs and benefits of all these multiple initiatives. How do all these pieces fit together? What has been their impact? Which ones should we support? What have we learned from decades of investment in both on the ground conservation actions and international processes?

David Humphreys' book does not answer all of these questions. But it does provide many useful insights into the way that efforts to achieve the sustainable management and conservation of forests have evolved over the past decades. The present generation of forest policy makers and activist campaigners are exceedingly prone to reinventing wheels. They are often surprisingly ignorant of what was attempted in the past and of what the outcomes were. The international processes and the field activities of aid agencies and conservation groups have become so complex and time consuming that many people become experts on a small subset of the range of issues involved. There is a serious problem of lack of historical perspective and of what Tony Blair refers to as 'joined up thinking.' Most people are focused on a tiny piece of the puzzle observed only over a very short time span.

David Humphreys' latest book provides a broad, historical perspective on international forest politics and policies. It should be read by all people who are concerned with the management and conservation of forests. David is one of the few people who has an overview of all the international processes concerning forests and he is a card-carrying 'joined up thinker.' I have been more involved than most in all of the forest processes over the past 30 years, but I still found the book full of thought-provoking comment and innovative perspectives. *Logjam* should be compulsory reading for all of those concerned in the current round of international debates on forests. But there is another important audience. The world of international negotiations has become so arcane that most practitioners on the ground have totally disconnected from it. I am constantly surprised during my field work throughout the world to find how little of the international polemic reaches conservation and forestry professionals on the ground.

This is a dangerous situation. If the international negotiators are not communicating with the field practitioners then one has to ask if they are negotiating about the right things. As this book constantly reminds us, the issues that are considered important change with amazing rapidity. In the short period of two decades covered by the book the perspectives on the big issues confronting forests have changed several times. In the 1980s it was logging, in the 1990s shifting cultivation and agricultural conversion were the villains, while in the new millennium governance and forest crime have taken centre stage. The impacts of climate change are also emerging as a major concern.

So it is probably too soon to reach a verdict on the impact of the international forest processes of recent decades. It is clear that no single initiative has provided a silver bullet solution for the global decline of forests. One could argue that all of these international processes have been futile and that they were motivated by a misconception about the levels at which rules should be made. Following this line of reasoning one might conclude that the forces of neoliberalism and globalization will inevitable lead to the victory of private greed over public good. But one could also argue that the proliferation of international forest activity constituted the birth pangs of a gradual movement towards global environmental governance – and given the ambition of this endeavour we should not be surprised that progress has been slow.

I prefer to think that we need to move forward on a broad front. That we need to keep forests in the international spotlight but we also need to keep fighting in the trenches at the front line to gradually solve problems, build institutions and establish rights and responsibilities at the local level. But above all I believe that we need to get all of these things to link up. Forests are important for their global and local impacts and it is not helpful to debate these independently. We need integrated, holistic thinking, and action, to conserve forests.

The need for holistic approaches to forest problems has been widely recognized for a long time. Even the international negotiators urge us to take ecosystem, landscape or ecoregional approaches. But few people actually do have the broader, inclusive overview of forest issues that we need. David Humphreys' book covers the full range of international initiatives and the most important developments in recent forest diplomacy. It makes a valuable contribution to our understanding of the issues and will help us, in the future, to reach judgements about how effective all this activity has been. It weaves together in a readable and authoritative way a comprehensive account of the actors and issues in international forestry of recent decades. All who are concerned about the future of the world's forests will gain by reading it. It is with great pleasure that I welcome this new title to the Earthscan Forest Library.

Jeffrey A. Sayer
Lally sur Blonay, Switzerland
July 2006

Preface

All books have a biographical dimension, even scholarly works. Our choice of research reflects our professional interests and those of our employers and funders. How we carry out research is an expression of our academic training and influences, such as the literatures we read and the colleagues with whom we associate. And more personal factors may also have a bearing. I came to the subject of forest politics through concern about the worldwide loss of forest cover. This led to my first book, *Forest Politics: The Evolution of International Cooperation,* which analysed international forest politics in the period from 1983 to 1995.[1] Three developments in my personal life since *Forest Politics* was written have had a bearing on how this present book has developed. I have met and married my life partner; we are the proud parents of two young children; and I have experienced some health difficulties. The combined effects of these life-changing developments prompted me to reconsider what I would like the impacts of my research to be and, more generally, to think further on the politics of scholarship and how the publicly funded academic may contribute to a world in which people behave more responsibly to nature and to other people.

This process of reflection continued throughout the researching and writing of this publication. During this time a number of scientific papers were published that offered predictions on the state of the global environment in the year 2050. In 2001 the Intergovernmental Panel on Climate Change (IPCC) predicted that by 2050 the globally averaged surface temperature would increase by 0.8°C to 2.6°C based on 1990 figures. During the 21st century, the IPCC projected that the rate of warming 'is very likely to be without precedent during at least the last 10,000 years, based on paleoclimate data.'[2] Of course, all predictive models contain uncertainties and such figures can be disputed; but the IPCC's projections form part of a wider trend in which scientific data point to significant and dangerous global environmental change. In 2004 a major study published in the journal *Nature* predicted that by 2050 a minimum of 15 per cent of species and a maximum of 37 per cent would be 'committed to extinction' as a result of climate change. These predictions, which are based on sample ecological regions covering 20 per cent of the globe, are expected to hold when extrapolated globally.[3] Also in 2004 the Tokyo-based United Nations University estimated that by 2050 two billion people would be vulnerable to floods as a result of climate change, rising sea levels, deforestation and population growth.[4] Some 520 million people are currently exposed to the risk of flooding, which, as we see in Chapter 1, can devastate entire communities.

In 2050 – the year on which these predictions focus – my young children will, give or take two years, be the same age I am now. What future will they inherit? What is my responsibility to them and to their generation? What should they expect from me? In many respects they have been the audience for whom I have written this book. I have imagined them looking over my shoulder, reading the drafts every bit as rigorously as those colleagues who commented on early drafts of this book, checking that I am doing what I should be doing as a parent and a publicly funded academic. After all, the burdens of adjusting to environmental degradation will fall on their generation, which has not caused the problem. As if to prove the point, in April 2006, as I finalized the manuscript, another major study was published in *Nature* which found that current trends of agricultural expansion and deforestation in the Amazon indicate that, by 2050, 40 per cent of the Amazon forests will be lost, with one quarter of the 382 mammalian species examined losing two-fifths of their Amazon forest ranges.[5]

Those like me who work in the public education sector and are funded from the public purse should contribute to public debate on matters of public importance. I researched this book with the intention of contributing to dialogue on what is the most critical public welfare issue of our age: global environmental degradation. But precisely what sort of contributions should the publicly funded social scientist make? Is our legitimate role solely to analyse and explain 'what happens' – to identify problems of public importance before stepping back so that others, such as policy-makers and politicians, can come forward with, hopefully, the right solutions? If that is the case then this book should, presumably, concentrate solely on trying to explain the processes that drive deforestation and on analysing the various international political processes that have addressed this problem.

Most, though by no means all, of my colleagues in the public education sector would argue that it is perfectly legitimate for publicly funded academics to go further than this and to address normative questions on the policies, political systems and governance structures that are necessary or desirable. In other words, academics may engage in advocacy. But this then leads to an important question: on whose behalf should we claim to speak? Given the growing role of corporate finance in funding higher education research, this is a fundamental question. There is an increasing risk that publicly funded academics, under pressure from employers with financial difficulties, can be co-opted into research that is driven not by some generic notion of the common good, but by the specific interests of private donors. For example, the corporate sector has become adept at funding scientific research in public education establishments that can later be patented and used in the commercial manufacture of products. Some corporations have lobbied successfully for the redirection of public funds for such ends. Others have funded 'policy-relevant' research on issues that promote corporate interests, such as the role of private sector investment in public service provision. I contend that scholars from the public education sector who carry out research on behalf of private sponsors and then engage in advocacy to promote the interests of such sponsors violate the impartiality and integrity of the public research enterprise.

So under what conditions should the publicly funded academic become an advocate? I would argue that academics should be free to engage in advocacy, indeed may have a public duty to do so, providing that one simple but crucial condition is met: that the advocacy in question is expressly intended to enhance the public interest through, for example, the provision of public goods or the promotion of human rights. This is the key criterion that must be satisfied if a publicly funded academic should seek to exert influence outside the academy. But it involves grappling with an essentially, and perpetually, contested concept: the public interest. No public body, even the most independent and well intentioned, can ever authoritatively determine what the public interest is. And there are other difficulties. The spatial scale at which we analyse 'the public' may vary from one community to another. How we view the public interest changes significantly when we consider the idea of a 'global public,' and changes further still if we stretch time to factor in the interests of generations who have yet to be born. Another complication, as we see throughout this publication, is that it is not always easy to separate the public domain from the private. Promoting the interests of a private group or organization may yield broader benefits that are in the public interest. Partnerships between public and private actors may be in the public interest, although very often they serve particular private interests at the expense of the public at large. Public officials and politicians may masquerade as servants of the public interest while promoting the interests of favoured corporations. Indeed, patron–client networks between forest industry corporations and politicians in the tropics help to drive deforestation, although such networks are by no means confined to tropical countries.

It is helpful here to consider Robert Cox's seminal distinction between problem-solving approaches and critical theory. A problem-solving approach seeks to address problems without challenging dominant actors, relationships and ideologies, all of which are seen as 'realities' that have to be accepted. A problem-solving approach would see environmental problems as 'managerial' issues to be addressed through more effective policies and strengthened environmental institutions. In distinction, a critical theory admits the possibility that a problem may be a defining feature of a social system itself. A critical theoretical view might see environmental problems not as unusual or accidental, but as more ordinary phenomena that are routinely generated by dominant power structures and which can be solved only through fundamental systemic change.[6] The distinction between the two approaches reminds us that advocacy on public issues may come in various guises, from contesting the fine details of policy to 'broad-brush' analysis that concentrates on 'the big picture.' If, as this book argues, the global loss of forest cover is a systemic problem, then an exclusive focus on the details of policy will, at best, lead only to incremental improvements in forest quality while ignoring the deep driving forces of deforestation.

Pondering these issues led me to engage with the growing literature on global public goods. While the public interest may vary from community to community, all publics have an interest in the conservation and integrity of environmental public goods. Forests provide a diversity of public goods, such as climate regulation and soil conservation, that benefit us all, but which can be owned by none. They also provide private goods, such as timber and fruits, that may be owned through acquiring property rights. Balancing the private and public goods that forests provide is the defining challenge for forest policy-makers. With respect to forests, the long-term global 'public interest,' which includes future generations, may be defined as the maximum provision of forest private goods but only to the extent that such provision does not erode forest public goods or otherwise degrade the integrity of the forest resource base. Deforestation takes place when forest private goods are harvested at the social cost of forest public good depletion. The continuing exploitation of forest private goods is enabled by a hegemonic neoliberal global economy that promotes private sector investment in forests, international trade of forest products and the voluntary regulation of the private sector. It is a central argument of this publication that not only does neoliberalism drive excessive exploitation of forests, it also establishes the parameters of policies that set out to arrest deforestation, often rendering them ineffective.

Chapter 1 outlines the concepts of global public goods and neoliberalism that provide the theoretical framework for this book. The intention of the next eight chapters is to assess all international forest policy processes from 1995 to 2006, beginning with the Intergovernmental Panel on Forests (Chapter 2). The same year that the Panel first met (1995), the World Commission on Forests and Sustainable Development was also created. This commission, the first of its kind to address sustainable development since the influential World Commission on Environment and Development, reported in 1999, after which it disbanded. Lacking high-level political support, it had virtually no impact on forests discourse or global governance (Chapter 3). During the latter half of the 1990s, the main international forest policy dialogue was confined to the Panel and its successor, the Intergovernmental Forum on Forests, which was discontinued after three years in 2000 (Chapter 4).

The politically thorny question of whether states should agree a forests convention coloured the deliberations of all these bodies. This question was first considered at the United Nations Conference on Environment and Development held in 1992, when hopes for a forests convention were shattered when tropical forest governments, wary of erosions of their sovereignty over their natural resources, blocked a convention. Since then the question has been revisited several times, each time without agreement, most recently by the United Nations Forum on Forests (UNFF). Created in 2001 to replace the Intergovernmental Forum on Forests, the UNFF is the most senior body to address forests within the United Nations system. In 2006, it decided to negotiate a non-legally binding instrument on forests, a decision that was a compromise between the pro- and anti-convention states (Chapter 5).

At present, it is far from clear that the UNFF or any other intergovernmental organization can play a meaningful role in reversing the loss of global forest cover. This recognition has led to the partial migration of rule-making authority on forests out of the UN system. Two chapters explore the experiments in new governance that have been initiated as a result. The first deals with voluntary certification and labelling. In 1993, a coalition of non-governmental organizations (NGOs) and environmentally enlightened forest businesses created a voluntary timber certification organization, the Forest Stewardship Council (FSC). Other forest owners and timber businesses in North America and Europe reacted by creating trade-dominated certification schemes. The ensuing and, as yet, unresolved 'certification wars' have seen these different schemes competing for legitimacy and political support (Chapter 6).

The second example of rule-making authority leaking from the UN system relates to illegal logging. The realization that the real problems of deforestation are on the ground within countries has led to the creation of a network of linked regional initiatives to promote forest law and governance reform. Between them, these initiatives cover most of the important timber-producing and consuming countries of Asia, Africa and Europe. The emphasis on governance reform within producing countries has been supplemented by a recognition among many developed countries that demand-side measures are necessary if the international trade in illegally logged timber is to be arrested. This has led to a European Union (EU) action plan under which the EU can negotiate voluntary partnership agreements with timber-producing countries that undertake to ensure that timber exported to the EU is harvested from licensed legal sources. However, the EU commitment to demand-side measures has not been matched by all major timber importing states, with the US opposed to any measure that can be construed as a trade restriction (Chapter 7).

Both timber certification schemes and the regional initiatives to tackle the trade in illegally logged timber rely entirely on voluntary measures. This is so that such measures will not be ruled as barriers to trade under World Trade Organization (WTO) rules. The WTO has established a neoliberal constitutional order under which free international trade and other neoliberal objectives trump public goods provision. Trade restrictions to ban the international trade of unsustainably managed or illegally harvested timber are inadmissible. Other forest conservation initiatives are delimited and enfeebled by neoliberalism. So, despite the many international forest policy initiatives on forests that this book considers, deforestation, especially in the tropics, continues largely unchecked.

The World Bank, another international organization that is associated with neoliberalism, first introduced a forest policy in 1978. Since then the World Bank's forests strategy has gone through various changes, including a ban on financing logging in primary tropical moist forests that was introduced in 1991 and then, controversially, lifted in 2002. Historically, the World Bank's forests strategy has included both conservationist measures, such as establishing protected areas and supporting strong certification standards, and neoliberal policies, such as opening forests for private sector investment and supporting voluntary private sector regulation. The Bank has introduced safeguard policies on financing forest projects, although these safeguards do not apply to adjustment lending, which accounts for one quarter of the Bank's lending portfolio. Central to the World Bank's forests strategy is an alliance with the World Wide Fund for Nature (WWF) (Chapter 8).

Despite the continuing absence of a forests convention, a distinct international regime on forests has emerged based on hard, soft and private international law. The main hard legal instrument of the regime is the Convention on Biological Diversity. The forests regime comprises a number of principles promoting economic, environmental and social objectives, such as forest conservation and protection, sustainable use of forests, the rights of forest indigenous peoples, and the equitable sharing of the benefits that accrue from exploiting forest biodiversity. However, the various regime principles stand alone from each other. The result is that the regime promotes both forest public good enhancement and an expansion in the trade of forest private goods (Chapter 9).

Chapter 10 develops this argument. Under neoliberalism, the trade in forest private goods has a stronger normative force in international law than public goods provision. While the WTO agreements promote the trade of private goods, environmental and human rights principles have been deliberately and systematically excluded. In order to comprehend the processes that drive deforestation, it is thus necessary to look not to forest-related organizations but to broader structures of global governance. Deforestation, it is argued, is both the result and a symptom of a crisis of public accountability. Publicly accountable bodies at all levels of international society – intergovernmental organizations, the state and public authorities at the sub-state level – lack the capacity, the political will or both to represent and regulate in the public interest and to hold business corporations to account for their socially and environmentally destructive activities. The reason for this is that corporations have colonized public authorities, exerting significant influence on public policy and successfully recasting the notion of the public interest in their own image. The public interest is best realized, argue corporate executives and the neoliberal politicians who support them, when businesses are free to pursue their interests with minimal state regulation and according to common international trade and investment rules. Chapter 10 dismisses both this argument and neoliberal policy responses such as corporate social responsibility. It then advocates a model for the reassertion of publicly accountable politics, in particular the democratic regulation of the business corporation. A multilevel system of democratic governance is proposed that would see publicly accountable authorities acting solely and exclusively in accordance with their original *raison d'être*: as guardians of the public interest. Under this governance system, final decision-making authority on local economies would rest with local communities, who are invariably the first to suffer from forest loss.

Acknowledgements

This book could not have been written without the help of many people. Colleagues in the Geography discipline of the Open University agreed to fund the research carried out from 2004 through to 2006. Felix Dodds and Toby Middleton arranged accreditations for attending United Nations negotiations. Peter Dauvergne, Lloyd Pettiford and Jeffrey Sayer provided invaluable feedback on the initial book proposal. Sir Martin Holdgate kindly handed over his personal archive from his time as co-chair of the Intergovernmental Panel on Forests. John Bazill, Jon Buckrell, David Cassells, Sasha Courville, Elena Kulikova, Sam Lawson, Jag Maini, Bill Mankin, Jan McAlpine, Manuel Rodríguez, Taghi Shamekhi and Gijs Van Tol gave detailed interviews. Ben Cashore, Debbie Davenport, Radoslav Dimitrov, Lars Gulbrandsen, John Hudson, Carole Saint-Laurent, Graeme Thompson, Dave Toke and Peter Willetts provided invaluable help, including reading and commenting on early draft chapters. Marcus Colchester generously provided a huge number of detailed and constructive comments that greatly improved the text. Thanks also to Michele Betsill, Andy Blowers, Grazia Borrini-Feyerabend, Duncan Brack, Brianna Cayo Cotter, Elisabeth Corell, Ken Creighton, Mike Dudley, Maurizio Ferrari, Peter Gondo, Anne Guest, Peter Hough, Petr Jehlicka, Libby Jones, Björn Lundgren, Peter Mayer, Diarmuid McAree, Américo M.S. Carvalho Mendes, the late David Pearce, Reidar Persson, Susan Saidi, Phil Sarre, Joe Smith, Roy Smith, David Sone, Viktor Teplyakov, Katharine Thoday, Mitzi Gurgel Valente da Costa, John Vogler, Juliette Williams and Stuart Wilson. Many thanks to the staff at the libraries of the Open University and the University of Kent at Canterbury. Jan Smith displayed her customary good humour and patience in styling and formatting the manuscript. At Earthscan Camille Adamson, Gudrun Freese, Andrea Service and Rob West competently guided the book towards publication. The usual author's disclaimers apply.

While researching this book I benefited immensely from some high-quality websites. Two, in particular, should be mentioned: www.illegal-logging.info (managed by Chatham House, London and financed by the UK Department for International Development); and http://forests.org/. Those who maintain such sites provide an invaluable public good by disseminating state of the art knowledge on deforestation and how it can be addressed.

The song lyrics at the start of the book are reprinted with the kind permission of Midnight Oil Enterprises Pty Ltd (with thanks to Rob Hirst and Arlene

Brooks). The extract from the statement of the UN Millennium Project Task Force on Environmental Sustainability to the United Nations Forum on Forests is reprinted with the kind permission of Don Melnick.

A big thank you to all those who rallied around in 2003. You know who you are.

A message for my son, Timothy, and daughter, Anna: I love you lots and I'm very proud of you. And, finally, many thanks to Colette, who is a constant source of love, support and inspiration. Like me, this book is dedicated to her.

Acronyms and Abbreviations

ABS	access and benefit-sharing
AFP	Asia Forest Partnership
AFPA	American Forest and Paper Association
AIDS	acquired immune deficiency syndrome
AMAN	Aliansi Masyarakat Adat Nusantara (Indigenous Peoples' Alliance of the Archipelago, Indonesia)
APEN	Asia Pacific Peoples' Environmental Network
ASEAN	Association of Southeast Asian Nations
BBC	British Broadcasting Corporation
BP	Bank Procedure (*of the* World Bank)
C+I	criteria and indicators (for sustainable forest management)
CAFTA	Central American Free Trade Agreement
CBD	Convention on Biological Diversity (1992)
CCD	Convention to Combat Desertification in Countries Experiencing Serious Drought and/or Desertification, Particularly in Africa (1994)
CDM	Clean Development Mechanism (*of the* FCCC)
CEC	Commission of the European Communities
CEO	chief executive officer
CEESP	Commission on Environmental, Economic and Social Policy (*of the* IUCN)
CEPI	Confederation of European Paper Industries
CERES	Coalition of Environmentally Responsible Economies
CFC	chlorofluorocarbon
CIFOR	Center for International Forestry Research
CITES	Convention on International Trade of Endangered Species of Wild Fauna and Flora (1973)
COMIFAC	Conference of Ministers in Charge of Forests in Central Africa
CPF	Collaborative Partnership on Forests (created 2001)
CSA	Canadian Standards Association
CSD	United Nations Commission on Sustainable Development
CSR	corporate social responsibility
°C	degrees Celsius
DG	Directorate-General (*of the* European Union)

DRC	Democratic Republic of the Congo
ECOSOC	United Nations Economic and Social Council
EIA	Environmental Investigation Agency
EMS	environmental management system
EU	European Union
FAO	Food and Agriculture Organization
FCCC	Framework Convention on Climate Change (1992)
FERN	Forests and the European Union Resource Network
FFCS	Finnish Forest Certification Scheme
FLEG	Forest Law Enforcement and Governance
FLEGT	Forest Law Enforcement, Governance and Trade (EU action plan)
FLO	Fairtrade Labelling Organizations International
FSC	Forest Stewardship Council
FTAA	Free Trade Area of the Americas
G7	Group of 7 Developed Countries (Canada, France, Germany, Italy, Japan, the UK and the US)
G8	Group of 8 Developed Countries (the G7 plus Russia)
G77	Group of 77 Developing Countries
GATT	General Agreement on Tariffs and Trade
GDP	gross domestic product
GEF	Global Environment Facility
GFPP	Global Forest Policy Project
GLOBE	Global Legislators Organization for a Balanced Environment
GM	genetically modified
GNP	gross national product
GRASP	Great Apes Survival Project
GURTS	genetic-use restriction technologies
HDI	Human Development Index
HIV	human immunodeficiency virus
IAF	International Accreditation Forum
IBAMA	Instituto Brasileiro do Meio Ambiente e dos Recursos Naturais Renováveis (Brazilian Institute of Environment and Renewable Natural Resources)
ICFPA	International Council of Forest and Paper Associations
ICRAF	International Centre for Research in Agroforestry (World Agroforestry Centre)
IFC	International Finance Corporation
IFF	Intergovernmental Forum on Forests (1997–2000)
IFIR	International Forest Industry Roundtable
IFOAM	International Federation of Organic Agriculture Movements
IIED	International Institute for Environment and Development
IIFB	International Indigenous Forum on Biodiversity
ILO	International Labour Organization
IMF	International Monetary Fund

IOAS	International Organic Accreditation Service
IPCC	Intergovernmental Panel on Climate Change
IPF	Intergovernmental Panel on Forests (1995–1997)
IPR	intellectual property right
ISEAL	International Social and Environmental Accreditation and Labelling
ISO	International Organization for Standardization
ITFF	Interagency Task Force on Forests (1995–2001)
ITTO	International Tropical Timber Organization
IUCN	World Conservation Union (*formerly the* International Union for the Conservation of Nature and Natural Resources)
IUFRO	International Union of Forest Research Organizations
JATAN	Japanese Tropical Forest Action Network
*l*CER	long-term certified emission reduction
LFCC	low forest cover countries
LMMC	like minded megadiverse country
MAC	Marine Aquarium Council
MAI	multilateral agreement on investment
MCPFE	Ministerial Conference on the Protection of Forests in Europe
MSC	Marine Stewardship Council
NAFTA	North American Free Trade Agreement
NFAP	national forestry action programme
NFP	national forest and land-use programme (*commonly abbreviated to* national forest programme)
NGO	non-governmental organization
NIEO	New International Economic Order
NOVIB	Nederlandse Organisatie voor Internationale Bijstand (Oxfam, the Netherlands)
ODA	official development assistance
OECD	Organisation for Economic Co-operation and Development
OED	Operations Evaluation Department (*of the* World Bank)
OP	operational policy (*of the* World Bank)
OP	operational programme (*of the* GEF)
PEFC	Programme for the Endorsement of Forest Certification schemes (*formerly* Pan-European Forest Certification scheme)
PFII	United Nations Permanent Forum on Indigenous Issues
PPM	production and processing method
PROFOR	Programme on Forests (*of the* World Bank)
RAN	Rainforest Action Network
SAI	Social Accountability International
SAP	structural adjustment programme
SBSTTA	Subsidiary Body on Scientific, Technical and Technological Advice
SFI	Sustainable Forestry Initiative (US)
SFM	sustainable forest management

SGS	Société Générale de Surveillance
TBT	Technical Barriers to Trade Agreement (*of the* WTO)
*t*CER	temporary certified emission reduction
TFAP	Tropical Forests Action Programme (*formerly* Tropical Forestry Action Plan)
TFD	The Forests Dialogue
TFRK	traditional forest-related knowledge
TGER	Theme on Governance, Equity and Rights (*of the* IUCN)
TRAFFIC	Trade Records Analysis of Fauna and Flora in Commerce
TRIPS	Agreement on Trade-Related Aspects of Intellectual Property Rights (*of the* WTO)
TTF	UK Timber Trade Federation
UK	United Kingdom
UN	United Nations
UNCED	United Nations Conference on Environment and Development (1992)
UNCTAD	United Nations Conference on Trade and Development
UNCTC	United Nations Centre for Transnational Corporations
UN-DESA	United Nations Department for Economic and Social Affairs
UNDP	United Nations Development Programme
UNEP	United Nations Environment Programme
UNESCO	United Nations Educational, Scientific and Cultural Organization
UNFF	United Nations Forum on Forests (*created* 2001)
UNGASS	United Nations General Assembly 19th Special Session (1997)
UPOV	Union Internationale pour la Protection des Obtentions Végétales (International Convention for the Protection of New Varieties of Plants)
US	United States
VPA	voluntary partnership agreement (*of the* EU's FLEGT action plan)
WALHI	Wahana Lingkungan Hidup Indonesia (Indonesian Environment Forum)
WBCSD	World Business Council for Sustainable Development
WCFSD	World Commission on Forests and Sustainable Development
WCPA	World Commission on Protected Areas
WHO	World Health Organization
WIPO	World Intellectual Property Rights Organization
WRM	World Rainforest Movement
WTO	World Trade Organization
WWF	World Wide Fund for Nature (World Wildlife Fund in Canada and the US)

1

Forests as Public Goods

Ships approaching the Venezuelan port of La Guaira are greeted by what, at first sight, appear to be small hillside holiday homes idyllically overlooking the Caribbean Sea. As the ships draw closer, it becomes apparent that the dwellings are, in fact, slums. For those who disembark and take the road from La Guaira to the capital Caracas, the scene is similar, with shanty towns clinging perilously to the mountainsides.[1] In December 1999, torrential rains led to huge mudslides, washing the shanties off the mountains into the sea or into rivers that had broken their banks. Along a 60 mile swathe of coastal Venezuela, some 30,000 people died, with thousands more losing their homes.

While unusually heavy rainfall precipitated this tragedy, a major contributory factor was the deforestation of the mountainsides on which the shanties perched. The trees had bound the soil to the mountains, protecting the valleys below from mudslides. Deforestation, which had taken place over decades as trees were felled to clear space for new shanties, had eroded this vital local public good. While the trees remained, everyone in the local populace benefited; but when the mudslides came no one in the vicinity escaped the consequences. The mudslides claimed prosperous houses and public infrastructure in the valleys; but it was the poor on the mountainsides who were the first to suffer and who bore the heaviest cost.[2] Lacking secure land tenure and with little where else to go, these people inhabited peripheral lands on the periphery of the global economy, excluded from more productive land after decades of private enclosure by more powerful interests.

This tragedy took place in one country. But it illustrates broader themes that are relevant to global forest governance, such as the destruction of local forest commons, insecure land tenure rights for the poor and the privatization of land. It also highlights the importance of forest public goods, and what happens – and to whom – when public goods are eroded. To help appreciate this, we begin this chapter by briefly introducing public goods theory. We then use this theory to develop a taxonomy of forest goods and services. In developing this taxonomy we consider the role of local communities, the modern state and business corporations in forest governance. We also examine the market-based ideology of neoliberalism and consider to what extent this ideology is congruent with the provision of environmental public goods.

Public goods theory

Forests provide a range of public and private goods.[3] The term 'good' is used here to denote non-tangible goods, usually referred to as services, in addition to tangible goods. Private goods, which can be bought and sold, meet two criteria: rivalry and excludability. A good is rival when consumption by one person will reduce what is left for others. An apple, once eaten, cannot be consumed by others. A good is excludable when its owner can prevent others from enjoying it. I own the laptop on which I am writing this chapter, and can exclude others from using it if I so wish. In distinction, public goods are non-rival, as one person's consumption does not affect what is left for others, and non-excludable, as no one can be prevented from enjoying the good. Public goods provided by human agency include footpaths, street lighting and flood control defences. The atmosphere, the ozone layer and the high seas are examples of global public goods that are given by nature. They existed prior to humans, but are now being degraded by human activity.[4]

In between the categories of public and private goods are goods that are neither purely private nor purely public. Goods that are excludable but non-rival are club goods. Individuals can gain entry to a club for payment of a fee or toll.[5] For some club goods there may be an element of rivalry among users; a tennis club may be subject to high demand during particular times. However, pure club goods are entirely non-rival and fully excludable. An example is satellite television. It is easy to exclude non-members from the 'club,' in this case through descrambler technology.

Two categories of goods are non-excludable but rival, in that one person's consumption affects the consumption of others. First, there are open access goods or *res nullius*, which means the property of no one. *Res nullius* goods are prone to depletion and exhaustion. Classic examples include fish stocks on the high seas and wilderness land that has not been claimed and which has no legal title. Second, there are impure public goods. An example here is the fire brigade. Any citizen may use the fire brigade, yet a fire engine cannot be in two places at once. So if there were several fires simultaneously, there would be rival consumption. Impure public goods – which also include the health service and school education – are liable to congestion, rationing and waiting lists when subject to high demand.

Table 1.1 shows the four categories of goods. The dualism of rival/non-rival is best seen as two poles on a continuum, as is that of excludable/non-excludable. The four cells of the table thus have permeable boundaries. Some public goods, such as the health service, are not innately non-excludable, and the decision that no citizen should be excluded from them is an essentially political one. Much depends on the social and political context, and whether a society or government considers a good sufficiently important that it should be collectively provided as a public good. Which cell a good is placed in will thus vary according to time and space.

Economists view the underprovision of public goods by markets as market failure.[6] Under conditions of market failure, resources are not allocated

Table 1.1 *Taxonomy of private and public goods*

	Rival	**Non-rival**
Excludable	***Quadrant 1*** *Private goods* Goods that can be bought and sold	***Quadrant 2*** *Club goods* Satellite TV Telephone networks Tennis clubs Private healthcare
Non-excludable	*Res nullius (open access)* Goods accessible to all that are subject to depletion and exhaustion, such as fish stocks and unsettled land *Impure public goods* Goods subject to congestion or rationing: Fire brigade School education National health service ***Quadrant 3***	*Public goods* Provided by nature: Atmosphere Ozone layer High seas Provided by human agency: Street lighting Footpaths Cultural heritage ***Quadrant 4***

efficiently and key consumer demands are unsatisfied. Because markets work best when goods are both rival and excludable, they will tend to undersupply, or not supply at all, goods with non-rival and/or non-excludable properties. So, whereas private goods can be supplied through the market, public goods require the exercise of agency by public authorities or publicly accountable bodies charged with acting *pro bono publico* (for the public good).

The spatial range at which the benefits of a public good are felt will vary. Where the benefits accrue at the country level, a national body may be best positioned to provide a public good. This may be the state – although the Keynesian notion that the centralized state is best placed to provide public goods is no longer accepted uncritically – or it may be another type of organization, such as a charitable foundation that acts *pro bono publico*. Other public goods have a global dimension, and their provision clearly depends on some form of global governance. Global public goods are essentially non-excludable and non-rival over a worldwide scale. Examples of global public goods provided by human agency are international air safety regulations, international disease eradication programmes and an international criminal justice system.

Of these examples, international air safety regulations are the purest as a global public good, as virtually every country has adopted the air safety regulations agreed by the International Civil Aviation Organization. Vaccination against a communicable disease is a private good in that it protects the vaccinated individual from the disease; but it also has a public good function since it reduces the risk of the disease spreading. At their most effective, vaccination programmes can completely eradicate a disease, as with smallpox. The international criminal justice system is an impure public good since it is easy to evade; some governments, notably the US, have refused to ratify the Rome Statute of the International Criminal Court and are therefore not bound by its provisions.[7]

Forest commons and forest public goods

We now apply the framework developed so far to yield a taxonomy of the goods that forests provide. Forests can be harvested to provide a diversity of private goods, including timber, fruits, nuts, berries, rattan and rubber. Another highly controversial example is bushmeat, namely the meat of forest wildlife, such as great apes, that some forest people consume as a traditional food. Forests also supply public goods, both pure and impure, at different spatial scales. The beneficiaries – the *publicum* – of a forest public good vary depending on the spatial range at which the benefits of the good are felt. Some benefits occur principally at the local or national level, such as watershed management and soil protection. Forests provide habitat for a range of flora and fauna. If one accepts the view that all humanity has the right to see a tiger, gorilla or toucan in the wild, then the habitat function of forests can be see as a global public good, even if not everyone takes advantage of the right. Charles Perrings and Madhav Gadgil argue that forests, along with other species-rich ecosystems, contribute to the global public good of biodiversity conservation, which maintains the global gene pool necessary for resilient and adaptable species and ecosystems.[8] Forests sequester carbon dioxide, an important greenhouse gas. They thus play a role in global climate stabilization – a global public good – so that humanity as a whole is the *publicum*.

Forests are shared, therefore, not in a spatial or ownership sense, but in the sense that they provide public goods for both proximate and distant users. No one can be excluded from these public goods, and no one will be able to escape the public bads that will result from severe forest depletion, such as global warming, biodiversity loss and soil degradation. As forests provide public goods at different spatial scales, thus benefiting different *publicums*, three public claims have been made on forests, ranging from the local to the global (see Box 1.1).

Effective local common regimes play an important role in conserving forest public goods. Common property regimes have been defined by Margaret McKean as 'institutional arrangements for the cooperative (shared, joint, collective) use, management, and sometimes ownership of natural resources.'[11] Garrett Hardin argued in his 1968 essay on the tragedy of the

Box 1.1 Three public claims to forests

It has sometimes been claimed that the world's forests should be seen as a global common, or a common heritage of humankind, as everyone derives benefits from them. This claim was ventured in the forest negotiations prior to the 1992 United Nations Conference on Environment and Development (UNCED). It was rejected by the Group of 77 Developing Countries (G77). The claim that forests are a global common is a flimsy one; it has no foundation in international law and has not been ventured in international forest diplomacy since the UNCED forest negotiations.

Second, forests can be seen as a sovereign resource of the state. Governments, especially from the developing world, assert their right to use forests in line with national development policy. Legally, this claim is the strongest of the three public claims, and it has a firm basis in international environmental law.[9] Furthermore, most forests in developing countries are legally administered by state agencies (see Table 1.2).

Third, indigenous peoples and local communities have asserted that forests should be seen as local commons and that local peoples are best placed to ensure their conservation and sustainable use. Local people, it is claimed, can best do this when they have secure land tenure rights and legal ownership of the forests. Many indigenous peoples and local community groups resist both state administration and for-profit privatization as ownership forms that exclude local people and degrade the commons.

These three claims are not mutually exclusive. In particular, indigenous peoples' groups have emphasized that 'State sovereignty does not and cannot preclude attention to and respect for indigenous peoples' internationally guaranteed rights.'[10] In this view, the principle of state sovereignty over natural resources encompasses respect for the traditional land rights of indigenous peoples.

commons that each individual user of a local common has a short-term interest in overexploiting the resource.[12] According to this view, common property resources should be seen as rival in consumption, but non-excludable, in that anyone can take what they want.

The arguments of Hardin's critics can be distilled as follows. Hardin was not writing about a genuine local common, where all members of the community respect the local rules and cooperate to monitor the access to the resource that outsiders may or may not enjoy. Hardin assumed that within a local common, access to the resource was open to all, whereas in fact use rights are agreed and enforced communally. Commoners tempted to free ride are likely to be caught in a well-functioning common property regime. Hardin's analysis is, however, relevant for open access resources that belong to no one and to resources where the owner is absent or disinterested. It also applies to well-managed commons that have been undermined by outside groups that are powerful enough to ignore the traditional rights of commoners and treat the resource as open access.[13] For example, commoners can easily be displaced by determined and unscrupulous agents from outside the forest, such as private corporations wishing to clear a forest for alternative land uses, such as agriculture. This has been a major problem in many tropical forests, including those of Brazilian Amazonia and Indonesia.

The fate of forests as commons has changed throughout history. In Europe, the commons were largely respected, except in times of war, until medieval times when the aristocracy and political elites organized the systematic and widespread displacement of commoners from common land, and the subsequent enclosure of this land by fencing.[14] In England, the 13th-century Forest Charter of Henry III gave forest access and use rights to royalty and the nobility, including the right to hunt.[15] Enclosure displaced commoners and placed outside the law those who continued to harvest forest goods from traditional lands. The legend of the mythic English folk hero Robin Hood relates the resentment of commoners of the royal forest laws that legalized the seizure of local commons by the monarchy and aristocracy for their personal use. When Robin commits the capital offence of killing a deer in the Royal Sherwood Forest, he becomes an outlaw. The tale strikes a populist note as Robin Hood plays a role in wealth distribution. By robbing from the rich and giving to the poor, he helps to right the historical wrong of land seizure from the poor by the aristocracy. This type of legend is not unique to England. The forest laws were codified in medieval France around this time,[16] and the 13th-century Norman French poem *Roman d'Eustache li Moine* relates the story of an outlaw who, like Robin Hood, found refuge in the forests.[17] The tales of Robin and Eustache illustrate that who is within the law and who is an outlaw depends on who writes the law.

Commons enclosure continued with the rise of the centralized state in the 16th and 17th centuries. James Scott sees the state as a modern project that attempts to render its territory and society legible and manipulable. The undermining of commons and customary ownership forms is, at least in part, an attempt to regularize complex and varied land tenure customs within standardized systems that enable easy recording, census and mapping.[18] However, centralized records and maps do not solely record land tenure and ownership. They first create, and then perpetuate, a system of more or less homogenous land registration, refashioning landownership to the advantage of those who accept the new administratively convenient ownership forms, which have the force of law, at the expense of those who wish to retain the earlier, often highly localized, customary ownership forms.[19] To Scott, the early modern state in Europe viewed forests as a revenue source. Trees with economic value became timber, while those without were considered 'underbrush.'[20] By conserving only economically valuable species, this instrumental, utilitarian and abstractionist logic eroded biodiversity in the forest and promoted the development of scientific forestry, which first emerged in Germany during the late 18th century as a product of the centralized state management of that period.[21] Early scientific forestry, based on geometrically neat seeding and planting patterns, systematic tree felling and a steady revenue stream, was subsequently adopted in other European countries.[22] This recasting of forests as an economic resource fuelled forest enclosure across Europe, further eroding the commons, which were considered unproductive in economic terms.

The practices of enclosure and scientific forestry thus led to a more regular, and more readily managed, ordering of forest space in Europe. These practices were later exported to the European colonies in Africa, Asia, Australia and the Americas.[23] In the process, the colonial authorities destroyed the centuries old and previously effective resource management regimes of the peoples whom they displaced. Two broad processes were at work. First, the colonial powers handed ownership of much forestland to imperial forest authorities, ignoring the traditional ownership claims and protests of indigenous peoples. Second, land that was not seized by the colonial authorities was regarded as *res nullius*, in line with the prevailing legal norms and property law of the colonial power. Migrants from the colonial country could lay legal claim to forestland, which was often seen as inhospitable and threatening jungle, by clearing it of trees and then settling it. European political thought helped to legitimize this process. John Locke's theory of the origin of private property holds that man (sic) has a 'natural right' to land with which he has 'mixed' the labour of his body, for example, by enclosure and tilling.[24] Locke's theory of individual rights and his individualist notion of property ownership clashes with the collectivist notions of rights and property that underlie local common property regimes. The practice of claiming legal title to land through forest clearance has continued in some countries since the end of the colonial era, including Costa Rica, Ecuador, Honduras and Panama.[25]

Enclosure during the medieval and colonial eras degraded and destroyed forest commons across the world. Forest degradation in the 20th and 21st centuries has rarely been due to poorly functioning common property regimes. It is invariably the result of the enclosure of commons by state and private interests, who overexploit the forests for economic gain and who have a totally different relationship to the forest than the commoners whom they displaced. As *The Ecologist* magazine has noted, tragedies of the commons are often 'tragedies of enclosure.'[26] They are also often tragedies of uncontrolled access to the forest by interests that ignore traditional land claims and treat forestland as *res nullius*. Commons vary according to excludability. Depending on how easy it is for commoners to exclude or to monitor resource use by outsiders, local common property regimes should be placed in either quadrant 1 or quadrant 3 of Table 1.1. Excludability depends on three factors. The first is how effectively a local community manages its resources. The more efficiently a collective of commoners monitors resource use, the better positioned it will be to regulate access from outsiders. Second, the spatial location and topographical features of a common have a direct bearing on excludability. A community managing a small forest common in a geographically remote mountainous area will find it easier to exclude outsiders compared to, say, a dispersed community managing a relatively large area of low-lying forest land surrounded by non-commoners. Third, the power of local commoners relative to those of political and economic agents from outside the forests is a key factor in excludability and, therefore, whether local resources are managed sustainably. Many commoners find it increasingly difficult to prevent outsiders, such as powerful timber corporations, from using and abusing common resource goods compared with earlier eras.

Encroachment by outsiders is a product of contemporary socioeconomic processes, such as government-led development programmes, international aid for forest-based industries and forest conversion. Often the modern state legitimizes interests from outside the forest by providing them with legal title to forestland, particularly when such interests may engage in economic activities that can be levied or taxed, thus contributing to the national exchequer. The result is that many incursions into the commons take place within national law, whereas many sustainably managed commons continue to have no legal status at all. The conflict between the communal ownership of local commons regimes and the legal title that is granted by the state remains central to forest politics. However, the two are not necessarily mutually exclusive: forest commons may enjoy legal title and where they do, the rights of commoners to manage their resources free of external encroachment is greatly enhanced.

While some government policies may, intentionally or otherwise, degrade forests, others can conserve forests and their public goods. One example is protected forest areas. Protected areas vary according to rivalry and excludability. A protected area is a club good when people must pay a fee to gain temporary access to the forest (quadrant 2). Where no fee is charged and members of the public enjoy free use, the forest in question is, in effect, a public good (quadrant 4). Where a protected forest area is not effectively secured, so that outsiders can enter the forest and exploit it for free, then that forest has a *res nullius* character (quadrant 3).

Forest governance is a multilevel affair. Just as local commons regimes and national forest policies can help to maintain forest public goods, so, too, can international regimes. A range of forest-related international legal conventions has been agreed, and one of the aims of this book is to consider how effective they have been in maintaining global forest public goods (see Chapter 9). Protecting global public goods requires effective funding mechanisms. In 1998, a background paper for the Intergovernmental Forum on Forests noted that 'If biodiversity conservation is a public good enjoyed by the world, it is appropriate to seek direct international sources of financing for conservation.'[27] But with the exception of the Global Environment Facility (GEF), which pays the incremental costs of policies that yield global environmental benefits consistent with the aims of international conventions (see Chapter 9), no multilateral mechanism is dedicated to global public goods financing. Most public goods financing is included in official development assistance (ODA) budgets. It can be argued that public goods financing and ODA should be separate areas of international public finance.[28]

Who owns the forests?

We have noted three public claims to the world's forests. But who legally owns them? Prior to the evolution of the modern state, most of the world's forests were either common property resources or open access regimes. Now most of the world's forests are controlled by a government agency on behalf of the state, which, we have seen, has tended to adopt a utilitarian approach to forests as revenue sources. However, in many countries there are unresolved disputes between local communities and the state. In many cases, the state permits community use of state forests, but without formally recognizing traditional land claims. Jim Sato has coined the term 'ambiguous lands' to describe forests and other geographical spaces that are legally administered by the state, but managed and used by local people.[29]

In 2002, Forest Trends and the Center for International Environmental Law surveyed forest ownership in 24 of the world's 30 most forested countries.[30] For six of these countries, the entire forest estate was government administered. Five of these countries were in the tropics (Burma, Cameroon, Central African Republic, Democratic Republic of the Congo and Gabon), with the sixth being Russia, the country with the world's largest expanse of temperate and boreal forest. In four tropical countries (Guyana, Indonesia, Sudan and Tanzania) and one temperate/boreal country (Canada) the government administers between 90 and 100 per cent of forest cover (see Table 1.2). On a worldwide scale, governments administer approximately 77 per cent of the world's forests.[31]

Figures for government administered forests have remained more or less static since the 1940s,[32] although there have been two parallel global pressures for the state to relinquish its dominance over forests. First, environmental and human rights non-governmental organizations (NGOs) have pressed for traditional claims to forests to be recognized. In Table 1.2, land for indigenous peoples and local communities is subdivided into, first, lands legally owned by the state and other public authorities, but reserved for indigenous and local peoples, and, second, lands that are legally owned by indigenous and local peoples under national law. Community ownership dominates over all other ownership forms in China, Mexico and Papua New Guinea. In Bolivia, approximately one third of state forestland is reserved for communities and indigenous groups. However, in ten countries indigenous peoples and local communities have neither any formal ownership rights to forestland, nor are any government administered forests reserved for such groups. These countries are Argentina, Burma, Cameroon, Central African Republic, Democratic Republic of the Congo, Gabon, Guyana, Japan, Russia and Sweden.

The second pressure comes from forest industries, which claim that forests are most effectively managed when under private ownership. Overall, there is a higher percentage of privately owned forests in developed countries than in tropical countries. For three countries, the figure exceeds 50 per cent (Sweden, Japan and the US). Only one developing country – Argentina – has more than 50 per cent of its forests under private ownership. The view that forests

Table 1.2 *Official forest ownership in 24 of the top 30 forested countries*

Country	Area in million hectares (percentage of country total)							
	Public				Private			
By descending area of forest cover as identified by FAO, 2001*	Administered by government		Reserved for community and indigenous groups		Community/indigenous		Individual/firm	
Russian Federation	886.5	(100)	0.0	(0)	0.0	(0.0)	0.0	(0.0)
Brazil	423.7	(77.0)	74.5	(13.0)	0.0	(0.0)	57.3	(10.0)
Canada	388.9	(93.2)	1.4	(0.3)	0.0	(0.0)	27.2	(6.5)
United States	110.0	(37.8)	17.1	(5.9)	0.0	(0.0)	164.1	(56.3)
China	58.2	(45.0)	0.0	(0.0)	70.3	(55.0)	0.0	(0.0)
Australia	410.3	(70.9)	0.0	(0.0)	53.5	(9.3)	114.6	(19.8)
Democratic Republic of Congo	109.2	(100)	0.0	(0.0)	0.0	(0.0)	0.0	(0.0)
Indonesia	104.0	(99.4)	0.6	(0.6)	0.0	(0.0)	0.0	(0.0)
Peru	n.d.		8.4	(1.2)	22.5	(33.0)	n.d.	
India	53.6	(76.1)	11.6	(16.5)	0.0	(0.0)	5.2	(7.4)
Sudan	40.6	(98.0)	0.8	(2.0)	0.0	(0.0)	0.0	(0.0)
Mexico	2.75	(5.0)	0.0	(0.0)	44.0	(80.0)	8.3	(15.0)
Bolivia	28.2	(53.2)	16.6	(31.3)	2.8	(5.3)	5.4	(10.2)
Colombia	n.d.		n.d.		24.5	(46.0)	n.d.	
Tanzania	38.5	(99.1)	0.4	(0.9)	0.0	(0.0)	0.0	(0.0)
Argentina	5.7	(20.5)	0.0	(0.0)	0.0	(0.0)	22.2	(79.5)
Burma (Myanmar)	27.1	(100)	0.0	(0.0)	0.0	(0.0)	0.0	(0.0)
Papua New Guinea	0.8	(3.0)	0.0	(0.0)	25.9	(97.0)	0.0	(0.0)
Sweden	6.1	(20.2)	0.0	(0.0)	0.0	(0.0)	24.1	(79.8)
Japan	10.5	(41.8)	0.0	(0.0)	0.0	(0.0)	14.6	(58.2)
Cameroon	22.8	(100)	0.0	(0.0)	0.0	(0.0)	0.0	(0.0)
Central African Republic	22.9	(100)	0.0	(0.0)	0.0	(0.0)	0.0	(0.0)
Gabon	21.0	(100)	0.0	(0.0)	0.0	(0.0)	0.0	(0.0)
Guyana	30.9	(91.7)	0.0	(0.0)	2.8	(8.3)	0.0	(0.0)
Totals	**2802.25**		**131.4**		**246.3**		**443.0**	

Notes: n.d. indicates no data.

*The authors of this report refer to the list of countries with the highest forest cover provided in: FAO (Food and Agricultural Organization) (2001) *FAO's Forest Resource Assessment – Forest Cover 2000*, Rome: FAO.

Source: White, Andy and Martin, Alejandra (2002) *Who Owns the World's Forests: Forest Tenure and Public Forests in Transition*, Washington DC: Forest Trends and Center for International Environmental Law, p.5

are better managed under private ownership has its basis in neoliberalism, to which we now turn.

Neoliberalism in the forest

Neoliberalism, the hegemonic ideology of our time, emphasizes the primacy of the individual, and holds that the collective common good will be maximized if people and firms are free to pursue their own interests in the marketplace. Instead of 'burdensome' state regulation – for example, on social and environmental standards – there should be voluntary private sector codes of conduct. To neoliberals, state ownership of the economy is inherently wasteful; private ownership is more efficient. Hence, financial and business interests have agitated for public spending cuts, tax cuts and the privatization of state-owned assets. But while neoliberalism urges a reduced role for regulation, it does not advocate an enfeebled state; state power is necessary to the neoliberal project since it can introduce market-based discipline to new areas.[33] Privatization is appealing to political interests in developed countries as it relieves the state of economic obligations, while promising greater efficiency.

The private forest sector encompasses a broad range of actors, from the small family-owned forest to one of neoliberalism's favoured agents, the transnational corporation. As we have seen, the modern state has promoted a distinct rationalism and instrumentality to forest use; this logic has been carried to a more extreme degree by forest industry transnational corporations. While corporations are effective in supplying private goods, their efficacy in public good provision is highly questionable. Corporations have one overwhelming responsibility, namely to maximize shareholder value. This is a fiduciary responsibility in private law in most countries, and if a corporation neglects to do this, for example, through environmental protection measures that do not increase the value of the corporation, then it can, in principle, be sued by its shareholders.[34] This legal responsibility to enhance shareholder value rationalizes both the internalization of monetary benefits and the systematic externalization of social and environmental costs.[35]

In line with neoliberal orthodoxy, many developed country governments and forest businesses have pressed for an enhanced role for the private sector in tropical forests, which, it is claimed, will be more sustainably managed if removed from administration by state agencies. However, developing country governments have been reluctant to cede forest ownership to the private sector. A key reason for this is that under international trade and investment rules, foreign businesses have the same rights to purchase forests as domestic businesses. Selling off public forests would work to the advantage of those countries whose corporations and financial sectors dominate transnational investment flows, and to the disadvantage of resource-rich countries with relatively weaker corporate and financial sectors. Tropical countries could thus lose ownership of their forests to private foreign interests, something that the governments of these countries, which view forests strategically as sovereign resources, are unwilling to countenance. But, while reluctant to transfer ownership rights to the private sector, many developing countries have been prepared to grant use rights to private companies. The usual instrument for transferring use rights is a concession allowing a company to extract from a public forest an agreed volume of timber. In many countries concessions have enabled widespread crime and corruption, with the politicians and public officials who grant concessions taking bribes from the companies involved, and subsequently ignoring violations of forest law and the extraction of timber in excess of that stipulated in the concession contract.

One forum through which developed governments have pushed for privatization is the Group of Eight Developed Countries (G8). In 1998, the G8 Action Programme on Forests committed its members to 'further examine ways of promoting private investment and partnerships in sustainable forest management' (see Chapter 7).[36] The world's largest timber and forest products corporations are from developed countries. In terms of annual revenue, seven of the top ten corporations are from G8 countries, while in terms of annual wood consumption, six are from the G8 (see Table 1.3). The remainder are from Scandinavian countries with significant private forest industries. So if the world's tropical forests were to undergo large-scale privatization, we could, under existing international investment rules, expect the main beneficiaries to include forest industry corporations from the developed world. The drive for privatization can thus be seen as an attempt by forest businesses and politicians in developed countries to pry open new markets in the tropics.

With governments in tropical countries reluctant to privatize their forests, governments and forest industries from developed countries have pressed for greater private sector access to tropical forests, particularly through logging concessions. And they have been very successful. Forest Trends and the Center for International Environmental Law have found that, for the 16 countries for which data were available, some 396 million hectares of publicly owned forest, or 23 per cent of the forest estate, has been allocated to private sector timber concessions.[37] So, while most forests are nominally state owned or government administered, the dominant actors in resource extraction are from the private sector (see Table 1.4). Corporations from the developed world have profited from this process, although they have not been the sole beneficiaries. As Peter Dauvergne has demonstrated, in many tropical countries the power of the state to grant concessions has fuelled the growth of local timber barons embedded within patron–client networks to national political elites, on whom the timber barons depend for timber licences, and to forest product enterprises in consumer countries, on whom they depend for retail outlets.[38]

A growing number of transnational logging corporations are based in tropical countries. The depletion of forests in Southeast Asia has led to timber corporations from that region migrating elsewhere in search of unlogged forests. Malaysian corporations have gained concessions in Belize, Brazil, Burma, Cameroon, Gabon, Guyana, Laos, Papua New Guinea, Russia, Solomon Islands and Zimbabwe.[39] Chinese companies are active in Brazil and Papua New Guinea.[40] Indonesian corporations have been responsible for the overlogging of

Table 1.3 *Ranking of top ten forest industry corporations by annual revenue and by annual wood consumption*

Rank	*Top ten forest products & paper corporations by annual revenue (2000)*	*Rank*	*Top ten wood processing companies by annual wood consumption (2000)*
1	International Paper (US)*	1	International Paper (US)*
2	Georgia-Pacific (US)*	2	Georgia-Pacific (US)*
3	Kimberly-Clark (US)*	3	Kimberly-Clark (US)*
4	Weyerhaeuser (US)*	4	Stora-Enso (Finland)
5	Stora Enso (Finland)	5	Smurfit-Stone (US)*
6	Oji Paper (Japan)*	6	Metsälitto (Finland)
7	UPM-Kymmene (Finland)	7	UPM-Kymmene (Finland)
8	Nippon Paper Industries (Japan)*	8	Abitibi (Canada)*
9	SCA-Svenska-Celluloso (Sweden)	9	Norske Skogindustrier (Norway)
10	Smurfit-Stone (US)*	10	Canfor (Canada)*

Note: * Indicates a corporation from G8.

Source: For top ten forest products and paper corporations by annual revenue; George Draffan, 'World's Largest Wood and Paper Products Corporations,' www.endgame.org/gtt-corps-ranked.html (citing Forbes Global 500) (accessed 22 April 2004). For top ten wood processing companies by annual wood consumption; WWF (World Wide Fund for Nature) (2001) *The Forest Industry in the 21st Century*, www.panda.org (accessed 21 April 2004)

vast forest areas in Suriname.[41] Corporations from China, Malaysia and Taiwan have profited from lucrative concessions in Cambodia, with some of these businesses engaged in illegal logging.[42] Indeed, the problem of illegal logging is one of the main causes of forest degradation in Southeast Asia, Central Africa and Latin America. In the absence of enforceable sustainable forest governance on the ground, many corporations have engaged in the blatant asset-stripping of forests both within and outside concession boundaries (see Chapter 7). The liberalization of capital flows between countries – another hallmark of the neoliberal global economy – enables unscrupulous corporations to launder the money gained from illegal logging.

Critics argue that increasing privatization and marketization of forests will mean that what ultimately counts in forest governance is financial returns for corporations, with certain tree species fitting better into the business worldview than others. One campaigner, Jamie Aviles, has argued that 'The eucalyptus is the perfect neoliberal tree. It grows quickly, turns a quick profit in the global market and destroys the Earth.'[43] Environmental and human rights groups campaigning on forest issues seek to counter neoliberalism by promoting a

Table 1.4 *Percentage of public forest under private concessions in 16 countries*

Africa	%	Americas	%	Southeast Asia	%
Central African Republic	71.4	Bolivia	10.2	Cambodia	64.3
Cameroon	37.1	Canada	56.6	Indonesia	60.0
Republic of Congo	79.2	Guatemala	4.8	Malaysia	57.7
Democratic Republic of Congo	36.4	Peru	1.7	Philippines	22.7
Equatorial Guinea	71.4	Suriname	22.4		
Gabon	56.7	Venezuela	5.9		

Source: White, Andy and Martin, Alejandra (2002) *Who Owns the World's Forests: Forest Tenure and Public Forests in Transition*, Washington DC: Forest Trends and Center for International Environmental Law, p.9

counterdiscourse that emphasizes respect for the traditional knowledge and customary land of local and indigenous peoples, and their right to participate in democratic and decentralized decision-making processes.[44] Forests, it is claimed, are most likely to be conserved and sustainably managed when under local community control, and are more likely to be degraded when under the control of large businesses (see Box 1.2).

The neoliberal policy prescription that forests should be privatized is resisted by environmental and social NGOs. Friends of the Earth International argues that privatization weakens democratic control of forests by removing them as an area of legitimate public concern. Privatized provision goes where there is economic demand; that is, needs or wants backed by money. Hence, those who have traditionally depended on the forest for sustenance but who have no money cannot make their needs felt in the marketplace, with the result that these needs will be neglected.[47] The Global Forest Coalition and Censat Agua Viva consider that the privatization of forests reflects a mercantilist approach to nature 'where the value of forest is limited to its commercial value and where its permanence depends on this value.'[48] Privatization reduces the access rights of local communities and caters to the demands of markets for immediate financial returns. Market mechanisms tend to generate only those forest private goods with monetary value that can be traded for profit. They cannot capture the full public goods value of forests and thus promote a reductionist view of forests.[49]

New environmental markets

Under neoliberalism, new private property rights and environmental markets have been promoted as policy responses to global environmental problems

Box 1.2 Forest workers as local communities

It is sometimes suggested that indigenous peoples and environmental groups have appropriated the concept of 'local communities' to the exclusion of other local interests with a stake in the forest, such as forest workers. When the campaigning Australian rock band Midnight Oil joined the international protests against the clearfelling of old-growth temperate rainforests in British Columbia, Canada, in 1993, they were harassed by timber industry workers, as the band's drummer, Rob Hirst, relates:

> *Take the angry crowd of loggers and their wives who rocked and hammered our car on our way to a dawn concert in 1993 at the Black Stump, a clear-felled site at pristine Clayoquot Sound on Vancouver Island. Our safety, along with the security of the local Native Canadians, the Douglas firs, the endangered deer-mouse and the delicate fungi and lichen, seemed of little concern to the blockading bushfolk ... holding up signs such as 'You're Barking Up the Wrong Tree' and screaming out friendly greetings like, 'Get the hell out of our community!'*[45]

The logging corporations involved at Clayoquot Sound, principally MacMillan Bloedel (now incorporated into Weyerhaeuser) and Interfor, were not local to the area but operated across Canada and some parts of the US. Timber workers depend for employment and livelihood on forests, which can meet the economic needs of local communities in perpetuity when sustainably managed. Many forest workers are committed to the forest as an environment and to its long-term sustainable management. However, clear-felling loggers, like the companies that employ them, are usually interested only in swift economic returns. They are not a forest community, if such a community is defined as a group of people grounded in a particular locality with a cultural attachment to forest spaces. Clear-fellers are a roving group, bound together as employees. They move swiftly from place to place once a site is clear-felled. Their attachment to the forest is as a source of income, and, at least in their working lives, clear-fellers have no affinity to forests as spaces that provide a diversity of social, cultural and economic goods. Their professional, if not personal, interest is in extracting for themselves a financial share of the private goods that forests provide at the expense of the public goods.

Across the world, confrontations between loggers and local communities happen on a daily basis. The balance of power on the ground between communities protecting their commons and loggers seeking income varies from place to place and over time. International publicity can tilt the balance of power in favour of local people. The long-running campaign to protect Clayoquot Sound, which included a blockade against logging companies, was eventually successful.[46] However, in many other cases, especially in the tropics, the logging companies, often protected by local politicians and security forces, prevail.

and public good provision. Unlike the *laissez faire* policies of the 19th century, which let the market work where it could, neoliberals agitate for the introduction of market forces and private property rights into areas where neither has hitherto existed.[50] Governments have been central to this process by opening up new spaces where the market can operate, thus creating new investment opportunities for business. The belief that public goods can be provided by the injection of market forces into new domains has proved a powerful one, with increasing emphasis in global environmental governance on 'new environmental markets' to provide global public goods. In order to create these new markets, it has first been necessary to introduce new property rights. Two examples are relevant to forests: intellectual property rights that permit the patenting of life forms, and tradable rights to emit carbon dioxide.

Biotechnology

Many of the properties of plants that are used in commercial medicines and crops were first identified by indigenous forest peoples, who shared the knowledge openly so that it was, in effect, a public good, freely available for all. However, in hundreds of cases corporations have patented such knowledge under US and European patent law. Patents, which are recognized under international law, particularly the Agreement on Trade-Related Aspects of Intellectual Property Rights (TRIPS), have been used to establish intellectual ownership of particular traits of plants and crops. Although the patenting of an entire species is not permitted, patenting of a particular trait is allowed on the basis that the isolation of a trait is an act of creativity that is eligible for patenting, like any other discovery. Patented knowledge is a club good: it is non-rival and excludable since it can only be used on payment of a royalty to the patent holder.

The patenting of traditional knowledge by corporations is a key conflict line that runs throughout global forest politics. The medicinal properties of plants can be harnessed for the commercial manufacture of drugs. Examples from African forests include *Ancistrocladus korupensis*, traits from which can be used to treat the human immunodeficiency viruses HIV-1 and HIV-2, and *Prunus africana*, traits from which have been used to treat prostate cancer (see Box 1.3).[51] The neem tree of India (*Azadirachta indica*) has often been referred to as the 'free tree' since it has so many uses, including cures for several human and animal ailments. Since 1985 more than a dozen patents have been taken out by US and Japanese corporations who have used traits of the neem for the manufacture of emulsions, toothpaste and a pesticide.[52]

Some new varieties of crops have also been produced using biotechnology. Corporations have taken out patents on these new varieties, many of which are produced using genetic-use restriction technologies (GURTS) so that the crops produce sterile seeds. Instead of saving their own seeds from the plants for the next planting season, farmers have to buy new seeds every year. What was once a free gift of nature, easily accessible for all, now has to be purchased anew for every planting season. The behaviour of biotechnology companies is

Box 1.3 Patents and medicines: HIV/AIDS and South Africa

Pharmaceutical corporations might claim that patenting is necessary to promote the research and development of drugs, and that this provides a public good through disease eradication and promoting public health. An argument against this is that patented drugs are more expensive than generic drugs. If patenting of traditional knowledge was abolished, any properly qualified commercial group would be able to use traditional knowledge to manufacture medicines commercially. Competition would drive down prices for consumers. Some antiretroviral drugs used to treat HIV/AIDS use properties from plants found in African forests. However, pharmaceutical corporations have patented and privatized some of this knowledge – and then charged high prices to sell HIV/AIDS medicines in Africa. Over one quarter of the adult population of Botswana, South Africa and Zimbabwe is HIV-positive. Due to the severity of the AIDS epidemic, the South African government overrode patents for several AIDS drugs during the 1990s so that generic versions could be produced at lower prices. Pharmaceutical companies responded by suing the South African government to protect their patents.[53] After an international campaign, the pharmaceutical companies dropped their action against the South African government and lowered their prices, although without conceding their patent rights. Patenting is first and foremost about enabling corporations to stake a monopoly claim to knowledge and to receive royalties for the use of this knowledge. The public goods case for the patenting of life forms is low. Like forest privatization, the expansion of intellectual property rights to cover life forms allows corporations to gain access to biological resources.

logical under the neoliberal market model; it is rational economic behaviour from utility-maximizing actors. Sterile seeds maximize returns on research and development costs.[54] But this has profound consequences for small farmers; their costs escalate, and many go out of business.[55]

The patenting of the traits of biological resources, including new species created using biotechnology, amounts to a form of market enclosure, part of what has been called the second enclosure movement. Whereas the first enclosure movement involved the seizure of common land, with enclosure taking the form of physical barriers to free movement, such as fences, the second enclosure movement involves the seizure of common knowledge, with enclosure taking the form of legal barriers to free use, such as patents.[56] Indian activists have been particularly vociferous in this area. To Vandana Shiva, the patenting of biodiversity and traditional knowledge is 'biopiracy.' Adapting the concept of *res nullius*, Shiva argues that the contemporary equivalent is *bio nullius*, which treats 'biodiversity knowledge as empty of prior creativity and prior rights, and hence available for "ownership" through the claim to "invention"'.[57] In much the same way as European colonialists treated land in the empires as *res nullius*, recognizing in law only those claims made under the imperial land registration system and, in the process, ignoring prior ownership, what matters under intellectual property rights law is not who first developed the knowledge, but who first patented it.

While indigenous peoples and community groups oppose the patenting of life forms by corporations, they are less united on what the alternative should be. Two views have made themselves heard. One view is that knowledge of biological resources should not be patentable and that it should revert to being a public good that is freely available to everyone. According to this view, corporations would still be able to use such knowledge to produce medicines and crops; but they would not be able to claim exclusive rights to that knowledge or charge royalties for its use.

A second view is that communities should play the corporations at their own game and take out patents themselves. They should also claim a share of the royalties from patents registered by corporations. In principle, this claim has been recognized in the Convention on Biological Diversity, which provides for the 'equitable sharing of the benefits' that arise from utilizing the 'knowledge, innovations and practices of indigenous and local communities.'[58] An objection, however, is that this would require recognizing a system of intellectual property rights that can be seen as unethical since it recognizes value in biological resources only when these resources are commercially owned. It is also unclear among which local groups the benefits should be shared and what share each should receive. This political and legal uncertainty has impeded the attention given to traditional forest-related knowledge in international forest negotiations (see Chapters 2, 4, 5 and 9).

Tradable emission permits

Trees take up carbon dioxide from the atmosphere through photosynthesis and store it as carbon. Since carbon dioxide is a major greenhouse gas, carbon sequestration in forests provides a global public good function, namely atmospheric regulation. No one can be excluded from the benefits of the atmosphere, and no individual's consumption affects anyone else's. In many respects, the atmosphere is the classic pure public good (quadrant 4 of our taxonomy in Table 1.1). However, its use as a common sink for atmospheric pollution also gives it a *res nullius* character (quadrant 3). Most carbon emissions come from industry and transport, although a major source is forest fire burning. Deforestation accounts for approximately 25 per cent of anthropogenic emissions of carbon dioxide.[59]

The 1997 Kyoto Protocol to the Framework Convention on Climate Change agreed the need for a global system of tradable emission permits for greenhouse gases, whereby low polluting states could sell part of their agreed quota of greenhouse gas emissions to high polluting states that exceed their quota. The Kyoto scheme has yet to be implemented, although the European

Union's Emissions Trading Scheme entered into effect in January 2005.[60] Tradable emissions schemes require the creation of a new property right: the right to emit greenhouse gas. Tradable emission permits are rival: the more emission permits that are held by one country, the less are available for others. Permits are also excludable since, under the terms of a permit system, no country is able to use another country's right to pollute. Tradable emission permits could be seen as a club good, as only a 'carbon club' of rich states would be able to afford to pollute. The euphemism of a rich man's club is appealing since rich states will be able to use their financial leverage to purchase pollution permits for themselves. But the club analogy can only be carried so far: club goods are, by definition, largely non-rival within the club, whereas the consumption of tradable pollution permits would be highly rival, as many rich states would compete with each other for the right to pollute. This combination of rivalry and excludability indicates that a tradable emission permit should be considered a private good (quadrant 1 in Table 1.1). While there is some contention on how the permits for an international tradable emission permits scheme should initially be allocated to countries, the secondary allocation would take place through the market. The concept of a tradable emission permit system is thus firmly grounded in neoliberal discourse.[61]

Concluding thoughts

We began this chapter by presenting a standard taxonomy of goods according to rivalry and excludability. The subsequent discussion enables us to apply this taxonomy to forests. Table 1.5 categorizes the various goods that forests provide. The allocation of new property rights has rendered excludable some goods that were previously non-excludable, such as knowledge of certain traits of forest species. Part of the problem of deforestation is that the overharvesting of private goods depletes forest public goods, the most obvious, and most destructive, example being unsustainable timber-felling. Another dimension to deforestation is that while forests provide many private goods, other private goods can be realized by forest clearance, which frees land for alternative uses. Widespread forest clearance for cattle pasture and soya bean cultivation takes place in Brazil, which over the last decade has become one of the world's leading suppliers of genetically modified-free soya to Europe.[62] Ironically, therefore, the market demand from consumers for a product perceived as environmentally responsible – namely, soya that does not contribute to contamination of the global gene pool through human-induced genetic change – has further increased the pressure on the forests and biodiversity of the Brazilian Amazon.

Table 1.5 *Taxonomy of private and public forest goods under neoliberalism*

	Rival	Non-rival
Excludable	***Quadrant 1*** *Private goods* Timber Nuts, berries, fruits, rubber Bushmeat Private forest land Local forest commons regimes: access by outsiders is regulated Tradable emission permits	***Quadrant 2*** *Club goods* Protected forest areas: access is regulated; a toll may be charged for entry Patents on the properties of forest species, including traditional forest-related knowledge – possibly with a share of the benefits being paid to host governments and traditional knowledge holders
Non-excludable	*Res nullius (open access)* Local forest commons that are undermined through unregulated access from outsiders Protected forest areas that are undermined through unregulated access from outsiders The atmosphere as a pollution sink ***Quadrant 3***	*Public goods* Biological diversity Carbon sequestration and atmospheric regulation Pollination Soil conservation Watershed management Sites of local cultural and spiritual value Protected forest areas under effective management with free access to the public Traditional forest-related knowledge ***Quadrant 4***

On a worldwide scale, forest public goods are being steadily depleted. The destruction of forests results in biodiversity loss, both directly by killing individual species of fauna and flora and indirectly through habitat loss. It causes soil erosion, the degradation of watersheds, the loss of places of local cultural and spiritual significance, and the destruction of other public goods. This begs the question of why, at a time when forest issues are receiving unprecedented international political attention, forest public goods continue to be seriously eroded? That is the question that this book explores. It does so by examining all of the main international political processes addressing forests and forest-related issues in the decade from 1995 to early 2006.

We now begin our exploration of these processes with the Intergovernmental Panel on Forests.

2

Intergovernmental Panel on Forests

Forest management first became an international issue in 1892 when, following a proposal for an international forest science research organ at the 1890 Congress of Agriculture and Forestry in Vienna, the International Union of Forest Research Organizations (IUFRO) was established.[1] In 1945, the Food and Agriculture Organization (FAO) was created with responsibility within the United Nations system for forests, which account for approximately 4 per cent of the FAO budget. In 1985, two major international initiatives were launched with a tropical-only focus, namely the International Tropical Timber Organization (ITTO), which remains the only international commodity organization with a conservation mandate, and the ill-fated Tropical Forestry Action Plan.[2] Despite the global environmental importance of forests, it was not until 1990 that negotiations were initiated that encompassed both tropical and non-tropical forests.

These negotiations took place prior to and at the United Nations Conference on Environment and Development (UNCED) held in Rio de Janeiro in June 1992. They are widely regarded as a failure. The developed countries of the North called for a global forests convention, which the Group of 77 Developing Countries (G77), led by Malaysia and India, resisted. One of the main points of contention was the proprietorial status of forests. Some delegates from developed countries intimated that forests should be seen as a global common as all humanity derives benefits from them. This was rejected by the developing countries, with a G77 spokesperson asserting that: 'We cannot accept the application of such concepts as "global commons" or the "common heritage of mankind" with regard to the territorial domain of developing countries.'[3] The G77 were suspicious of the interest of developed governments in tropical forests and insisted, successfully, that the UNCED recognize forests as a sovereign national resource of the state. Another point of conflict centred on finance, with the G77 making it clear that if tropical countries were to agree to conserve their forests, then the developed North would have to pay compensation for the opportunity cost foregone from forest development. While the fractious UNCED forest negotiations did not produce a convention, they did agree the first pieces of soft law on forests, namely Chapter 11, 'Combating deforestation,' of Agenda 21 and the non-legally binding Forest Principles.[4] Even so, by the end of the negotiations there was considerable mistrust and wariness between developed and developing countries on forest issues.[5]

In the same year that the UNCED was held, negotiations began for a second International Tropical Timber Agreement. Like the UNCED forest negotiations, these negotiations, which eventually led to the International Tropical Timber Agreement of 1994, were fractious, with the question of a forests convention hovering in the background.[6] With suspicions continuing to linger in global forest politics, the first attempts at bridge-building between developed and developing countries were initiated. An organizing committee to establish a World Commission on Forests and Sustainable Development was formed (see Chapter 3). Meanwhile, a second unrelated process took place in support of the UN Commission on Sustainable Development (CSD).

The creation of the Intergovernmental Panel on Forests

Created in 1993, the CSD was due to consider forests for the first time in 1995. It was clear that some preparatory work was needed if the CSD's work on forests was to be successful. The initiative that broke the impasse between developed and developing countries came from two unlikely protagonists: Canada, the country which, over the last decade, has argued most strongly and persistently in favour of a forests convention; and Malaysia, which vociferously opposed a convention at the UNCED.

During 1993, John Bell, the Canadian high commissioner to Malaysia, had a series of discussions with the Malaysian government on issues of mutual interest to Canada and Malaysia, including forests. These discussions would provide the spark that lit a new period of international political cooperation on forests. The task of strengthening cooperation on forests between the two countries fell to Jag Maini, the assistant forest minister of Canada, and Amha Bin Buang of the Malaysian forestry ministry. Maini and Bin Buang organized an Intergovernmental Working Group on Forests, co-sponsored by the Canadian and Malaysian governments, which met twice in 1994. The group was not a negotiating forum and did not seek to reach consensus on forest issues.[7] Instead, it served as a trust- and confidence-building process, facilitating dialogue and generating possible options on some key issues for consideration by the CSD (see Box 2.1).

The Canadian–Malaysian initiative led to support for the creation of a CSD subsidiary body on forests. The UK argued in a paper circulated within the European Union (EU) in January 1995 that it was important to 'ensure that the international debate is carried on in a pragmatic and non-confrontational way. One answer would be the creation of an intergovernmental panel or working group to take forward important issues.'[8] The EU subsequently proposed a CSD forests panel at the CSD's Intersessional Working Group on Sectoral Issues, held in New York in late February 1995. The intention of this body was to agree recommendations on the six issues due for consideration at the CSD's

Box 2.1 Agenda of the Intergovernmental Working Group on Forests

The Intergovernmental Working Group on Forests agreed an agenda of seven issues:

- forests conservation, enhancing forest cover and the role of forests in meeting basic human needs;
- criteria and indicators for sustainable forest management;
- trade and the environment;
- approaches to mobilizing financial resources and technology transfer;
- institutional linkages;
- participation and transparency in forest management;
- comprehensive cross-sectoral integration, including land-use planning and management and the influence of policies external to the traditional forest sector.

Source: UN document E/CN.17/1995/26, Annex, 'Report of the second meeting of the Intergovernmental Working Group on Forests, held at Ottawa/Hull, Canada from 10 to 14 October 1994,' pp.4–5

1995 session, of which forests was one.[9] There was considerable support from developed countries for an intergovernmental panel on forests reporting to the CSD, particularly from Australia, Finland, Germany, Norway and the UK.[10] There was less explicit support from developing countries. However, as the consensus among the developed states was for a temporary two-year body that would incur no significant costs to developing countries, and which could have potential benefits, no opposition was expressed. Some NGOs were cautious about the need for a panel. Ian Fry of Greenpeace cautioned against a new international institution before the Rio conventions had had time to prove themselves.[11] The intersessional group concluded by recommending that the CSD establish an intergovernmental panel on forests.

The next major event on the international forest policy circuit was the 12th session of FAO's Committee on Forestry in Rome, followed immediately by a ministerial meeting on forestry, also convened by the FAO, in March 1995. In hosting the ministerial meeting, the first of its kind, the FAO appears to have been trying to carve out a political leadership role for itself. However, the decision by the CSD's Intersessional Working Group on Sectoral Issues had firmly tilted the political balance away from the FAO and towards the CSD. Some delegates gently warned the FAO not to stray too far into the political arena. Sweden, for example, argued that the FAO Forestry Department 'should concentrate in areas where it has comparative advantage,' namely information-gathering, capacity-building and assistance to national forest action programmes.[12] The Committee on Forestry agreed that the FAO should support an intergovernmental panel on forests 'if such a panel is created.'[13]

During the ministerial meeting, the EU made clear that its eventual aim was to secure a legally binding instrument on forests, in other words a forests convention.[14] But with developing countries well represented, there was no agreement on this question. The Rome Statement on Forestry issued by the forest ministers merely noted that 'concerning the discussion on the controversial idea of a legally binding instrument on forests, the way forward should be based on consensus-building in a step-by-step process.'[15] (This language mirrors almost exactly the wording of the FAO's European Commission on Forestry that two months earlier had met as part of the preparations for the Rome ministerial, agreeing on the need to 'set up a step-by-step, non-confrontational process to discuss future legally binding instruments.'[16])

The 1995 session of the CSD subsequently agreed to create the Intergovernmental Panel on Forests (IPF) as an 'open-ended' body. This meant that while the IPF's formal membership was the 53 members of the CSD, other UN members attending as observer states would have the status and privileges of full members. The agenda of the IPF reflected both the UNCED forests negotiations and the work of the Canadian–Malaysian process (see Table 2.1). NGO lobbying also contributed to the agenda in two respects.

The first example is an issue on which forest NGOs have long campaigned, namely the contribution that traditional forest-related knowledge (TFRK) can make to sustainable forest management. The attention given within the UN system to indigenous peoples and traditional knowledge is arguably the most significant campaigning achievement of forest NGOs over the last two decades. Having successfully lobbied for inclusion of language on indigenous knowledge in the Forest Principles, Chapter 11 of Agenda 21 and the Convention on Biological Diversity (CBD),[17] NGOs argued for TFRK to be placed on the IPF agenda.[18] They succeeded because, as Kristin Rosendal points out, while TFRK has been promoted primarily by NGOs, as an issue it ranks highly with the G77, which asserts the importance of protecting the intellectual property rights of tropical forest countries over their genetic inheritance. Brazil and Ecuador have been especially active on this issue.[19] The G77 countries see access to TFRK as bargaining leverage for securing increased financial returns from their forest resources.

Second, NGOs have long campaigned for action to address the causes of deforestation.[20] Shortly before the IPF was created, two NGO networks – the Japan Tropical Forest Action Network (JATAN) and the Asia Pacific Peoples' Environmental Network (APEN) – stressed to the FAO that in international forest policy discussions there 'seems to be almost no debate on the underlying causes of massive forest ecosystem destruction' and that the trend was to 'ignore the underlying factors and obstacles to healthy forest management.'[21] NGO lobbying on this issue finally bore fruit when the CSD included it in the IPF's agenda. As with TFRK, the inclusion of this issue cannot be attributed to a single NGO source. Instead, it was the result of persistent campaigning by numerous forest NGOs over several years.

Table 2.1 *Intergovernmental Panel on Forests: Abbreviated work programme*

	Intergovernmental Panel on Forests agenda item	***Lead agency***
Programme area I: Implementation of UNCED forest decisions		
I.a	Progress through national forest and land-use programmes	FAO
I.b	Underlying causes of deforestation and forest degradation	UNEP
I.c	Traditional forest-related knowledge	CBD
I.d	Fragile ecosystems affected by desertification and drought	FAO
I.e	Impact of airborne pollution on forests	FAO
I.f	Needs and requirements of countries with low forest cover	UNEP
Programme area II: Financial assistance and technology transfer		
II.a	Financial assistance	UNDP
II.b	Technology transfer	FAO
Programme area III: Scientific research, forest assessment and criteria and indicators		
III.a	Assessment of multiple benefits of all types of forests	FAO
III.b	Forest research	CIFOR
III.c	Methodologies for the proper valuation of the multiple benefits of forests	World Bank
III.d	Criteria and indicators for sustainable forest management	FAO
Programme area IV		
IV	Trade and environment in relation to forests, including voluntary certification and labelling schemes	ITTO
Programme area V		
V	International organizations and multilateral institutions and instruments, including the identification of any gaps, areas requiring enhancement and areas of duplication	UN-DESA

CBD Convention on Biological Diversity
CIFOR Center for International Forestry Research
FAO Food and Agriculture Organization
ITTO International Tropical Timber Organization
UNCED United Nations Conference on Environment and Development
UNDP United Nations Development Programme
UN-DESA United Nations Department for Economic and Social Affairs
UNEP United Nations Environment Programme

Source: ITFF (1999) *The Interagency Task Force on Forests*, Rome: FAO

The Panel met on four occasions: September 1995 (New York); March 1996 (Geneva); September 1996 (Geneva); and February 1997 (New York). Sir Martin Holdgate (UK) was elected as the 'Northern' co-chair of the IPF for all four sessions.[22] For the first and second sessions, the 'Southern' co-chair was N. R. Krishnan of India. He was succeeded for the third and fourth sessions by Manuel Rodríguez, the former forest minister of Colombia.[23] Jag Maini was appointed as the head and coordinator of the IPF secretariat.[24] For the next seven years, he would be the most senior international civil servant working on the UN's forest policy dialogue.

The working modalities of the Intergovernmental Panel on Forests

The Panel's first session was purely organizational, with substantive business carried out in the remaining three sessions. It was clear from the outset that the IPF would not be able to deal with all the issues agreed at the CSD in its three substantive sessions, which totalled just six weeks. At the first session a number of governments proposed that intersessional initiatives be held between the formal IPF sessions.[25] These would not constitute a part of the IPF programme in any formal sense, but would feed into the process by, for example, providing draft texts for negotiation. Co-chair Holdgate noted that some developing country delegations were suspicious that developed countries were trying to use intersessional initiatives 'to take over the programme of the Panel and dictate its results.' He concluded that there was 'clear evidence that many delegations are more concerned to limit the analysis of forest issues than to stimulate it.'[26] This suspicion reflected the delicate political consensus on forests at this time. Despite this fragility, it was recognized that the informal nature of the Canadian–Malaysian initiative had contributed to confidence-building, and that similar initiatives might broker consensus on some technically complex and politically contentious issues during the IPF process. The developing countries thus agreed to intersessional initiatives, but insisted that they should be open to all countries (see Table 2.2).

It was agreed that an interagency group would be created, comprising international organizations with a forest-related mandate to support the Panel. The idea, which was the brainchild of Joké Waller-Hunter, a member of the UN secretariat,[27] received support from several delegations, including Finland, New Zealand and the US.[28] Subsequently, the Interagency Task Force on Forests (ITFF), comprising eight international organizations, was created (see Table 2.3). ITFF member organizations assumed responsibility as lead agencies for the IPF's agenda (see Table 2.1). They produced various analytical documents for the Panel. NGOs supported the creation of the ITFF, although they pressed unsuccessfully for it to be opened to participation from civil society and indigenous groups.

The ITFF was an innovation within the UN system. Although it is commonplace for international institutions to cooperate, the ITFF was the first

Table 2.2 *Intersessional initiatives in support of the Intergovernmental Panel on Forests*

Name of initiative	*Type*	*Venue*	*Organizers/ sponsors*
Certification and labelling of products from sustainably managed forests	International conference	Brisbane	Australia
Financial mechanisms and sources of finance for sustainable forestry	International workshop	Pretoria	Denmark, South Africa, United Nations Development Programme
Implementing the Forest Principles: promotion of national forest and land-use programmes (Feldafing Initiative)	Expert consultation	Feldafing, Germany	Germany
Rehabilitation of degraded forest ecosystems	Expert meetings	Lisbon	Cape Verde, Portugal, Senegal, European Community, FAO
Overview of international organizations, institutions and instruments related to forests	Expert meeting	Geneva	Switzerland, Peru
Trade, labelling of timber and certification of sustainable forest management	Expert meeting	Bonn	Germany, Indonesia
Criteria and indicators for sustainable forest management	Intergovernmental seminar	Helsinki	Finland
Long-term trends and prospects in wood supply and demand for wood, and implications for sustainable forest management	Study		Norway

Table 2.2 *Intersessional initiatives in support of the Intergovernmental Panel on Forests (Continued)*

Name of initiative	*Type*	*Venue*	*Organizers/ sponsors*
Sustainable forestry and land use: the process of consensus-building	Expert meeting	Stockholm	Sweden, Uganda
Integrated application of sustainable forest management practices	International workshop	Kochi, Japan	Japan, Canada, Malaysia, Mexico, FAO, ITTO
Conservation and sustainable management of forests	International meeting of indigenous and other forest-dependent peoples	Leticia, Colombia	Colombia, Denmark, International Alliance of the Indigenous–Tribal Peoples of the Tropical Forest, Indigenous Council for Amazon Basin

Source: UN forests website, www.un.org/esa/forests/gov-ipf.html (accessed 15 March 2004)

time that an interagency body was formed in support of an intergovernmental organization. This reflected both the multifaceted complexity of forest use as an international issue and the extensive range of international organizations with a mandate on forests. Intended as a 'high-level, informal, flexible and effective mechanism.'[29] the ITFF brought together some very different international institutions, with different organizational cultures. It functioned reasonably effectively, given that it had no independent budget or project management role. Similar bodies have since been established on disaster reduction, the role of sport in development and peace, and gender and trade.[30]

Prior to the Panel's first session, the German delegation suggested that the IPF 'should particularly aim at specific recommendations for activities ... which should be as brief and comprehensive as possible.'[31] The IPF adopted this suggestion and produced a series of *proposals for action*, namely suggestions and policy recommendations for consideration by governments and other actors. The first drafts of the proposals came from a range of sources, including the IPF secretariat, ITFF member organizations and intersessional initiatives. Some NGOs, which were granted the same access rights at the IPF as they enjoy at the CSD,[32] also submitted draft proposals.

Table 2.3 *Member organizations of the Interagency Task Force on Forests*

Name	Type of organization
Food and Agriculture Organization*	UN specialized agency
International Tropical Timber Organization	International treaty
Convention on Biological Diversity	International treaty
United Nations Development Programme	UN programme
United Nations Environment Programme	UN programme
World Bank	International financial institution
UN Department of Economics and Social Affairs	Department of the UN secretariat
Center for International Forestry Research	International research centre

Note: * Chaired the ITFF in its capacity as the UN task manager on forests.

The following sections analyse the IPF's work on programme areas I to IV, the bulk of which was accomplished during the second and third sessions. Attention then turns to the negotiations on the desirability of a forests convention, which dominated deliberations under programme area V during the Panel's fourth session in 1997.

Implementation of forest-related decisions of the United Nations Conference on Environment and Development

Programme area I related to the implementation of the UNCED's Forest Principles and Chapter 11 of Agenda 21. The IPF helped to establish the concept of national forest and land-use programmes (usually abbreviated to national forest programmes, or NFPs) as the commonly accepted national-level policy vehicle for implementing sustainable forest management. The legitimacy of national programmes had suffered during the 1980s when national forestry action programmes (NFAPs) created under the auspices of the Tropical Forestry Action Plan (TFAP, renamed in 1990 the Tropical Forests Action Programme) were criticized for being too dependent on the priorities of donors, ignoring the needs of indigenous peoples, contributing, in some cases, to further deforestation and ignoring the underlying causes of deforestation. The NFP concept developed by the IPF took into account these criticisms[33] and stressed that NFPs should be holistic, intersectoral and iterative programmes that recognize and respect the customary and traditional rights of indigenous people, local communities and other actors.[34] Table 2.4 details the conceptual differences between the different types of national programme elaborated by the TFAP and the IPF.

On the underlying causes of deforestation, many developing countries reiterated points they had made during the UNCED negotiations by asserting poverty, excessive consumption patterns in the North and high levels of external indebtedness as causes. Despite some resistance from the EU, all of these points survived the negotiations and appear in the final report. The G77 reasserted the principle of common but differentiated responsibilities, a conceptual expression that asserts that the rich developed countries bear a disproportionate share of the blame for global environmental problems. This principle was negotiated into the Framework Convention on Climate Change, although its operational meaning is unclear. It was blocked by the developed countries in the negotiations for the Forest Principles and Chapter 11 of Agenda 21. At the IPF, the G77 had more success, and the principle survived the final round of negotiations.[35]

It was noted in discussion that while deforestation and forest degradation pose serious problems in some regions, not all changes in forest cover are necessarily harmful. Much depends upon national circumstances and development plans.[36] The IPF produced and recommended the adoption of a 'diagnostic framework' as a conceptual tool to enable actors to identify the relationship between the direct and underlying causes of deforestation (see Table 2.5). It was recommended that the framework should be developed voluntarily and that its use should not be a basis for aid conditionality. NGOs were unsatisfied with the IPF's work on this subject, and after the IPF was replaced by the Intergovernmental Forum on Forests, NGOs held an intersessional initiative on the underlying causes of deforestation (see Chapter 4).

Table 2.4 *Conceptual comparison between national forestry action programmes (TFAP) and national forest and land-use programmes (IPF)*

National forestry action programme (TFAP)	National forest and land-use programme (IPF)
Main objective	
Slow the rate of deforestation in developing countries	Enhance sustainable forest management in all countries
Planning ideas	
Technocratic	Deliberative and consensus oriented
No iterative long-term planning	Iterative long-term planning
Participatory in implementation only	Participatory in both formulation and implementation
Intersectoral interpreted solely as the agriculture–forestry interface	Intersectoral between all sectors

Source: Pülzl, Helga and Rametsteiner, Ewald (2002) 'Grounding international modes of governance into National Forest Programmes,' *Forest Policy and Economics*, Vol. 4, No. 4, pp.259–279

Table 2.5 *Diagnostic framework: Relationships between selected direct and underlying causes of deforestation and forest degradation*

	Underlying causes							
Direct causes	***1***	***2***	***3***	***4***	***5***	***6***	***7***	***8***
Replacement:								
By commercial plantations	X					X	X	
Planned agricultural expansion	X	X				X	X	
Pasture expansion	X	X				X		
Spontaneous colonization		X	X	X		X	X	X
New infrastructure						X		
Shifting agriculture			X	X				X
Modification:								
Timber harvesting damage	X		X		X		X	
Overgrazing			X		X			
Overcutting for fuel			X		X			
Excessive burning				X	X			
Pests or diseases					X			
Industrial pollution					X		X	

The column headings for underlying causes are:
1 Economic and market distortions
2 Policy distortions, particularly inducements for unsustainable exploitation and land speculation
3 Insecurity of tenure or lack of clear property rights
4 Lack of livelihood opportunities
5 Government failures or deficiencies in intervention or enforcement
6 Infrastructural, industrial or communications developments
7 New technologies
8 Population pressures causing land hunger

Source: UN document E/CN.17/IPF/1996/2, 'Intergovernmental Panel on Forests, Programme Element I.2, underlying causes of deforestation and forest degradation,' 13 February 1996, Table 4, p.22

Indigenous peoples are now recognized as distinct actors in environmental politics. For example, Principle 22 of the Rio Declaration on Environment and Development states that 'Indigenous people and their communities and other local communities have a vital role in environmental management and development because of their knowledge and traditional practices.'[37] In support of the IPF, indigenous peoples' groups, in cooperation with the governments of Colombia and Denmark, organized an International Meeting of Indigenous and Other Forest-dependent Peoples on the Management, Conservation and Sustainable Development of all Types of Forests in Leticia, Colombia.

This resulted in the Leticia Declaration, which reiterated many of the positions that indigenous peoples groups had advocated throughout the IPF process, including the demand that use of traditional forest-related knowledge 'should not be made without the prior informed consent of the Peoples concerned.'[38] The principle of prior informed consent is one on which indigenous peoples' groups have long campaigned (see Chapter 9). The final report of the IPF subsequently stated that:

> *Governments and others who wish to use TFRK should acknowledge, however, that it cannot be taken from people, especially indigenous people, forest owners, forest dwellers and local communities, without their prior informed consent.*[39]

The principle was thus adopted by the IPF, although its formulation was weaker than that of the Leticia Declaration. The Leticia emphasis on 'Peoples' (upper case, plural) has been lost. Instead, the IPF refers to 'people' (lower case, singular), which denotes a lower status for indigenous peoples in international law (see Box 5.2 in Chapter 5). Furthermore, the emphasis on indigenous peoples has been broadened, with the IPF stressing that:

> *TFRK should be broadly defined to include not only knowledge of forest resources, but also knowledge of other issues that are considered relevant by countries based on their individual circumstances.*[40]

This emphasis can be interpreted to admit agencies that have promoted forest loss, such as industrial timber companies. This broadening of the concept to include the knowledge of actors from outside the forests thus weakened the original proposal.[41] Much discussion centred on who should benefit from TFRK. The Panel eventually adopted a formulation agreed in the Convention on Biological Diversity, namely that the 'effective protection of TFRK requires the fair and equitable sharing of benefits among all interested parties.'[42] The work of the CBD on access and benefit-sharing has had a direct bearing on the work of UN forests fora (see Chapters 4, 5 and 9).

The IPF noted that airborne pollution affects forest health in many parts of the world and recommended that preventative (as opposed to adaptive) policies should be adopted to tackle this problem. Beyond this, the Panel's conclusions and proposals on pollution were weak, emphasizing the need to 'continue monitoring and evaluating the impact of airborne pollution on forest health' and for international cooperation.[43]

Some delegates used the negotiations on the needs of countries with low forest cover to press their views on plantations. What can be seen as a 'plantation coalition' took shape with Australia, Chile, China, the EU, New Zealand, South Africa and Uganda all noting that plantations can relieve the pressure on national forests and conserve biological diversity.[44] The Panel's outputs on countries with low forest cover were also weak, although its successor, the Intergovernmental Forum on Forests (IFF), had more success in grappling with this issue (see Chapter 4).

Financial assistance and technology transfer

The twin issues of financial aid and technology transfer had dominated the UNCED forest negotiations. Recognizing the contentiousness of finance and technology, and how they could deflect attention from other issues, some delegations, such as Norway, pushed for them to be considered only late on in the IPF process.[45] However, they were raised continually by developing countries throughout the negotiations.

The developed countries had shifted their strategy on these issues since the UNCED. From the start of the IPF process, the EU, the US and Japan sought to broaden the range of financial issues under consideration. Unwilling to be placed continually in the position of refusing G77 demands for new official development assistance (ODA) commitments, the developed countries emphasized that financial and technological assistance could be raised from many different sources, in particular, from private capital flows, which, it was noted, are increasing at a faster rate than public funding. Japan emphasized the importance of a predictable political climate and investor-friendly markets for private sector investment. The developed countries thus tried to appear constructive on finance, while refusing to concede any new transfers. They emphasized that the developing countries should make more efficient use of existing mechanisms and sources. The response of the G77 was that the financial question was not one of efficiency, but of sufficiency, with the G77 claiming that ODA had declined since the 1990s.

This was rejected by the EU and the US. While statistically it is difficult to arrive at unambiguous figures for ODA, the data compiled by the UN secretariat support the G77 position. At the CSD's fifth session in 1997, where the IPF's report was tabled, a report from a United Nations High Level Advisory Board on Sustainable Development noted that:

> *It is disturbing that official development assistance (ODA) has fallen in real terms from over [US]$60 billion per year (1994 $) in the early 1990s to about [US]$55 billion in 1995. Despite the concomitant increases in private capital flows, which have been concentrated in a relatively few countries, ODA remains an essential element for sustainable development.*[46]

Some G77 countries, including the Philippines,[47] argued that private sector investors are not always motivated by environmental considerations. Colombia stated that private sector resources would not pay for the environmental services that forests provide.[48] The G77 insisted that there remains an important role for international public finance. The IPF was not the only forum where this debate took place. At the CSD session of 1996, India expressed concern at the emphasis given to the private sector during negotiations on environmentally sound technologies.[49] Tanzania, speaking for the G77, stated at the IPF that:

> *... environmentally sound technologies should be made available to developing countries at affordable terms and without the stringency of intellectual property rights. It is regretted that at present there is no internationally agreed mechanism*

> *for the transfer of technology from the developed countries apart from the commercial exchanges mainly through the private sector, which most developing countries ... cannot afford.*[50]

The private sector also featured in another area of the IPF's work, namely the possibility of a code of conduct to regulate private sector forest businesses. The negotiating positions on this issue can be summarized crudely thus: NGOs and many of the G77 countries favoured a binding code of conduct, while the developed countries, unwilling to increase the regulatory burden on forest industries, favoured only a voluntary code.

The question of a generic code of conduct to regulate transnational corporations has been on the international agenda since the 1970s, when the United Nations Centre for Transnational Corporations (UNCTC) was given the task of drafting a code.[51] NGOs lobbied hard for a code to be agreed at the UNCED, and they received some support from the G77 and China. However, the idea of a binding code, which runs counter to the free market and anti-regulatory ethos of neoliberalism, was blocked at the UNCED by developed countries. The US, responding to pressure from a newly formed corporate front group, the Business Council for Sustainable Development, was strongly opposed.[52] One year after the UNCED, the UNCTC was closed and its activities downgraded in the UN system with transfer to the United Nations Conference on Trade and Development (UNCTAD) in Geneva,[53] a move that was widely seen as a triumph for the neoliberal worldview and a defeat for those who favoured international regulation of business.

However, the defeat of a UN code has not, and as those who have resisted calls for corporate regulation would doubtless wish, led to a closure of the debate. It has, however, shifted the political space within which the debate is played out. First, some NGOs and civil society groups have been prepared to work with corporations in developing codes of conduct, such as the Coalition of Environmentally Responsible Economies (CERES) Principles. The discourse by which this voluntaristic approach to business regulation is known is corporate social responsibility, or CSR (see Chapter 10). Second, many NGOs have worked with forest-based businesses to promote market-based certification schemes, such as the Forest Stewardship Council (FSC) (see Chapter 6). Both CSR and market-based forest certification schemes are voluntary forms of governance. They have arisen due to the reluctance of many governments to regulate private sector activities that degrade the environment.

At the IPF, NGOs cooperated to press for a code of conduct to regulate forest businesses. The World Wide Fund for Nature (WWF), noting that transnational corporations have 'enormous economic power' and are involved 'in all aspects of production, from extracting the primary resource to high street retailing,' called for the IPF to develop 'realistic codes of practice.'[54] The World Conservation Union (IUCN) hosted an IPF workshop, attended by government delegates and the recently created World Business Council for Sustainable Development (WBCSD).[55] The IUCN recommended that:

> *The IPF should encourage or facilitate the development of a corporate code of conduct which would contain voluntary guidelines or principles for corporations operating timber concessions. This code would be developed by industry groups and a representative set of stakeholders.*[56]

With some G77 countries supporting the proposal, language on a code of conduct subsequently appeared in the draft text under negotiation by the Panel. However, by the end of the Panel's third session, all language on a code had been placed in square brackets, indicating disagreement between states (see Box 2.2). The main point of contention was whether any code of conduct should be voluntary:

> *[Formulation of [voluntary] codes of conduct [in cooperation with or] by the private sector should be further examined.] [In this context, voluntary codes of sustainable forest management to guide investments, concessions and forest management should be developed in cooperation with the private sector and all other major groups, including indigenous people and local communities.]*[57]

Bracketed text also remained on other issues. To prevent the IPF's fourth session from being bogged down in negotiating fine detail, co-chairs Holdgate and Rodríguez redrafted the entire text, aiming to leave language with which all delegates could, hopefully, agree.[59] The co-chairs' draft was then used as a basis for negotiation at the fourth session. This new draft contained only six sets of square brackets, including '[voluntary] codes that will guide investments, concessions and sustainable forest management.'[60] At the fourth session, delegates removed the brackets around 'voluntary' after the US insisted that it could only agree to mention of a code if the word was retained. By placing the primary responsibility for developing codes of conduct on the private sector, the agreed proposal for action has a distinct neoliberal flavour. The proposal:

> *Urged all countries, within their respective legal frameworks, to encourage efforts by the private sector to formulate, in consultation with interested parties, and implement voluntary codes of conduct aimed at promoting sustainable forest management through private sector actions, including through management practices, technology transfer, education and investment.*[61]

The use of the word 'urged' indicates that the proposal is among the strongest that the UN system can deliver,[62] although this was countered by the word 'voluntary,' which deprived the proposal of any substantive content. The proposal had something for everyone: developed countries were satisfied that any codes would be voluntary and formulated by the private sector, with other actors merely consulted; developing countries were pleased at the mention of technology transfer; while NGOs were pleased to have revived the idea of corporate regulation after the UNCED defeat on this issue. However, the proposal was a non-binding exhortation to take voluntary action; it proved to be another false dawn in the struggle for corporate accountability.

Box 2.2 Text negotiation in the United Nations system

Outside the Security Council and, for some issues, the General Assembly, votes are unusual in UN fora, and the aim of negotiation is to agree text consensually. Under consensual negotiation procedures, it takes all to say yes and only one to say no. Delegations signal their disagreement by inserting square brackets around words, articles or paragraphs with which they disagree. Generally speaking, the more square brackets that exist in a document, the greater the disagreement between delegates. Where disagreement is especially acute, the draft text will be littered with square brackets, proposals and counterproposals.

Where the time remaining for negotiation is limited, recourse may be made to a chair's draft. The intention is to provide a draft that all delegates can accept as a basis, if not for agreement, then at least for further negotiation. When preparing a new draft a chair or co-chair will work through the text, removing contentious words and formulations and replacing them with new text that is intended to faithfully and impartially reflect the interests of all actors. They may combine different proposals, standardize wording (for example, where there are different formulations for the same concept or idea), bring together different references to a concept or idea in one paragraph, and provide clarity by deleting redundant words. It is a job that calls for skills in diplomacy and draftsmanship, the respect of all (or at least the majority) of delegates, the ability to consider different angles and competing points of view, and a thorough grasp of the issues.

A chair's draft is necessary when negotiations are stalling and an intervention is needed to broker agreement, and when there is insufficient time to agree text by negotiation. Both of these factors applied after the IPF's third session. The draft prepared by co-chairs Sir Martin Holdgate and Manuel Rodríguez proved to be decisive, with almost all of the text surviving the final round of negotiations and appearing in the Intergovernmental Panel on Forests' final report.[58]

The subject of debt-for-nature swaps was raised several times in the negotiations. This idea had emerged during the 1980s when NGOs including WWF and Conservation International brokered deals between tropical forest governments and the banks that held their debt, the result being relief from some debt in return for a commitment to conserve an agreed area of tropical forest. Brazil, Colombia and Zimbabwe stated at the IPF that, while they agreed with the idea of debt-for-nature swaps, they opposed debt-relief packages that included the imposition of policy conditionalities – so-called 'debt-for-policy swaps.'[63]

Developing countries achieved no progress on finance and technology transfer. Their repeated calls for a global forest fund were ignored or rejected by the developed countries, as were calls for the transfer of environmentally sound technologies. Developed country delegations appended 'as mutually agreed' to claims from the G77 for technology transfer 'on concessional and preferential terms,' thus ruling out mandatory transfers outside the market. In this respect, the negotiations mirrored the UNCED forest negotiations. However, the failure of the G77 to achieve progress on technology transfer at the IPF was not entirely surprising: not only was the early 1990s a period when hopes for a code of conduct to regulate transnational corporations finally evaporated, it was also a time when efforts to negotiate a code of conduct on technology transfer were thwarted (see Box 2.3).

Box 2.3 The proposed code of conduct on technology transfer

In 1977, United Nations General Assembly Resolution 32/88 established the United Nations Conference on an International Code of Conduct on the Transfer of Technology under the auspices of the UNCTAD. From 1978 to 1985, six sessions of the conference were held. There were no negotiations for the next six years. In 1991, UNCTAD debated the draft code with the intention of bringing a fresh approach to bear on the negotiations. It was agreed that the views of governments on the best way to proceed should be solicited. Only ten governments replied, including the US, which stated that:

> *The Government of the United States is of the opinion that there is no basis to believe that there is a 'convergence of views' on the outstanding issues in the draft code of conduct, nor is there likely to be. Therefore, the Government opposes resumption of negotiations on the draft code of conduct on the transfer of technology.*

Since then, there has been no further progress towards the completion of the code.

Source: UN document A/47/636, 'International code of conduct on the transfer of technology,' 6 November 1992

Scientific research and forest assessment

Before the IPF's second session, it transpired that much of the UN analytical document for the Panel's work on 'methodologies for the proper valuation of the multiple benefits of forests,' which had been prepared by the FAO, had been plagiarized from a report prepared by the International Institute for Environment and Development (IIED) for the UK Overseas Development Administration. This came to light when, ironically, the Overseas Development

Administration asked the IIED to comment on the document as part of the UK delegation's preparations for the IPF's second session.[64] While it says much about the working practices of the FAO staff member or consultant who prepared the document that they adopted, without acknowledgement, material from another source, of deeper interest is that this reveals the limits of genuine intellectual debate on environmental valuation. It provides evidence of a knowledge-based network that shares agreement on the methodologies for environmental valuation in general, and forest valuation in particular.

This network is seen by Marie-Claude Smouts as an epistemic community. Peter Haas has defined epistemic communities as 'transnational networks of knowledge-based communities that are both politically empowered through their claims to exercise authoritative knowledge and motivated by shared causal and principled beliefs.'[65] The concept of epistemic communities in international environmental politics is often applied to science professionals; for example, international cooperation on ozone depletion and climate change was catalysed by a transnational network of scientists who identified the causes of the problem and who had influence over policy-makers. There is no equivalent scientific-based epistemic community on the causes of deforestation; analysis of numerous case studies has revealed no clear causal patterns to deforestation over time and space.[66]

However, Smouts argues that an epistemic community on forest valuation has emerged that has its origins in economics institutes in the US and Europe. The discourse of this community is grounded in neoclassical economics. In the neoclassical view, environmental problems arise from the failure of markets to properly value environmental goods and services. Hence, so it is argued, if a resource is to be conserved, then it should be given economic value. Forest resources without such value will suffer degradation and conversion to other land uses. Given this, the argument continues, the solution is to devise methodologies that provide such values, which can then be internalized into market prices.

The FAO/IIED analytical document on valuation faithfully reflected this belief system. It included the shadow pricing method, whereby prices for environmental goods and services are imputed by an analyst or analysts; hedonic pricing, whereby surrogate markets are used to arrive at prices for environmental goods and services; and contingent valuation, whereby prices for environmental goods and services are gauged by, for example, conducting a survey of a group of residents who are asked hypothetical questions about how much they would be willing to pay for, say, the conservation of a given area of forest.

The authors of the document note some disadvantages of these approaches, although the emphasis is more on the need for caution when using the methodologies in certain contexts, rather than a foundational critique. For example, for contingent valuation it is noted that the method is 'sensitive to numerous sources of bias in survey design and implementation.'[67] This assumes that contingent valuation is theoretically sound in principle, provided that due care is taken in designing the survey and in implementing the results.

However, a more fundamental critique is that contingent valuation presumes that people have money. Those who do not will not register in a contingent valuation survey – however well designed – since they will have a low (or no) willingness to pay. The methodology is thus biased against the poor.

There was virtually no support among IPF delegates for language on stronger, or compulsory, valuation techniques. Belarus, Ecuador, the G77, New Zealand and the US stated that valuation should be compatible with national accounts and carried out in accordance with national priorities.[68] These emphases on national accounts and priorities can be read as pre-emptive moves against possible international standards for accounting that would include the economic costs of environmental degradation. The US was particularly vocal on this subject and stated that valuation should be a 'neutral tool' that should not be used as 'a means of advocacy.'[69] The US interventions on this subject exposed the higher prioritization by the US of economic values over environmental values. The international consensus on neoclassical environmental valuation methodologies was apparent when few IPF delegates expressed reservations with the FAO/IIED analytical document, with some delegates seemingly fearing that more rigorous methodologies would increase forest industry operating costs.

Trade and environment in relation to forest products and services

Trade and environment was one of the programme areas where the IPF made least progress. Lack of information on the international and domestic trade in timber and non-timber forest products was used as an excuse to avoid substantive deliberations. A proposal tabled at the IPF's third session for the 'Trade and Environment Committee of the World Trade Organization to continue its work to ensure that trade and environment are mutually supportive, including in the area of forest products and services' was not adopted.[70] There were two reasons for this. First, the proposal fell prey to G77 suspicions that environmental considerations might be used to introduce trade restrictions. Second, the question of whether or not a forests convention should be negotiated was unresolved when the Panel debated the trade–environment relationship, and some pro-convention states were unwilling to refer to the World Trade Organization (WTO) an issue that could come under the auspices of a forests convention.

Table 2.6 presents a tabulated summary of the proposals for action agreed by the IPF.

The forests convention debate

At the IPF's fourth session, delegations turned their attention for the first time since Rio to the question of whether negotiations should be launched

Table 2.6 *The proposals for action of the Intergovernmental Panel on Forests*

IPF programme area		Proposals for action at the national forest and land-use programme level	Proposals for action requiring action only at the international level
I Implementation of forest-related decisions of the UNCED			
I.a	Progress through national forest and land-use programmes	6	1
I.b	Underlying causes of deforestation and forest degradation	3	4
I.c	Traditional forest-related knowledge	4	2
I.d	Fragile ecosystems affected by desertification and drought	2	1
I.e	Impact of airborne pollution on forests	3	1
I.f	Needs and requirements of developing and other countries with low forest cover	4	1
II International cooperation in financial assistance and technology transfer			
II.a	Financial assistance	5	5
II.b	Technology transfer, capacity-building and information	4	3
III Scientific research, forest assessment and criteria and indicators for sustainable forest management			
III.a	Assessment of the multiple benefits of all types of forests	2	4
III.b	Forest research	3	3
III.c	Methodologies for the proper valuation of the multiple benefits of forests	1	1
III.d	Criteria and indicators for sustainable forest management	1	4
IV Trade and environment in relation to forest products and services		10	3
V International organizations and multilateral institutions and instruments		2	3
Totals		**50**	**36**

Note: The final report of the IPF has 149 paragraphs, most of which contained proposals for action. These paragraphs vary in terms of content and there is considerable overlap between them. The following source screened the original proposals to yield 86 discrete proposals.

Source: FAO and UNDP (1999) *Practitioner's Guide to the Implementation of the IPF Proposals for Action*, Eschborn: Gesellschaft für Technische Zusammenarbeit (GTZ)

for a global forests convention.[71] Before we examine the IPF deliberations on this subject, we first consider some of the cases for and against a forests convention.[72]

An argument in favour is that a forests convention would strengthen existing multilateral environmental agreements. Since forests play a role in climate regulation, a forests convention would strengthen the Framework Convention on Climate Change. Similarly, as most of the world's biodiversity is found in tropical forests, a forests convention would support the Convention on Biological Diversity. It would thus provide the third component of a triumvirate of mutually reinforcing environmental regimes. Second, with the international legal provisions on forests scattered among several international instruments, the result is a fragmented and opaque coverage of forests in international law, resulting in an *ad hoc* regulatory environment and political uncertainties. These problems would be eliminated, so it is claimed, if all forest-related provisions were rationalized under a single legal cover. Third, a convention would demonstrate high-level political commitment to tackling deforestation and would provide strategic and focused leadership.

Two more cynical explanations can be forwarded that have their roots in domestic politics. First, a government department with responsibility for forests may advocate a convention in order to raise the status of forest management as a national issue, or even to raise the domestic profile of the department itself: bureaucracies with responsibility for an international instrument tend to have a higher status than those that do not. Second, a forests convention could be presented to domestic constituencies as evidence that the issue is being addressed internationally. Assuming that this is the sole reason, a multilateral environmental convention would, presumably, contain only general principles and would avoid specific obligations.

However, there are some persuasive arguments against a convention. It can be argued that a convention would lead to political complications and 'turf wars' with other legal instruments. It could be unclear, for example, whether the Convention on Biological Diversity or a forests convention would be the lead organization for forest biodiversity and protected areas. Far from providing a more rationalized and harmonized treatment of forests in international law, a convention could, by adding another layer of international regulation, lead to further legal uncertainties and complications. Hence it is often claimed that with most of the world's biodiversity being found in tropical forests, the most pragmatic choice for an instrument on forests is not a convention, but a forest protocol to the Convention on Biological Diversity.

In addition to these generalized arguments, there are arguments that individual government delegations may make based on their interests. In international negotiations, no delegation is obliged to say how it views its interests. The onus is on each delegation to attempt to judge the interests of other delegations from the statements they make and the positions they adopt. But this is a process fraught with difficulties. First, a negotiator may be bluffing. Second, if a negotiator merely presents the interests of her own delegation, she is not providing any good reason why other delegations should agree with her.

Skilled negotiators will, therefore, seek to garner political support by framing their statements to appeal to the interests of other actors, both opponents and potential allies who may be wavering, in order to build a winning coalition.

Despite the complications of gauging the interests of individual delegations at forest negotiations, we can posit five propositions on why governments may favour a forests convention and five propositions on why they may be opposed. These propositions are grouped under forest management standards, finance and technology transfer, sovereignty, forest industry and intergenerational equity (see Table 2.7).

During the UNCED forest negotiations, developing countries, represented by the G77, strongly opposed a convention. Developed countries – G7 and the European Community – argued in favour. At the IPF, the EU continued to support a convention[73] and worked behind the scenes to this end. Co-chair Rodríguez was lobbied to support a convention by EU delegations, especially Germany and the Netherlands. Some EU delegates intimated to Rodríguez that new financial resources would be available if a convention was agreed, but that developing countries should not expect this without a convention. During the third IPF session, some EU delegates treated Rodríguez to dinner at the Geneva Hilton, where this message was conveyed particularly strongly.[74]

The US, which had supported a convention during the UNCED negotiations, changed position to oppose such an instrument at the IPF. The influence of the corporate sector in the US explains this change. Whereas the first term of the Clinton administration saw a policy shift in favour of environmental protection, with the administration signing the CBD, Clinton's second term witnessed business reassert its opposition to environmental regulation. US business, including the influential American Forest and Paper Association, was supported by a Republican-dominated Congress. The Clinton administration was criticized by the American pharmaceutical industry for signing the CBD, which can be interpreted as a restriction on corporate access to tropical forest biodiversity.[75] At the same time, the US energy sector was failing to make progress on stabilizing carbon dioxide emissions as called for under the Framework Convention on Climate Change. Overall, US domestic politics had moved against additional international environmental commitments.

The Canadian forest industry, unlike its US counterpart, supported a convention. The Canadian Pulp and Paper Association intervened at the Panel to say that an element of a convention could be 'the promotion of worldwide trade in forest products.'[76] Canadian support for a convention needs to be understood within the context of the long-running US–Canada softwood lumber dispute. The US has claimed that Canada unfairly subsidizes its timber exports to the US through stumpage subsidies, thus giving its industry an unfair trade advantage over the US. One of the elements Canada would be looking for in a convention would be international rules on tariffs and trade barriers against timber exports. (After the 1996 US–Canada Softwood Lumber Agreement expired in 2001, the US raised tariffs against softwood lumber exports from Canada to the US.)[77] Russia was another temperate forest power that supported a forests convention.

Table 2.7 *Some arguments for and against a forests convention*

	Arguments for	***Arguments against***
Forest management standards	Some states with high forest management standards may favour a convention as an instrument that will raise other countries' management standards up to their own	Some states with weak forest management standards may oppose a convention that aims to raise standards since such an instrument may impose additional costs on forest industries
Finance and technology transfer	Some states may favour a convention since it may provide a route for increased flows of finance and technology, including possibly opening an additional window on the Global Environment Facility	Some developed states may oppose a convention because they do not wish to commit themselves to additional transfers of finance and technology
Sovereignty	Some states may support a convention in order to gain a measure of control over the forest policies of other states	Some states may oppose a convention, which could infringe their sovereign rights to exploit their natural resources in line with national development policies
Forest industry	Some states with a large forest industry sector may favour a convention as a mechanism to promote the international trade in forest products	Some states may oppose a convention as a form of international regulation that would impose additional costs on the forest industry
Intergenerational equity	Some states may favour a convention to promote long-term forests conservation for future generations	Some states may oppose a convention with a conservationist ethos since such an instrument may threaten key economic and political constituents

The debate had also shifted within the G77, with many African countries now supporting a convention. According to one NGO observer, the pro-convention French and Canadian delegations were able to exert some influence on the Francophone African countries, which by the end of the IPF process showed more support for a convention than other African states.[78]

Malaysia had been the strongest voice against a convention at Rio when its delegation was led by Ting Wen Lian, the Malaysian permanent representative to the FAO in Rome. But at the IPF, Malaysia now advocated a convention that established a relationship between forest conservation and financial assistance. As with the US, the shift in the Malaysian position can be explained by domestic politics. After UNCED, the lead government agency on forests was switched from the Ministry of Foreign Affairs to the Ministry of Primary Resources. This, according to Ans Kolk, led to more cooperative behaviour from Malaysia in international forest negotiations.[79] It helps to explain the Canadian–Malaysian initiative that led to the creation of the IPF. This change meant that Ting Wen Lian, whom some developed country delegations had found difficult to negotiate with, no longer headed the Malaysian delegation. It also enabled the views of Malaysian industry, in particular the Malaysian Timber Council, which now supported a forests convention, to come to the fore. By the mid 1990s, many Malaysian timber corporations had established logging operations outside Southeast Asia, and it seems that the Malaysian forest industry was looking for a convention to establish rights for forest businesses. Other developing countries that supported a convention included Costa Rica and Papua New Guinea. However, most Latin American countries, especially Brazil, Colombia and Peru, remained opposed to a convention. From these positions, the G77, which sought to speak with one voice, synthesized the common line that it was too early to commence negotiations for a convention, although the desirability of a convention should be reassessed later.

At Rio, NGOs had been divided on this issue, although most international NGOs offered cautious support for a convention, providing that it contained strong provisions on conservation and indigenous peoples' rights. At the IPF, however, almost all NGOs opposed a convention. As Bill Mankin of the Global Forest Policy Project relates, the shift by major tropical timber producers such as Malaysia in favour of a convention, along with support for a convention from the Russian government, the Canadian government and Canadian business, led NGO campaigners to conclude that a convention would promote the forest industry rather than forest conservation, and that it would be unlikely to contain the elements that would make it worthy of NGO support.[80] NGOs subsequently issued an international citizens' declaration against a forests convention. This stated that a convention 'could formalize unacceptably weak forest management standards, thereby giving a global "green light" to unsustainable forest practices.' It 'will be dominated and driven by powerful timber and commercial trade interests, and fail to address the predatory and unethical behaviour of an increasing number of transnational industrial timber corporations.'[81] The shift by the NGO community can also be explained by disillusionment with intergovernmental initiatives and fears that a convention would reinforce a global governance structure that gave authority to states at the exclusion of local communities. NGOs that adhered to this position included Greenpeace International, Friends of the Earth, the World Rainforest Movement and the World Wide Fund for Nature.[82]

The Environmental Investigation Agency (EIA) was the only major environmental NGO to support a convention at the Panel. The EIA shared the concerns of other NGOs on corporate practices, but differed from them in concluding that a convention could regulate the global timber industry.[83] However, given the difficulties that had surrounded the Panel's consideration on a code of conduct for forests businesses, it was politically unrealistic to expect states to agree to a forests convention that would regulate trade interests.

With Brazil, the country with the world's largest share of tropical forests, and the US, the country with the world's largest timber industry, firmly opposed to a convention, there was no chance that the pro-convention lobby would gather the critical mass of support that it needed. The fourth and final session of the Panel agreed on the need to continue the international dialogue on forests, but beyond this there was no consensus. The Panel generated three possible options: to continue intergovernmental dialogue within existing fora (effectively the *status quo ante* the Panel); to establish an Intergovernmental Forum on Forests that, like the IPF, would report to the CSD (effectively the *status quo*); and to launch negotiations for a legally binding instrument on forests. The Panel passed these options to the fifth session of the CSD, held in April 1997, which in turn passed the matter to the United Nations General Assembly 19th Special Session (UNGASS) to review the implementation of Agenda 21. This was held in New York in June 1997, where the pro-convention states finally conceded defeat. The outcome was an agreement to establish an Intergovernmental Forum on Forests for three years.

The Intergovernmental Panel on Forests in context

The IPF has the legacy of being the first truly global forum where government delegates could exchange views on forest policy and forest politics.[84] It was initially a fragile and delicate process, with suspicions lingering from the divisive UNCED forest negotiations. However, the Panel was able to build on the confidence-building Canadian–Malaysian initiative and add to the body of soft international law on forests by negotiating the IPF proposals for action, although, and as their name makes clear, these proposals are not legally binding and are merely recommendatory in nature. The Panel firmly established the concept of national forest programmes in international forest discourse and it addressed indigenous peoples and traditional knowledge for the first time in an international forest forum. On two issues, the IPF initiated a political dialogue that would be continued by the IFF, namely the underlying causes of deforestation and the needs of low forest cover countries.

However, there were no innovations on issues where progress could only be achieved if the foundational assumptions of neoliberalism were challenged. The role of private sector investment as a surrogate for public sector finance, the insistence that technology transfer should take place only through the market, and the emphasis that any code of conduct for business should be voluntary are all hallmarks of neoliberalism. The Panel's conclusions on forest valuation merely reiterated the assumptions of neoclassical economics. The Panel gave no sustained attention to public goods provision, forest conservation and expanding global forest cover.

The forests convention negotiations were much less contentious than those at Rio. Even so, these negotiations completely overshadowed the final stages of the IPF to the extent that there was no political space available for innovative solutions short of a convention. In his closing statement as co-chair, Manuel Rodríguez noted a lack of global solidarity and the poor commitment to forest issues from the developed world.[85] While the polarization between developed and developing countries had blurred since Rio, it remained discernible. But although some old wounds were reopened during the negotiations on the thorny convention issue, as well as those on finance and technology, the best testament to the IPF was that the cooperative spirit generated in 1994 still existed when the Panel's two-year life span expired. Delegates had recognized that there was value in continuing the international dialogue by agreeing to create the Intergovernmental Forum on Forests. We turn to this in Chapter 4. But first we consider the fate of a political process that emerged parallel to the IPF: the World Commission on Forests and Sustainable Development.

3

World Commission on Forests and Sustainable Development

For those who coveted a forests convention, the 1992 United Nations Conference on Environment and Development (UNCED) had been a failure. After the UNCED, some leading personalities in international forest policy began exploring the possibility of a world commission on forests that, it was intended, would reinvigorate forest policy discourse and generate a consensus for a convention. Eventually, in 1995, a World Commission on Forests and Sustainable Development (WCFSD) was convened. Here we examine the background to the creation of this commission, which involved some high-level political jockeying involving a former US president and the UN secretary-general. We consider the relationship between the commission, the Intergovernmental Panel on Forests (IPF) and its successor, the Intergovernmental Forum on Forests (IFF). We consider why the commission failed to build a consensus for a convention and provide an assessment of its overall contribution to global forests discourse. In order to set the commission in context, we first provide a brief overview of the roles that world commissions have played in global politics.

World commissions: An historical overview

The world commission is a fairly recent innovation in global politics and dates back to the 1960s. World commissions set out to examine a serious problem of global dimensions that, for various reasons, is being ignored or inadequately addressed by the international political system. These problems usually have a humanitarian and public goods dimension. Issues with which world commissions have grappled include international development, aid, refugees, famine, the arms race and environmental degradation. The typical commission comprises an elitist 'eminent persons' membership of about 20 to 30, drawn from prominent politicians, diplomats, scientists, academics and others with relevant expertise. The membership is usually balanced between the world's main regions.

World commissions are independent; they represent no particular organization, group or interest. However, occasionally an international organization will play an entrepreneurial role in establishing a commission by suggesting terms of reference, inviting a chair or co-chairs to form the commission, and pledging financial support. Examples include the Commission on International Development, the chair of which, Lester Pearson, was appointed by the World Bank, and the World Commission on Environment and Development, whose chair, Gro Harem Brundtland, was appointed by the UN secretary-general following a General Assembly resolution.[1] Once a world commission is established, however, there will be no attempt to interfere in its independence from the sponsoring organization.

The rationale of the world commission is that its membership will be able to engage in grand strategic thinking on how to address a major global problem without the shackles of loyalty to any particular government, institution or group. World commissions typically issue recommendations and policy proposals. Although they have no power to implement their recommendations, the stature and reputation of a commission's membership provides a certain moral authority that can lead to recommendations being adopted by those with political and economic power. One way of viewing a world commission is as a temporary, but high-profile, international think tank.

Most world commissions exist for three to four years and issue just one report. An exception is the Independent Commission on International Humanitarian Issues, which issued several reports during the 1980s on famine, street children, desertification, terrorism, modern warfare, refugees and other humanitarian issues. Two reports prepared for this commission are of relevance to us. *The Vanishing Forest: The Human Consequences of Deforestation* (1986) issued no policy recommendations, but did provide a useful overview of the social and ecological consequences of tropical deforestation.[2] *Indigenous Peoples: A Global Quest for Justice* (1987) focused on two types of rights for indigenous peoples: the right to land, and the right to self-determination. It urged that these rights be strengthened and promoted at the national level and in intergovernmental organizations, financial institutions, transnational corporations and at the International Court of Justice.[3]

Table 3.1 details all of the world commissions of the last four decades that fit the model described above. Not included, but of interest to our study of global forests governance, is the World Commission on Protected Areas. This is a IUCN network with a membership of 1000 or so leading protected area specialists (see Chapter 9). It should be noted that the term 'world commission' can be freely appropriated by any group or organization. Table 3.1 does not, therefore, include groups such as the World Commission for Peace and Human Rights Council, which is best seen as a pressure group that reports on human rights violations and humanitarian assistance,[4] and the World Commission on Global Consciousness and Spirituality, which was inaugurated at the World Congress of Philosophy in 1998.[5]

The organizing committee: 1992–1994

The World Commission on Forests and Sustainable Development (WCFSD) had a difficult gestation period that was dominated by three political issues: mandate, membership and relationship with the United Nations.

Table 3.1 *Major world commissions, 1968–2002*

Name of commission **Year created**	**Chairs** **Sponsoring institution (where appropriate)**	**Report(s), year published** **Main recommendations**
Commission on International Development (Pearson Commission) 1968	*Chair:* Lester B. Pearson (Canada) World Bank	*Partners in Development*, 1969[6] Recommended • A reversal in declining public aid for moral reasons (the fortunate have a duty to help those in need) and for enlightened self-interest (the world community is interdependent) • Developed countries to increase public aid equivalent to 0.7% of GNP by 1975
Independent Commission on International Development Issues (Brandt Commission, or North–South Commission) 1978	*Chair:* Willi Brandt (West Germany)	*North–South: A Programme for Survival*, 1980[7] *Common Crisis, North–South: Cooperation for World Recovery*, 1983[8] In order to abolish global hunger these two reports recommended: • Large-scale transfer of funds from North to South • Reform of the international monetary system to bring about more stable exchange rates • Commodity price reform to enable countries from the South to boost their export earnings
Independent Commission on Disarmament and Security Issues (Palme Commission) 1981	*Chair:* Olof Palme (Sweden)	*Common Security: A Blueprint for Survival*, 1982[9] Recommended a series of measures to achieve universal and complete disarmament, including 'zones of peace' and 'nuclear-free zones'

Table 3.1 *Major world commissions, 1968–2002 (Continued)*

Name of commission Year created	***Chairs Sponsoring institution (where appropriate)***	***Report(s), year published Main recommendations***
Independent Commission on International Humanitarian Issues 1983	*Inaugural co-chairs:* Prince El Hassan Bin Talal (Jordan) Sadruddin Aga Khan (Iran)	Several reports on subjects that included famine, desertification and refugees. Two reports relevant to forests: *The Vanishing Forest: The Human Consequences of Deforestation*, 1986 (This report issued no policy recommendations.)[10] *Indigenous Peoples: A Global Quest for Justice*, 1987 Recommended the strengthening of indigenous peoples' rights to land and self-determination at the national and international levels[11]
World Commission on Environment and Development (Brundtland Commission) 1983	*Chair:* Gro Harem Brundtland (Norway) *Vice-chair:* Mansour Khalid (Sudan) United Nations General Assembly	*Our Common Future*, 1987[12] • The UN should convene a conference on environment and development (subsequently held in Rio in 1992) • Proposed a set of legal principles for environmental protection and sustainable development
World Commission on Culture and Development 1992	*President:* Javier Pérez de Cuéllar (Peru) United Nations Cultural Organization (UNESCO)	*Our Creative Diversity*, 1995[13] Emphasized the importance of cultural pluralism including: • The role of culture in economic development • How certain types of economic development can erode cultural diversity

Table 3.1 *Major world commissions, 1968–2002 (Continued)*

Name of commission Year created	***Chairs Sponsoring institution (where appropriate)***	***Report(s), year published Main recommendations***
Commission on Global Governance 1992	*Co-chairs:* Ingvar Carlsson (Sweden) Shirdath Ramphal (Guyana)	*Our Global Neighbourhood*, 1995[14] Recommended strengthening of the UN, including: • A UN army • An economic security council • UN authority over the global commons • A parliamentary body • A court of criminal justice • Global taxation.
Independent World Commission on the Oceans 1995	*Chair:* Mario Soares (Portugal) Nine vice-chairs	*The Ocean: Our Future*, 1998[15] Recommended: • A United Nations conference on ocean affairs • A world ocean affairs observatory to monitor 'ocean governance'
World Commission on Forests and Sustainable Development 1995	*Co-chairs:* Ola Ullsten (Sweden) Emil Salim (Indonesia) InterAction Council of Former Heads of State and Government	*Our Forests: Our Future*, 1999[16] Proposed: • A Forest Security Council of the most important forested countries • ForesTrust International, a citizens' association for the defence of the public interest in forests • An international forest capital index
World Commission on Dams 1998	*Chair:* Kader Asmal (South Africa) *Vice-chair:* Lakshmi Chand Jain (India)	*Dams and Development: A New Framework for Decision-Making*, 2000[17] Proposed: • A set of 'strategic priorities' for the sustainable development of water and energy • Criteria and guidelines for applying the strategic priorities

Table 3.1 *Major world commissions, 1968–2002 (Continued)*

Name of commission Year created	**Chairs Sponsoring institution (where appropriate)**	**Report(s), year published Main recommendations**
World Commission on the Social Dimensions of Globalization 2002	*Co-chairs:* Tarja Halonen (Finland) Benjamin William Mkapa (Tanzania) International Labour Organization	*A Fair Globalization: Creating opportunities for all*, 2004[18] Proposed: • Fairer rules for international trade, finance, investment and migration • Core labour standards and social protection in the global economy • Policy coherence initiatives between international organizations

Sources: Endnotes 6 to 18 for Chapter 3

In October 1991, when it was clear that no forests convention would be agreed at the UNCED in Rio, a meeting on international forest policy was held at the Woods Hole Research Center in Massachusetts. Attended by some 30 academics, scientists, policy-makers and NGO representatives, this concluded that a global forests convention was 'a prerequisite to reversing current global trends toward biotic impoverishment and to preserving forests for present and future generations.'[19] It proposed that an independent world commission on forests be established with the goal of creating a consensus for a convention. This was the first time that a proposal for a world commission on forests was publicly aired.[20] A supporter of the Woods Hole initiative was the former Swedish prime minister Ola Ullsten. During the early 1990s, Ullsten was Sweden's ambassador to Rome, in which capacity he represented Swedish interests at the Food and Agriculture Organization (FAO). He was one of the co-authors of the independent review of the FAO's Tropical Forestry Action Plan that in 1990 recommended a convention on forests.[21]

In March 1992, just three months before the UNCED, the director of the Woods Hole Research Center, George Woodwell, attempted to solicit support for a commission on forests from former US President Jimmy Carter. He wrote to Carter:

> *Our initiative is the suggestion that an International Commission on the Conservation and Utilization of Forests be established immediately following the Brazil conference.* The Commission would have the responsibility for developing support for an international treaty on forests *[emphasis in original]*.[22]

Carter declined to play a role in the initiative, although both he and the Carter Center that bears his name were later involved in debates on the commission's membership.

One month after the UNCED, Ullsten chaired the first meeting of the organizing committee for a World Commission on Forests and Sustainable Development in Rome.[23] Those involved in the organizing committee included George Woodwell, Jim MacNeil (secretary-general of the World Commission on Environment and Development) and Maurice Strong (secretary-general of the UNCED). The organizing committee worked on the mandate and agenda of the WCFSD and debated who should be invited to serve as commissioners.

Throughout 1992 and 1993, the organizing committee was lobbied by the NGO community. Bill Mankin of the Global Forest Policy Project (GFPP) argued that the plan to create a commission to build a consensus for a convention should be abandoned: 'the WCFSD would not only prejudice its own objectivity, but could even doom itself to political failure if it began with a pre-ordained outcome so widely opposed by developing countries.'[24] Mankin also urged that the conventional 'eminent persons' membership of previous world commissions should not be adopted, and that of the 25 planned commissioners, five should be representatives of forest user groups, such as indigenous peoples, small-scale farmers, non-timber extractivists and other forest dependent communities. Without the representation of forest user groups, Mankin argued, the WCFSD would have no legitimacy or credibility.[25] In April 1993 he circulated an open notice to all forest activists and NGOs, noting that the WCFSD organizing committee was 'somewhat resistant' to forest user group representation, but suggesting that NGOs and civil society groups consider a process to agree representatives to the commission, should the organizing committee prove receptive to this idea at a later date.[26]

Aware that the Woods Hole Research Center had approached Jimmy Carter as a possible commissioner, the GFPP approached the Carter Center with concerns about the planned membership of the WCFSD. Subsequently, the Carter Center wrote to George Woodwell, noting that the creation of a world commission would be 'a unique opportunity to ensure more direct and intense local involvement by *actually appointing to the Commission representatives of forest user groups* [emphasis in original],'[27] and that including forest users 'would generate even more excitement about the Commission and would lend it great legitimacy and support at the grassroots level and throughout the developing world (as well as with the environmental movement in this country).'[28]

Woodwell's reply to the Carter Center two weeks later stated that:

> *You are not alone in suggesting that there be strong representation on the Commission for the direct users of forests ... I am concerned that if we bow specifically to every interest advanced, we shall have to balance various exploitative interests against the public interest when the purpose of the Commission is to examine the public interest in its broadest context.*[29]

The organizing committee eventually decided to adhere to the 'eminent persons' format. Only one NGO representative served on the commission, namely Yolanda Kakabadse of Ecuador, although she was invited as the incumbent president of the IUCN, and not as an NGO representative. No forest user group representatives served as commissioners.

The relationship between the UN and the WCFSD took up considerable time in the organizing committee. There were two separate but inter-related questions. First, should the organizing committee seek for the WCFSD a status similar to that of the World Commission on Environment and Development, which had been called for by the General Assembly and whose chair was appointed by the UN secretary-general.[30] Second, what relationship should the WCFSD have with UN institutions with a forest-related mandate, such as the CSD. The CSD, which has a rotating agenda, was scheduled to consider forests for the first time at its 1995 session (where, as we saw in Chapter 2, the Intergovernmental Panel on Forests was created).

The CSD thus posed a 'threat' to the WCFSD before it was even launched. According to one campaigner, the organizing committee wanted the CSD to wait for the WCFSD to complete its work before it discussed forests.[31] But most NGOs wanted no delay in the CSD's consideration of forests. They saw more promise in the CSD process, which was open to NGO observers, than the WCFSD. Five US-based NGOs wrote in June 1993 to the UN secretary-general, Boutros Boutros-Ghali, stating their concern that 'the CSD may be asked to delay its work on forests to allow the world commission to complete its work.'[32] They argued that:

> *Although the findings of the proposed world commission on forests will no doubt be helpful to the CSD in its work, the CSD cannot afford to wait for these findings before beginning its own work on forests ... we strongly believe the CSD should be free to begin its work on forests without deferral to other bodies.*[33]

One week later, Jimmy Carter wrote to Boutros Boutros-Ghali. He mentioned 'the potential for conflict' between the WCFSD and the CSD and added: 'I would hope that, under your leadership, appropriate roles and responsibilities could be set forth for both the WCFSD and the UN CSD which enable each to support and complement, not compete with, the other.'[34] Carter also noted that he wished to see the WCFSD include forest user groups in its membership. He offered the offices of the Carter Center in identifying potential commissioners.

Boutros-Ghali's reply to Carter avoided both the relationship between the CSD and the WCFSD, and the question of whether forest user groups should serve as commissioners. Significantly, he offered no support for the WCFSD and stated that the view among states is that the CSD 'is the appropriate forum, and has the status and flexibility required, to deal effectively with issues of forest [sic] and sustainable development.'[35] Although Boutros-Ghali had met with the organizing committee,[36] he did not give the formal approval of the office of the UN secretary-general to the WCFSD. The organizing committee thus had to turn elsewhere for legitimacy. Eventually, support was given from the InterAction Council of Former Heads of State and Government, a fairly obscure body of ex-political leaders. It was, however, the only source to provide the WCFSD with any external legitimacy.

There are a number of reasons why there was no support within the UN system for the WCFSD. First, after the arduous UNCED negotiations, forests was simply too politically contentious an issue for agreement to create a UN-sanctioned world commission. Second, and as we have seen, the organizing committee faced opposition from NGOs, some of whose concerns were shared by a former US president. Third, the initiative was regarded with suspicion in tropical forest countries, where many governments were unwilling to leave the international forests agenda in the hands of what was essentially an unelected group. The dominant view in developing countries was that any international dialogue on forests should be under the control of an intergovernmental forum. This helps to explain the creation of the IPF.[37]

By 1995, the organizing committee had raised sufficient money to formally launch the commission.[38] At this stage, the InterAction Council of Former Heads of State and Government invited Ullsten to co-chair the WCFSD. Emil Salim, a former environment minister of Indonesia, was invited to be the other co-chair.[39] Salim had been a commissioner on the Brundtland Commission and was involved in the WCFSD organizing committee. However, although the InterAction Council sent the invitations, the WCFSD was, to all intents and purposes, a self-appointed commission.[40] The original idea that the commission would aim to build a consensus for a forests convention had now been dropped. The mandate of the WCFSD was to:

- *Increase awareness of the dual function of world forests in preserving the natural environment and contributing to economic development*
- *Broaden the consensus on the data, science and policy aspects of forest conservation and management*
- *Build consensus between North and South on forest matters with emphasis on international cooperation.*[41]

Just two months after the CSD agreed to create the IPF, the WCFSD finally held its inaugural meeting in June 1995.

The work of the World Commission on Forests and Sustainable Development: 1995–1999

Even before the WCFSD was launched, it was being marginalized by the moves to create the IPF. Indeed, this was precisely the intention of some governments, principally (although not exclusively) from the developing world, which supported the creation of the IPF because they did not wish to surrender the international forest policy dialogue to a body over which they exercised no control. Government support for the WCFSD was lukewarm. The governments of Canada, the Netherlands and Sweden voiced some support; but there was little support from other governments.

The WCFSD sought to build links between itself, the IPF and the IPF's replacement, the IFF, in three ways, none of which were particularly successful. The first was to identify potential areas of collaboration between the two processes. Three areas were identified: participation and benefit-sharing; the underlying causes of deforestation; and trade and the environment.[42] However, while these were all items on the IPF's agenda, no coordination mechanisms were created, such as a joint committee of the IPF/IFF and WCFSD secretariats, to ensure complementarity of work programmes. Second, the commission appointed IPF co-chair Manuel Rodríguez as a commissioner. Rodríguez believes that he was appointed because the WCFSD's organizers hoped that he would be able to ensure that the IPF was favourably disposed towards the work of the commission. Rodríguez acted as a reporting link between the IPF and the WCFSD, although it is his assessment that the two processes were 'very separate' and there was 'no real interaction between them.'[43] Third, some commissioners attended IPF and IFF meetings, where they made verbal interventions, distributed documents on the work of the WCFSD and hosted side events. However, these activities had minimal impact. For example, a summary of the commission's final report[44] was circulated only at the final session of the IFF, far too late to influence the negotiations (see Chapter 4).

The criticisms that the organizing committee had attracted for failing to consult with forest user groups appear to have had an impact. From its inception, the WCFSD aimed to operate in an open and participative manner. Five public hearings were held – in Jakarta, Indonesia; Winnipeg, Canada; San José, Costa Rica; Yaoundé, Cameroon; and St Petersburg, Russia – to seek citizens' perspectives on forest-related issues. Some 2000 people attended the five public hearings, including small farmers, private woodland owners, local forest communities, indigenous communities and representatives from forest-based transnational corporations and trade organizations. Manuel Rodríguez recalls that the Yaoundé meeting was one of the first opportunities for people from the Congo Basin countries to talk freely and publicly on forest conflicts.[45] However, a downside to the public hearings was noted by another commissioner, the British environmental economist David Pearce: the appellation 'world commission' and the high political status of some commissioners gave the impression to some people attending the public hearings that the commission had political power, whereas it did not.[46] A further problem, noted by Marcus Colchester of the Forest Peoples Programme, was that poor record-keeping at the public hearings led to much valuable testimony being lost.[47] While the WCFSD can claim to have operated the most inclusive global process on forests in history, it remained essentially a top-down process, albeit one that spent considerable time and effort consulting, rather than a genuine 'stakeholder commission' in which forest user groups had full representation.

In addition to the public hearings, the commission held eight closed session meetings attended solely by the commissioners and select invitees. It created three working panels on the sustainable, equitable use and management of forests; trade and the environment; and financial mechanisms and international agreements and institutions.

The findings of the World Commission on Forests and Sustainable Development

The report of the WCFSD, published in 1999, reflects the broad range of views that fed into its work and the disparities between them. The commission was an elite-driven process, yet through the public hearings and extensive consultations it had a clear grassroots element. The result is an uneasy juxtaposition of different ideological positions. The discourse that runs throughout the report is a composite of the orthodox and the innovative, with recommendations to governments and international institutions, as well as a distinct radical edge.

The commission's report can be seen as an attempt to offer an alternative worldview to neoliberalism. The emphasis on the 'public interest' and the fate of communities whose livelihoods have been destroyed by deforestation are recurring themes that serve as an antidote to the emphasis on the private sector that has dominated recent intergovernmental forest processes. The relationship between poverty and forest degradation is particularly prominent. While local communities provide custodial services that play a vital role in forest conservation, the rights of such communities are not always valued. The theme of economic development was far less prominent than in the report of the Brundtland Commission. The WCFSD paid lip service to the concept of sustainable development, but the concept was in no sense central to its work. Corruption and the unsustainable practices of forest-based industries were strongly criticized.[48] The WCFSD stressed the public goods functions of forests, particularly ecological and social functions, more than it stressed industrial functions. The commission both reflected and reinforced NGO efforts to shift the forest conservation agenda towards increased public and community involvement in forests and the elimination of short-term exploitation for profit. There was considerable discussion within the commission on how to balance the concept of the 'public interest' with the rights of indigenous peoples and local communities. As Marcus Colchester of the Forest Peoples Programme noted, the concept of the public interest would pose a risk to communities and indigenous peoples if it were to be interpreted according to majority rule principles or defined by an elite 'in the name of the people.'[49] The WCFSD compromised by highlighting both the public interest and communities, but it did not explore the possibility of a tension between the interests of a local community and those of the broader public.

The emphasis on the public interest led the commission to consider how the public goods functions of forests should be quantified and measured. The solution was to propose a forest capital index scheme under which countries would record on a scale of 0 to 100 the changing state of their forests according to various indicators, such as surface area, species diversity and standing biomass. In distinction to neoclassical economics, which concentrates primarily on measuring the flow of commercial goods and services in markets and gross national product (GNP), a forest capital index would keep a tally of the stock value of forests as a natural capital asset, and it would enable changes to this value to be tracked over time.[50]

But while the commission challenged neoliberalism, it also reflected many neoliberal assumptions, with parts of the report reflecting the neoclassical view on markets for environmental services: 'Market mechanisms need to be broadened to capture the full economic value of forests.'[51] There is considerable emphasis in the commission's report on the need to take into account the monetary valuation of forest goods and services and how missing markets and the absence of prices for forest goods and services can lead to the overexploitation of forests.[52] However, the commission also acknowledged that part of the driving force of forest destruction was market forces, and the emphasis on the individual utility maximizer that informs the neoclassical paradigm is implicitly questioned: 'The public interest goes beyond the material interests of a single group and ... exceeds the sum of all group interests;'[53] forest management cannot 'be left only to private interests, as the public interest becomes more pressing as human demands increase.'[54]

The tension between what economists claim can be achieved through the monetary valuation of forest goods and new environmental markets, and the failure of economic methodologies to generate policies that maintain the capital stock values of forests, is unresolved in the commission's report. The fundamental problem is that markets, even markets for environmental goods and services, always tend to maximize short-term gains for individual private actors, whereas the public interest requires maximizing long-term collective gains. Although the commission appears to have recognized this problem, it had no success in reconciling these two different imperatives.

Even the proposal for a forest capital index, which has not subsequently been adopted, stopped well short of a radical innovation, as there was no indication of how the index would become an integral feature of forest economy decision-making. The mere existence of a global index does not necessarily mean that it will be taken into account by decision-makers. For example, the Human Development Index (HDI) of the United Nations Development Programme provides a benchmark by which changes in human and social conditions can be tracked, but it does not *necessarily* factor into national and international economic decision-making. This is because there is no feedback loop by which the HDI can register on decision-makers in industry or government, in the same way that, for example, profitability and shareholder value registers in the decisions of commercial executives and the rate of inflation registers with government macro-economic policy-makers. There is no reason to suppose that a forest capital index, which in the absence of such a feedback loop would be decoupled from mainstream economic decision-making, would fare any better than the HDI.

A forest capital index would certainly enhance our knowledge of global forest loss. However, global trends in forest degradation and deforestation have previously been widely documented, and this has not solved the crisis.

The problem is not so much the absence of a forest capital index, but more the failure of existing knowledge about the loss of forests to register in economic decision-making, in part due to the neoclassical paradigm which does not recognize human need unless it takes the form of economic demand; that is, demand backed by money. The fundamental disconnect between knowledge on the loss of environmental capital and the economic decisions that cause this loss cannot simply be tackled with a new index.

Other ideological divisions also appeared within the commission. A comparison between an internal discussion draft produced by the WCFSD in 1997 and the final report of 1999 reveals that some of the more radical initial ideas were filtered out and replaced by proposals that are more orthodox when viewed from a neoliberal perspective. For example, language that was critical of business was deleted. The 1997 internal draft noted the financial and political power behind many large-scale forest industrial and infrastructure projects, observing that given:

> *... their [industrial interests'] refusal to abide by socially and environmentally acceptable codes of conduct, it is not surprising that environmental activist movements such as Oil Watch and Mine Watch have sprung up whose main objective is to mobilize civil society opposition and, in extreme cases, to encourage boycotts of a company's products or services.*[55]

The internal draft also included a section headed 'Penalizing companies that fail to adopt responsible forest stewardship,' which recommended that where there is 'obvious disregard for sustainable forest management practices, coercive measures will be necessary.'[56]

But by the time the final report was published, strong criticisms of business, including mentions of boycotts and 'coercive measures,' had been replaced with language on voluntarism, with the commission citing with approval a voluntary code of conduct developed by the global forest industry, noting that:

> *... it is encouraging that some corporations are beginning to realise that long-term success depends upon effective performance in integrating profitability with environmental and social needs and standards... Such corporations need encouragement.*[57]

Cooperation between the World Bank and forest corporations is cited as one way of deepening and expanding corporate commitment, and is considered 'a recognition of the important role the corporate sector has to play in achieving sustainable forest management.'[58] Some of the commission's policy recommendations fit comfortably within neoliberal discourse, such as working with and 'encouraging' business. Compulsory public oversight and sanctioning and fining environmentally destructive businesses is eschewed. The commission thus drew back from a fundamental radicalization of the political discourse on forests. Had its membership been more constituency based rather than led by establishment figures, it is likely that its conclusions would have been more critical and forceful.

The commission proposed ten resolutions addressed to forest stakeholders. Each resolution was accompanied by various proposed actions. The first resolution takes the form of a bland statement on the severity of the crisis. Some of the resolutions reiterate proposals made previously in other institutions and are hardly new, such as the need for poverty alleviation (Resolution 2), education, research and training (Resolutions 8 and 9), and political leadership (Resolution 10). Other resolutions follow more directly from the commission's work on securing the public interest in forests. The unresolved debates within the commission are apparent, with separate but unrelated resolutions on prices for the goods and services that forests provide (Resolution 4) and the capital stock value of forests (Resolution 6) (see Box 3.1).

The institutional recommendations of the World Commission on Forests and Sustainable Development

Some politicians looked to the WCFSD as a track that could lead to a forests convention, while others were more suspicious and preferred to support the IPF. Some commissioners wanted the commission to openly endorse a convention. Prominent among them was Hemmo Muntingh, a Dutch member of the European Parliament. Muntingh was a major player in the European Parliament's 1990 decision to recommend a forests convention.[59] He was a founding member of the Global Legislators Organization for a Balanced Environment (GLOBE), which, during the early 1990s, commissioned a draft of a forests convention that was intended to feed into the UNCED forest negotiations.[60] Just before the UNCED, a European Parliament committee on the environment, for which Muntingh was rapporteur, drew from the GLOBE draft when it, too, recommended a forests convention.[61] Muntingh supported Ullsten and some other commissioners in arguing that the WCFSD should recommend a convention.

However, a majority of commissioners argued against this.[62] Prominent among them was Angela Cropper, a biodiversity specialist from Trinidad and Tobago,[63] David Pearce and the IPF co-chair Manuel Rodríguez.[64] The commission was gently advised not to support a convention by a study from the IIED, which concluded that 'it is *premature* to negotiate an intergovernmental convention on forests if such negotiation is based on current intergovernmental mechanisms and does not benefit from a searching review of other instruments' [emphasis in original].[65]

The WCFSD's recommendation on a forests convention was carefully crafted to paper over the substantial differences between the commissioners on this question. The commission, it was stated, supports 'use of international law in the service of societal goals,' and:

Box 3.1 The ten resolutions of the World Commission on Forests and Sustainable Development

The WCFSD recommended:

1. that radical and urgent attention should be given to arrest the decline in forests, since forests, their products and services are threatened.
2. that conserving and sustainably managing the world's forests should go hand in hand with the priority of reducing poverty and sustaining the livelihoods of millions of poor people and numerous communities who depend heavily upon forests.
3. that the public interest in a stable and secure environment must become paramount in decisions about the use and management of forest lands.
4. that we must get to the roots of the crisis by bringing prices and policies in line to better reflect the range of values and uses of forests.
5. that the threshold for responsible forest stewardship should be raised to reflect the new responsibilities of forest management to integrate economic, environmental and social considerations and to make the public interest paramount.
6. that new measures must be created to keep track of the value of the capital stock of forests and to create a basis for compensation to countries for the ecological services of forests.
7. that participatory planning for the use of landscapes, not just forests, should be instituted to ensure conservation objectives.
8. that the information base about forests should be enhanced and made more directly useful and applicable to policy-makers and in education programmes for the public.
9. that research and training should be adapted and accelerated to support the new responsibilities of forest management.
10. that additional avenues for political and policy leadership should be exposed to accelerate progress towards solutions.

Source: WCFSD (1997) *Our Forests, Our Future*, Cambridge: Cambridge University Press, pp.154–164

> *... commitments which have the force of international law would underpin the efforts that are required, providing that they address the fundamental issues, that they identify appropriate responses, that they secure the means for their implementation, including financing, and that they are pursued with diligence and commitment... The commission therefore considers that under these conditions a global forests convention would be a useful instrument.*[66]

This conclusion had no impact at all on the forests convention negotiations then unfolding at the Intergovernmental Forum on Forests (see Chapter 4).

The commission recommended the establishment of a new institution, ForesTrust International, to act as 'a citizens' force for discharge of the public trust and defence of the public interest in forests.'[67] It was envisaged that ForesTrust International would have various components, including an ombudsman with the power to investigate abuses of power and to pass judgements on forest conflicts, and ForestWatch, with responsibility for monitoring forests worldwide. The ForestWatch idea has taken root, although not as a result of the WCFSD: in 1997, the World Resources Institute created Global Forest Watch, a network of NGOs that aims to promote transparency and accountability in the world's forests by tracking the actors involved in forest development and by monitoring the degree to which they adhere to national and local laws and regulations.[68] The World Resources Institute established Global Forest Watch before the commission issued its report.

The WCFSD also recommended the creation of a Forest Security Council that would 'take up the mantle of leadership on behalf of their own, and the world's, forests and citizens.'[69] Twenty-five countries with significant forest cover or prominent forest industries, and which, it was therefore argued, had an 'extraordinary duty of care for forests on behalf of the world's people and the planet',[70] were proposed as Forest Security Council members (see Box 3.2). These countries were selected to provide a balanced reflection of the world's forest types with representation from the main forested regions. While the proposal for a Forest Security Council had its own logic, only a very few governments, of which Canada was one, expressed interest in the idea,[71] and the commission did not come close to gathering the critical mass of political support that it needed to launch the institution.

Box 3.2 World Commission on Forests and Sustainable Development: Proposed membership of a Forest Security Council

Asia and Oceania	Australia, China, India, Indonesia, Malaysia and Japan
Africa	Democratic Republic of Congo, Ethiopia, Ghana and Nigeria
Latin America and the Caribbean	Bolivia, Brazil, Chile, Colombia, Ecuador, Peru and Venezuela
Europe	Russia, Finland, France, Germany and Sweden
North America	Canada, Mexico and the US

Source: World Commission on Forests and Sustainable Development (1997) *Our Forests, Our Future*, Cambridge: Cambridge University Press, p.147.

Parallel lines

It was always clear that the long-term contribution of the commission would depend on the extent to which its findings could influence those actors with the political and economic power to affect changes in forest governance, such as governments, forest corporations and international organizations. It is now clear that the WCFSD has achieved minimal influence on these actors. The commission tried to broaden the global conversation on forests, and while its efforts to rehabilitate the public goods value of forests were commendable, they found little support among those governments pursuing neoliberal trade policies, which emphasize the primacy of individual values over collective values, of private and commodifiable values over public goods, and of trade and the global market over environmental conservation. Despite diluting or deleting some of the more radical recommendations that appeared in its early reports, the commission remained at odds with the entrenched discourse of neoliberalism that has dominated the global economy over the last three decades.

In this respect, there was a noticeable difference between the WCFSD and the Brundtland Commission. The latter started with the premise that economic growth was necessary, especially for developing countries. It argued that economic growth and environmental protection were mutually reinforcing objectives and that international trade would help to promote sustainable development.[72] This overall approach fitted neatly with neoliberalism and made the recommendations of the Brundtland Commission palatable to government and economic elites from both the developed and developing worlds.[73] In distinction, the WCFSD focused more on the social dimensions of the global forest crisis and was noticeably more cynical about economic growth than the Brundtland Commission. This element was regarded with suspicion by developing world governments. The commission's emphasis on the need to resolve forest-related conflicts in favour of the landless poor ultimately offered no material alternative to governments wishing to exploit their forests for economic gain.

The commission made a serious effort to broaden the discourse on forests and to place public concerns centre stage in international forest policy. However, it never enjoyed support or commitment from the governments of major forested states. Nor was there support from major timber consuming countries, such as Japan and China. Although Canada, the Netherlands and Sweden offered some support, this was insufficient to provide any high-level political momentum. Unable to garner support within the UN and eclipsed by the IPF and IFF, the commission faced an uphill battle before it held its first meeting. Although it made a respectable contribution to network-building and knowledge dissemination, its aims were far more ambitious. It is now clear that its overall impact on mainstream forest discourse has been negligible.

As we have seen, the commission's hearings occurred contemporaneously with the IPF and IFF. The work of the IPF and IFF had very little impact on the WCFSD, while the WCFSD's emphasis on corruption, corporate responsibility and the empowerment of civil society found little support in the IPF and IFF. The WCFSD process and the IPF/IFF process existed virtually in isolation from each other. Each had different approaches to dialogue and debate. The WCFSD was best suited to dealing with big questions and grand ideas in a way that the IPF and IFF, which used consensual decision-making procedures during detailed textual negotiations, were not. Logically, big ideas should precede attention to detail, and in this respect it would have made sense for the WCFSD's report to have been considered in full by a body within the UN system. However, the suggestion that the UN should delay its work on forests until the WCFSD had reported faced strong opposition, and there was simply no political agreement for the WCFSD to have any status within the UN system. The WCFSD and the IPF/IFF process were parallel initiatives and, like parallel lines, they never met.

4

Intergovernmental Forum on Forests

Introduction

The Intergovernmental Forum on Forests (IFF) occupied the same position within the UN system as the IPF, namely as a subsidiary body of the Commission on Sustainable Development (CSD). Like the IPF, the IFF was an open-ended body, meaning any UN member could partake in its work, and like the IPF it spent considerable time generating proposals for action. Indeed, to all intents and purposes, it was the IPF with a new name and revised mandate. It met four times: October 1997 (New York); August–September 1998 (Geneva); May 1999 (Geneva); and January–February 2000 (New York). It reported to the 2000 session of the CSD.

This chapter provides an analytical overview of the IFF. We examine an NGO initiative on the underlying causes of deforestation, and the IFF's deliberations on trade, financial resources and technology transfer. We pay particular attention to explaining the IFF's lack of progress on traditional forest-related knowledge, an issue that falls within the purview of several institutions. We argue that the IFF's work on valuation and economic incentives was heavily influenced by the World Bank. The chapter concludes with the IFF's deliberations on whether negotiations should be launched for a convention on forests.

The working modalities of the Intergovernmental Forum on Forests

The member organizations of the Interagency Task Force on Forests (ITFF), initially formed to support the IPF, served as lead agencies for individual items of the IFF work programme (see Table 4.1). The IFF had three categories of work. Category I was devoted to reviewing the implementation of the Panel's proposals for action. Category II concentrated on matters left pending from the IPF, such as the trade–environment relationship. Category III addressed institutions and organizations.

Several intersessional initiatives were held in support of the IFF (see Table 4.2). Examples included research and information systems, the needs

Table 4.1 *Intergovernmental Forum on Forests: Abbreviated work programme*

Intergovernmental Forum on Forests agenda item	***Lead agency***
Category I: Implementation of the Intergovernmental Panel on Forests proposals for action	
I.a Promoting and facilitating implementation	Various*
I.b Monitoring progress in implementation	Various*
Category II: Matters left pending from the IPF	
II.a Financial resources	UNDP
II.b Trade and environment	ITTO
II.c Transfer of environmentally sound technologies	FAO
II.d Issues that need further clarification	
i. Underlying causes of deforestation	UNEP
ii. Traditional forest-related knowledge	CBD
iii. Forest conservation and protected areas	UNEP
iv. Forest research	CIFOR
v. Valuation of forest goods and services	World Bank
vi. Economic instruments, tax policies and land tenure	World Bank
vii. Supply and demand of forest goods and services	FAO
viii. Forest cover in environmentally critical areas	FAO
II.e Forest-related work of international and regional organizations and under existing instruments	UN-DESA
Category III: International arrangement and mechanisms	–

Note: * The lead agency for Category I issues varied according to the proposals for action under consideration.

Source: ITFF (1999) *The Interagency Task Force on Forests*, Rome: FAO

of low forest cover countries and the Six Country Initiative on national forest programmes. Some intersessionals that had been created during the IPF process reported to the IFF. One example, which we consider later, is the NGO initiative on the underlying causes of deforestation.

The co-chairs throughout the IFF process were Bagher Asadi of Iran and Ilkka Ristimäki of Finland.

Implementation of the Intergovernmental Panel on Forests proposals for action

The decision to make implementation of the IPF's proposals for action a part of the IFF's mandate was welcomed by environmental and social NGOs.

Table 4.2 *Intersessional initiatives in support of the Intergovernmental Forum on Forests*

Name of initiative	***Type***	***Venue***	***Organizers/ sponsors***
Putting the IPF proposals for action into practice at the national level: Six Country Initiative on national forest and land-use programmes	National studies and workshop	Baden-Baden, Germany	Finland, Germany, Honduras, Indonesia, Uganda, UK, IFF Secretariat, UNDP, FAO
Research and information systems in forestry	International consultation	Ort/ Gmunden, Austria	Indonesia, Austria, IUFRO, CIFOR, FAO, IFF Secretariat
Underlying causes of deforestation and forest degradation	Regional consultations and international workshop	San José, Costa Rica	NGOs, Costa Rica, UNEP
International arrangements and mechanisms to promote the management, conservation and sustainable development of all types of forests	Regional consultations and expert meetings	San José, Costa Rica, and Ottawa, Canada	Costa Rica, Canada
South Pacific sub-regional workshop on IFF issues	Workshop	Nadi, Fiji	Australia with New Zealand and Fiji
Forest conservation and protected areas	Expert meeting	San Juan, Puerto Rico	US, Brazil
The global outlook for plantations	Study		Australia with Brazil, Chile, China, Japan, Fiji, Indonesia, New Zealand, Korea, South Africa, UK, US, CIFOR, FAO, IFF Secretariat

Table 4.2 *Intersessional initiatives in support of the Intergovernmental Forum on Forests (Continued)*

Name of initiative	*Type*	*Venue*	*Organizers/ sponsors*
The role of planted forests	Expert meeting	Santiago, Chile	Chile, Denmark, New Zealand, Portugal
Special needs and requirements of developing countries with low forest cover and unique types of forests (Tehran Process)	International meeting	Tehran, Iran	Iran, UNEP, FAO

Source: UN forests website, www.un.org/esa/forests/gov-iff.html (accessed 15 March 2004)

However, states agreed only to voluntary reporting. As with the IPF's review of the implementation of the forest-related decisions of the UNCED, there was no systematic national reporting and peer review of implementation. Those countries that did present national reports did so at satellite meetings, rather than as part of the formal IFF process.

Nevertheless, useful work was carried out by the German-led Six Country Initiative, which analysed how the IPF proposals could best be translated into action in different national contexts. In addition to Germany, the participants were Finland, Honduras, Indonesia, Uganda and the UK. The initiative concluded that while not all of the IPF proposals for action were relevant for all countries, they could be used as a 'checklist' in a process of assessment, interpretation and planning to improve and monitor the implementation of national forest programmes.[1] The initiative 'screened' the IFF proposals for action, eliminating repetition and distilling the final IPF text to yield 86 separate proposals (see Table 2.6 in Chapter 2).

NGOs were deeply dissatisfied with the IFF's record on implementation. The NGO Forests and the European Union Resource Network (FERN) reported that in Europe 'the wider issues addressed by the IPF, including air pollution, underlying causes, trade and consumption, and internalizing externalities, are hardly being tackled.'[2] An NGO survey of 17 countries reported implementation problems, with many governments claiming that existing forest policy already fulfilled the IPF proposals; it recommended that governments create formal mechanisms to implement the proposals, including establishing focal points and lead government agencies.[3] NGO campaigners at the IFF argued that implementation and national reporting should be kept separate from further political negotiations.

Matters left pending and issues requiring clarification from the Intergovernmental Panel on Forests

Most of the IFF's 120 proposals for action addressed the 12 Category II issues on which the IPF had failed to conclude its work due to time pressure or political conflicts (see Table 4.1). As with the IPF, some progress was made on the more technical issues, although trade, finance and technology transfer again proved contentious.

Trade and the environment

There was discussion on whether the IFF should consider the possibility of a forest products agreement, although it was eventually decided not to discuss this as the World Trade Organization (WTO) was taking an interest in forest products at this time (see Chapter 9). The G77, keen to secure new markets for their forest products, pressed for language on further trade liberalization in forest products. There was no agreement on a draft proposal to remove 'remaining and emerging trade restrictions which constrain market access, particularly for value added forest products.'[4] Those proposals on trade that were agreed were often weakly worded, a typical example being the proposal that:

> *Urged countries, including trade partners, to contribute to achieving trade on wood and non-wood products and services from sustainably managed forests, and implement policies and actions, in particular avoiding policies that have adverse effects, either on trade or environment.*[5]

In a statement on behalf of the G77 at the end of the IFF process, Ositadinma Anaedu of Nigeria noted that international markets were dominated by the developed countries, and trade liberalization remained a key concern for developing countries.[6]

Underlying causes of deforestation

NGOs played the leading role on this issue, carrying out research and hosting a series of meetings that culminated with a global workshop hosted by the government of Costa Rica with support from the United Nations Environment Programme. The final report of the initiative, prepared by the Washington-based Biodiversity Action Network, was presented to the IFF's third session.[7] A comparison between text that NGOs proposed in this report with language adopted by the IFF reveals similar findings as those on the impact on the IPF of the Leticia Declaration on Indigenous Peoples; some text proposed by the NGOs was adopted in amended form, but much was not adopted at all (see Chapter 2). Some of the causes of deforestation reported by the NGOs appear in the IFF's report, although the text proposed by NGOs was substantially modifed. The NGO report stated that: '*The non-recognition of the territorial rights of indigenous and other traditional peoples*, and the resulting invasion of these territories by external actors, was often highlighted as an underlying cause' [emphasis in original].[8] The IFF adopted a weaker formulation, noting the role in deforestation of 'inadequate recognition of the rights and needs of forest-dependent indigenous and local communities within national laws and jurisdiction.'[9] Other NGO proposals were not adopted by the IFF. The NGOs emphasized as underlying causes '*Government-led colonization processes into the forests,* stemming from *inequitable land tenure patterns*' [emphasis in original].[10] This criticism of government-led policies is not reflected in the IFF's report, not surprisingly given that most developing countries will countenance no erosion of sovereignty over natural resources.

NGO proposals that ran counter to neoliberal discourse were also blocked. Government delegates screened the NGO text for language that challenged neoliberal precepts, and such language was either not admitted to, or filtered out of, the negotiations. For example, NGOs lobbied unsuccessfully for the IFF to recognize as an underlying cause '*the privatization of forests for the benefit of large-scale private or corporate landowners*' [emphasis in original].[11] Instead the IFF took an opposite view by supporting further private sector investment in forest projects: 'The concept of an international investment promotion entity to mobilize private sector investment in SFM (viz. sustainable forest management) deserves further consideration.'[12] The negotiations thus reinforced the existing consensus that private sector investment in forests should be promoted. An IFF analytical document hardened this consensus by explicitly viewing the world's forests as an important area for business investment:

> *... the essential role of the private sector in the productive, entrepreneurial aspects of sustainable forest management needs to be recognized, reinforced and promoted. Mobilizing private sector resources has thus been identified as a key component of a global sustainable forest management financing strategy. To accomplish such mobilization, many of the barriers to investment in sustainable forest management activities and operations must be removed or mitigated.*[13]

Most NGOs wish to limit private sector involvement in forests, especially the tropics, arguing, with considerable justification, that corporate involvement has usurped local commons regimes, undermined livelihoods and led to severe forest degradation. But many governments, especially developed governments, argued for an enhanced role for the private sector, with barriers being dismantled to 'investment opportunities.' This benign view of the private sector is certainly not typical of all governments. However, those governments that express reservations about the private sector are in the minority, so that language promoting private forest investment invariably survives the negotiation process. The IFF's final report calls on governments and international organizations to:

> *Encourage private investments in sustainable forest management by providing a stable and transparent investment environment within an adequate regulatory framework that also encourages the reinvestment of forest revenues into sustainable forest management.*[14]

Private sector investors have tended to concentrate their attention on countries with abundant forest resources, principally in the tropics. Countries with limited profit-making potential, such as low forest cover countries, have attracted limited interest. Overall, the case that private sector investment can be a force for sustainable forest management is far from proven.

Financial resources and technology transfer

The G77 again pressed its claims for increased financial flows and the transfer of environmentally sound technologies, and again the developed countries made no concessions. One of the most ambitious proposals on finance came from India, which proposed a global forests facility structured along similar lines to the Global Environment Facility (GEF).[15] However, this proposal, as well as others from the G77 for a global forests fund, was blocked by developed states. The IFF recognized that 'developed countries should fulfil the commitments they have undertaken to reach the accepted United Nations target of allocating 0.7 per cent of gross national product to ODA (viz official development assistance) as soon as possible.'[16] This language is strikingly similar to that agreed in the UNCED outputs.

The negotiations on environmentally sound technologies – which include, for example, technology that enables the harvesting of timber and its extraction from the forest while minimizing damage to other trees – were dominated by differing views on property rights. In line with neoliberalism, developed countries emphasized that technology transfer should take place using market-based mechanisms that respect the intellectual property rights of patent holders. Meanwhile, developing countries argued for concessional terms. The agreed text reflects these conflicts when noting the need to promote the:

> *... transfer of environmentally sound technologies and the corresponding know-how, in particular to developing countries, on favourable terms, including concessional and preferential terms, as mutually agreed, taking into account the need to protect intellectual property rights as well as the special needs of developing countries.*[17]

The G77 insisted that the IFF proposals on technology transfer also mention the benefits arising from the utilization of genetic resources.[18] In doing so, the G77 made it clear that it wished to establish an issue linkage between intellectual property rights to technology and those to genetic resources. By linking the two sets of rights, the G77 appears to have been trying to create the conditions for a bargain encompassing the transfer of environmentally sound technology from developed to developing countries and the transfer of genetic resources from developing to developed countries. So far, no such package has been agreed. First, developed countries have refused to accept that the issues are linked. Second, the issue of patent rights to biological resources has proved politically charged. In order to understand why this is so, it is necessary first to consider one of the IFF's Category II issues: traditional forest-related knowledge.

Traditional forest-related knowledge

Traditional forest-related knowledge (TFRK) forms part of a broader system of knowledge, namely traditional ecological knowledge (that is, the knowledge of their environments held by indigenous peoples and local communities). The practice of traditional ecological knowledge dates back centuries to hunter-gatherer cultures, although the use of the term has only become widespread since the 1980s.[19]

The global politics of traditional ecological knowledge embrace distributive questions, such as who should benefit from the use and commercial application of such knowledge, and the form that these benefits should take? Such questions do not simply apply to TFRK, and are relevant also to the knowledge developed and held by traditional agricultural and fishing communities. As a political issue, traditional ecological knowledge is contested in other institutions, notably the CBD and the WTO. A temporary low-level forests institution such as the IFF stood little chance of making substantive progress on an issue area with such a strong non-forests dimension.

Another factor that prevented the IFF from making progress on TFRK is that traditional forest-related knowledge and traditional ecological knowledge are ambiguous terms with no commonly accepted definition.[20] This lack of definitional clarity has prompted political manoeuvring in international negotiations, with different actors pursuing different definitions according to their self-interest. We have seen that at the IPF governments agreed a broad definition of TFRK that included actors from outside the forest, such as forest businesses. NGOs had favoured a narrower definition centred on forest peoples (see Chapter 2). A further dimension to the definitional problem is whether traditional ecological knowledge exists in a 'pure' form. Analytically, there is a clear distinction between traditional ecological knowledge and Western knowledge systems, such as scientific forestry. In practice, TFRK may become enmeshed with other knowledge systems such as scientific knowledge (see Box 4.1). Numerous forest research institutes and businesses in Europe and North America have benefited from knowledge developed and acquired by local forest communities on, for example, the medicinal properties of plants. Similarly, in many places in Africa and Asia, the ecological knowledge yielded from research by colonial foresters was acquired by local communities during the 19th and 20th centuries. This knowledge has since been passed on from generation to generation, so that today, in many localities, it is difficult to separate the traditional knowledge from the colonial. According to one view, the notion of traditional knowledge is too narrow, and it makes more sense to talk, instead, of local knowledge, which is rooted in a particular ecology and culture, but which was not necessarily entirely generated by traditional communities.[21]

So while in principle there is a clear distinction between scientific knowledge and traditional ecological knowledge, the two different knowledge systems have drawn from each other. Western science has informed traditional ecological knowledge, and *vice versa*. The two different knowledge systems can 'know' the same things. However, the key differences between the two are not so much epistemological – in other words, how we know what we know – but political. Several political questions arise. Should traditional ecological knowledge be communally owned as a public good, or can it be patented, with the rights to the knowledge being held by actors outside the forest, or even in another country? If the rights can be patented, which actors or groups should legitimately own the intellectual property rights? How should the monetary benefits from the commercial application of traditional knowledge be shared? If the local community should receive a share of these benefits, then how should the 'local community' be defined, and how large should the share be? Finally, for many indigenous peoples' organizations, the key political questions do not solely concern the rights to traditional knowledge, but also the practice of such knowledge, which, in turn, leads to land rights issues, such as access and tenure, as well as to the issue of self-determination.

Property rights are thus absolutely central to resource use in the developing world. Those who holds property rights – either physical ownership of the land or the intellectual property rights to the particular uses to which biological resources can be put – control resource extraction and use.[24] There are three principal claimants to the intellectual property rights of TFRK:

1. traditional communities and indigenous peoples who initially developed TFRK (some of these groups see patenting as a form of biopiracy; see Chapter 1);
2. the private sector interests that commercially develop TFRK and which are granted rights to knowledge through a patent or other type of property right;
3. the government of the state where the resource is physically located (the claim that the state should receive a share of the benefits from utilizing TFRK is made especially by tropical forest governments).

At the time of writing (June 2006) parties to the CBD are negotiating a benefit-sharing regime that could recognize all three claims (see Chapter 9). However, contemporary international law on this area is messy. On balance, it favours commercial patent holders, who have successfully pressed their claims that corporations should receive the rights to any 'innovative' uses of biological resources. But there have also been numerous legal declarations affirming that the state has sovereignty over natural resource use. Meanwhile, International Labour Organization (ILO) Convention 169 of 1989 stipulates that indigenous and tribal peoples 'shall have the right to decide their own priorities for the process of development as it affects their lives, beliefs, institutions and spiritual

Box 4.1 Scientific knowledge and traditional ecological knowledge

The analytical distinction between Western scientific knowledge and traditional ecological knowledge can be summarized thus. Western science is based on the Enlightenment view that there is a separation between values and facts, and between the observer and the observed. The values of the scientist should not interfere with the process of collecting data, testing evidence and establishing facts. This approach to the scientific method is known as positivism. The 17th-century French philosopher René Descartes argued that the biological world of plants, trees and animals can be studied and understood mechanistically in much the same way as, say, the movements of the planets. Descartes extended this view to human beings: the human body works mechanistically and is separate from the soul. This view – that the human mind and human values exist separately from nature – has been termed Cartesian dualism, after Descartes. The job of the scientist is to understand systems, such as the solar system or a natural system, by reductionism; knowledge is gained by reducing a system into discrete parts and analysing how these parts work. This knowledge can then be distilled into generalizations and principles that, it is claimed, apply irrespective of cultural context, time or space.[22]

In distinction, the idea of traditional ecological knowledge rests on very different philosophical foundations. The positivist notion that human values and beliefs exist separately from the world is rejected. In distinction to the reductionism of Western science, traditional knowledge is seen as holistic and embracing a native ecological wisdom in which humans are an integral part of ecosystems. Whereas the scientific positivist-reductionist paradigm has promoted an instrumentalist and utilitarian approach to nature, which, its critics claim, treats nature as a resource to be extracted and used elsewhere, traditional ecological knowledge promotes localized uses of nature and long-term nature conservation. With traditional ecological knowledge the manager of nature is also the user of the goods and services that nature produces, whereas Western science promotes a separation between the manager and the user. Western science, therefore, is disembedded from space and place, whereas traditional knowledges, which acknowledge no separation between nature and culture, are embedded in and attached to place and the local environment. Whereas Western science has been used to 'organize' nature for commercial purposes that can be seen as exploitative of the environment of others, traditional knowledges stress a harmonious and non-exploitative relationship between humanity and nature.[23]

well-being and the lands they occupy or otherwise use.'[25] This can be interpreted as a recognition that indigenous peoples have the right to determine how their natural resources are used.

Political contention over the intellectual property rights (IPRs) to biological resources has seen different groups of states negotiating different legal declarations in different international institutions. A given state will favour those institutions which, it is judged, will yield international law that best promotes its interests. The result is different legal instruments providing different, often competing, rights, leading to some fundamental, and still unresolved, contradictions in international law on intellectual property rights.

The history of this area dates to the seed wars of the 1960s on the development of new varieties of plants and crops through the crossbreeding of seeds. At the risk of oversimplification, the seed wars saw, on the one side, communal farmers, usually in developing countries, asserting *farmers' rights* – that is, the rights of farmers to continue using seeds for free – and, on the other side, commercial plant breeders, often in developed countries, asserting *breeders' rights* – namely, the right to benefit commercially from the scientific breeding and large-scale production of seeds. The seed wars were a precursor to the contemporary conflict between the view that biological diversity is a public good or common resource, and commercial claims that species can be privately owned and developed for profit.

The opening salvo in the seed wars came in 1961 when the first international agreement on plant breeders' rights, the International Convention for the Protection of New Varieties of Plants (UPOV)[26] was negotiated. The convention allowed a plant breeder who has developed a variety of seed that is 'new, distinct, uniform and stable' to apply for a breeder's right to that seed.[27] This right extended to the subsequent production, reproduction, sale, marketing and trade of the variety.[28] Most of the signatories of the UPOV convention, which relates to the commercial breeding and sale of new plant varieties, were developed countries.[29] The plant breeders' rights that the convention granted virtually amounted to a form of patenting, although without the legal status of a patent, and they were opposed by developing countries, many of which, two decades later, successfully negotiated within the FAO the 1983 International Undertaking on Plant Genetic Resources. This can be seen as a developing country antidote to the pro-business emphasis of the UPOV convention.

The 1983 international undertaking, which was non-legally binding, defined farmers' rights as 'rights arising from the past, present and future contributions of farmers in conserving, improving and making available plant genetic resources, particularly those in the centres of origin/diversity.'[30] It defined plant genetic resources as 'a common heritage of mankind to be preserved, and to be freely available for use, for the benefit of present and future generations.'[31] The notion of common heritage was opposed by UPOV signatory states, which viewed it as a 'strike at the heart of free enterprise.'[32] The precise relationship between breeders' rights and farmers' rights has never been defined, and the two have an uneasy co-existence. The tension between the UPOV convention and the FAO undertaking was still unresolved when the IFF considered TFRK (see Box 4.2).

Two further layers of international law relevant to TFRK were agreed during the 1990s. The first is the 1992 Convention on Biological Diversity, which, in Article 8(j), provides for the equitable sharing of the benefits that arise from the use of traditional knowledge. In principle, the concept of benefit-sharing can reconcile the various claims to the benefits from using biodiversity made by states, transnational corporations, indigenous peoples, farmers and so on. However, it should be emphasized that some developing states, notably Brazil, have championed benefit-sharing not in the interests

Box 4.2 International Treaty on Plant Genetic Resources, 2001

One year after the IFF was wound up, the FAO agreed the International Treaty on Plant Genetic Resources of 2001 to replace the non-legally binding International Undertaking on Plant Genetic Resources of 1983. There are some differences between the 2001 treaty, which entered into legal effect in June 2004, and the undertaking. In 1983, when the undertaking was agreed, the collective response of developing countries to the patenting and commercialization of genetic resources was that such resources were a common heritage of mankind, freely available for use by all.

During the UNCED negotiations, there was a strategic shift by developing countries, which, realizing the bargaining leverage that biodiversity provided, now gave up the common heritage argument and instead asserted that developing countries were entitled to a share of the commercial benefits arising from the use of their genetic resources. Hence, the Convention on Biological Diversity does not mention the concept of common heritage, and neither does the 2001 International Treaty on Plant Genetic Resources. Both the CBD and the 2001 treaty aim to promote a multilateral system of access and benefit-sharing. The 2001 treaty is the first legally binding instrument to recognize farmers' rights (Article 9), and it aims to move beyond the tension between the International Convention for the Protection of New Varieties of Plants (UPOV) and the 1983 undertaking. FAO Director-General Jacques Diouf remarked that the entry into legal effect of the treaty was 'the start of a new era. The treaty brings countries, farmers and plant breeders together and offers a multilateral approach for accessing genetic resources and sharing their benefits.'[33] But while the treaty provides a clear endorsement of farmers' rights, it is ambiguous on the status of breeders' rights. One environmental activist has claimed that the treaty 'leaves the door open to private property rights: patents and plant breeders' rights.'[34] It remains to be seen whether the treaty can reconcile the competing rights claims made by communal farmers and commercial plant breeders.

of indigenous peoples, but because they wish any benefits from the use of traditional knowledge to accrue to national governments rather than the local level (see Chapters 5 and 9). Given that the practical application of Article 8(j) has yet to be agreed, there was no prospect of the IFF making any progress on the issue.

The second layer of international law is the Agreement on Trade-Related Aspects of Intellectual Property Rights (TRIPS), which forms part of the single undertaking agreed during the Uruguay Round negotiations that created the WTO. The single undertaking means that TRIPS applies to all WTO members, and its provisions are subject to the integrated WTO dispute settlement procedures.[35] The migration of IPRs into the WTO is significant, given that previously the sole relevant international organization handling IPRs was the World Intellectual Property Organization (WIPO). The interest of the General Agreement on Tariffs and Trade (GATT)/WTO system in IPRs can be explained thus: WIPO was favoured by many developing countries since it left each country free to pursue its own legislation on intellectual property, subject to the restriction that national law should not discriminate between foreigners and a country's own nationals. The WIPO system was thus non-discriminatory within countries, but allowed for differences between countries. This did not suit the interests of Western-based corporations, who sought a unified international system of IPRs that included rights to biodiversity and biotechnology. The biotechnology, food and pharmaceutical industries subsequently successfully lobbied developed countries, in particular the US, to shift IPR negotiations into the Uruguay Round negotiations as a 'trade-related' issue.[36]

A key TRIPS clause that relates to traditional forest-related knowledge is Article 27.3(b), which allows states to exclude from patentability 'plants and animals other than micro-organisms, and essentially biological processes for the production of plants or animals other than non-biological and microbiological processes.' However, what at first sight seems an exemption for biological resources is then qualified: states 'shall provide for the protection of plant varieties either by patents or by an effective *sui generis* system or by any combination thereof.'[37] The phrase *sui generis* refers to an alternative to patenting that provides some guarantee of the rights of those who develop new plant varieties. One possible model is the rights system of the UPOV convention.[38] When TRIPS was negotiated, it was agreed that the exemptions in Article 27.3(b) should be reviewed in 1999. At the time of writing (June 2006), this review has not been concluded (see Box 4.3).

TRIPS makes no mention of communally held traditional knowledge. It reflects the interests of those actors that were instrumental in its negotiation, namely developed states and corporations seeking to promote the commodification of nature and the private ownership of biological resources. This makes TRIPS highly unpopular with social and environmental groups. For example, in 1997, the year that the IFF was created, 45 civil society organizations issued the Thammasat Resolution, calling for a moratorium on bioprospecting, the revision of TRIPS to allow countries to exclude life forms and biodiversity,

Box 4.3 Agreement on Trade-Related Aspects of Intellectual Property Rights Article 27.3(b) and the Convention on Biological Diversity

The WTO's Doha Ministerial Declaration of 2001 instructed the TRIPS Council 'to examine, *inter alia*, the relationship between the TRIPS Agreement and the Convention on Biological Diversity, the protection of traditional knowledge and folklore, and other relevant new developments.'[39] The Doha Declaration thus broadened the review to be carried out by the TRIPS Council of Article 27.3(b). This was the result of lobbying by the developing countries, most of which favour a *sui generis* system (the details of which are still to be decided) for the protection of TFRK, under the auspices of the CBD. The mention in the Doha Declaration of the CBD has not been welcomed by biotechnology and agriculture corporations, which wish to ensure that the intellectual property rights of knowledge on biological resources remain a 'trade-related' issue firmly under the purview of the WTO.

and the 'primacy of international agreements on biodiversity, such as the CBD and FAO instruments, over TRIPS and other trade regimes.'[40] The IFF was not mentioned, suggesting that the NGOs promoting the resolution did not consider it a forum that would have any influence on the politics of traditional knowledge.

Like the NGOs, many developing countries are suspicious of TRIPS. The concepts of farmers' rights and TFRK were initially promoted by citizens' groups and NGOs resisting the expansion of capitalist markets into nature. These groups have received support from developing world governments not because these governments oppose commodification, but because they oppose the terms of commodification of the TRIPS regime, which allows most of the benefits to flow to transnational corporations. Most developing countries consider that their interests will best be realized through developing the concept of benefit-sharing, as elaborated in the CBD and the International Treaty on Plant Genetic Resources of 2001.

Whereas the international politics of the property rights of nature were initially played out between the UPOV convention and the FAO, today the main axis of this issue can be found in the struggle for ascendancy between TRIPS, the CBD and the WIPO, the latter of which, since 2000, has acquired a renewed relevance to this area through the negotiation of a Substantive Patent Law Treaty (see Chapter 9).[41] Different states have gravitated towards different international institutions, depending on which institution they consider will best promote their interests. The resulting incoherence in international law on IPRs and traditional ecological knowledge, with different legal instruments stipulating different provisions, symbolizes one of the central struggles of our age: who wins and who loses in the politics of natural resource use.

Against this background the IFF's deliberations were inconsequential, merely reflecting external legal and political uncertainties. The text negotiated at the IFF makes several references to the role of the CBD in protecting TFRK, including the CBD's role in fair and equitable benefit-sharing.[42] There is no mention of TRIPS. In this respect, the IFF's outputs suited the interests of the G77 more than the developed states, although not too much significance should be attached to this: most delegates were aware that the real politics of this issue were being played out elsewhere. While the IFF had a mandate to consider TFRK, it simply did not have the political authority to handle such a complex and contentious issue. It was thus no surprise when the IFF reported that it had 'discussed but could not conclude the debate or reach consensus' on a proposal to encourage countries to enable local communities, indigenous peoples and forest-dependent groups to realize the benefits from TFRK, including through 'due recognition to the use of traditional forest-related knowledge in patent applications for technologies.'[43]

Valuation and economic incentives

We have argued that the work of the IPF on forest valuation and economic incentives reflected the premises and assumptions of neoliberalism (see Chapter 2). Similar dynamics were at play in the IFF. The analytical documents on these issues were prepared by the World Bank, which advocates neoliberal policy responses, such as decreasing state spending, privatization and an enhanced role for market-based solutions. The World Bank has had a longstanding interest in new environmental markets.[44] This was reflected in the documents it prepared for the IFF, which noted five areas in which new markets could provide incentives to conserve forest goods and services.

At least two of these areas are politically controversial. First, the Bank proposed the use of carbon sequestration markets. This is a contentious idea, with opponents claiming even before the Kyoto Protocol entered into legal effect that tradable emission permits would 'privatize' the atmosphere, grant pollution rights to rich states and corporations, and lead to the planting of 'carbon forests' in which the non-carbon goods and services of forests are diminished (see Chapters 1 and 9). Second, the Bank noted the role of markets for biological diversity. Like carbon sequestration markets, biodiversity markets are possible only because property rights have been extended into an area where they did not previously exist.[45] The Bank noted the role of bioprospecting as 'an emerging source of forest-based revenue,' with the royalties from genetic compounds being 'particularly relevant for specific forest areas rich in biodiversity.'[46] However, the distributive question of who should receive these revenues and royalties was not considered in the World Bank's analytical documents for the IFF.

Of the remaining three areas, one is an area where the role of markets is currently unproven, namely hydrological services. The remaining two areas are hardly new: non-timber forest products, such as nuts, fruits and berries, where markets have great potential for conservation; and ecotourism, which can assist conservation efforts, although ecotourist projects have proved divisive when they undermine local rights and livelihoods.

While the World Bank's analytical papers to the IFF have a logic that fits comfortably with neoliberalism, the Bank also elaborated some other policy options. The Bank stated that the cultural, spiritual and ethical values of forests 'should not be contested in principle, even if they cannot be denominated in monetary terms'; valuation alone will not attain conservation and is no substitute for balanced political decisions.[47] The World Bank emphasized that well-designed policies will include 'a mix of regulatory and economic instruments.'[48] Although the Bank argued for biodiversity markets, it also envisaged a role for international public finance to conserve the public goods value of biodiversity.[49] However, the main emphasis of the World Bank documents was on expanding new markets, market-based solutions, using incentives to modify market-based behaviour, and internalizing environmental values within the market. In this respect, the Bank reiterated orthodox neoclassical valuation models. In several cases, language from the documents was used as a basis for negotiating IFF proposals for action. And in many cases wording proposed by the Bank survived the negotiations virtually intact. Two such examples are now provided. In Table 4.3, the first column provides excerpts from a World Bank analytical document, while the second is language from the IFF's final report.

It can be seen that the IFF made only minor changes to the World Bank's wording. In the first example, a provision was added during the negotiations that data should be collected for non-wood materials, presumably meaning plastics, metals, concrete and so on. Minor amendments have been made in the second example: 'benefits' has been added to 'costs' and the word 'incremental' has been added. Nonetheless, to all intents and purposes the language in the second column was not negotiated by delegates: it was fed into the negotiation process by the World Bank and emerged at the end of the negotiations largely unscathed.

It might be argued that this does not really matter; the examples do not deal with politically substantive issues and relate only to uncontroversial matters. At negotiations, delegations have insufficient time to negotiate everything *ab initio*. Advance preparation by organizations such as the World Bank facilitates negotiation by providing an initial draft. Delegates could have altered the text proposed by the Bank, but exercised their right not to do so.

Some responses should be made to these points. First, the examples given have been selected because they are glaring and obvious. There are many other examples where the influence of the World Bank is considerably more subtle, nuanced and difficult to track. Second, what is happening here says a great deal about how international declarations are agreed, by whom and for whom. Public international law, even soft international law such as the IFF proposals, should be agreed by states and not drafted by the secretariat of an international organization. Third, at international negotiations, especially those taking place in institutions with a finite lifespan such as the IFF, time constraints frequently preclude meaningful negotiation on all paragraphs. It is easy for text to slip through unnoticed by some delegations. Fourth, if an organization such as the

Table 4.3 *Language proposed by the World Bank that appears in the proposals for action of the Intergovernmental Forum on Forests*

Language in analytical documents prepared by the World Bank	**Final text negotiated by government delegates to the Intergovernmental Forum on Forests**
'Urge governments to improve collection of quantitative data to enumerate and develop physical accounts for the range of forest outputs and services, including inventories of timber and other goods and services, and impacts of changes in forest use on the environment'	'Urged Governments to improve collection of quantitative data to enumerate and develop physical accounts of the full range of forest goods and services, including inventories of timber and other goods and services, and impacts of changes in forest use on the environment. This should also be done for substitute non-wood materials'
Source: UN document E/CN.17/IFF/1999/12, para 44(a)	*Source:* UN document E/CN.17/2000/14, para 107(a)
'Develop an approach for the identification of the costs of sustainable forest management (a supply price), which can be used for a cost-efficient use of scarce investment funds for forest values.'	'... to develop approaches for the identification of the costs and benefits, including incremental costs and benefits, of sustainable forest management, which can be employed for a cost-efficient use of investment funds for forests.'
Source: UN document E/CN.17/IFF/1999/12, para 48(b)	*Source:* UN document E/CN.17/2000/14, para 107(c)

World Bank is going to feed into UN negotiations language that reiterates an existing orthodoxy, in this case on valuation and economic incentives, and if this language is changed only moderately during negotiation so that the agreed output produces nothing new and merely reiterates conventional thinking, then what precisely is the point of the negotiations?

This is not the first example we have noted of text on economic valuation produced by one actor being uncritically adopted by another. Scholarship from the International Institute for Environment and Development (IIED) on valuation was copied into an IPF analytical document (see Chapter 2). This, we suggested, indicated the hegemony of environmental valuation methods among international forest policy-makers and illustrated how limited genuine intellectual debate is on this subject.

The IPF and IFF examples took place at different stages of what we can see as an international text agreement process. The IPF example is one where an international organization, namely the FAO, adopted text from outside the UN system, whereas the examples presented above are of text from an international organization being formally endorsed in UN negotiations. The IFF examples have thus taken place further 'downstream' than the IPF example. When considered together as part of an international text agreement process, these different examples not only raise fundamental questions on the efficacy of UN forest negotiations; they also say a great deal about how neoliberal thinking is perpetuated and reinforced throughout the international political system. Ideas become hegemonic when they are passed on, unaltered and unfiltered, from one institution to another. With respect to valuation and economic incentives, the IFF, like the IPF, produced nothing new or original. It merely reproduced deeply embedded ideas and concepts derived from neoclassical economics and neoliberal thinking.

To Michel Foucault, language and discourse is central to the exercise of power in politics. The production of discourse and the production of power are not separate processes; they are inextricably interwoven. The power of political actors is, in large measure, derived from discourse. A dominant discourse confers legitimacy on certain social practices while delegitimizing others. The more dominant a discourse is, the more it 'makes sense' to think within it. Ideas grounded within a dominant discourse will appear realistic, feasible and pragmatic. Proposal that runs counter to such a discourse will seem controversial, radical, impractical and maybe even heretical.[50]

Indeed in intergovernmental negotiations, a form of self-censorship often takes hold. Many delegates will refrain from making innovative proposals if they believe that such proposals will not survive the negotiation process. So in the IPF and IFF proposals, no challenges to neoliberal orthodoxy emerged from the negotiations, whereas statements that reiterated that orthodoxy were adopted. For example, a proposal from Canada that 'policies that distort the efficient operation of markets may contribute to the unsustainable management of forests' was accepted.[51] There is another example: the G77, US and EU collaborated to negotiate the following text:

> *Most recent outlook studies have reached the general conclusion that, at the global level, wood fibre supply will be broadly matched with demand without likely price increases; but at the national level some countries may experience shortages and possibly price increases.*[52]

We shall leave aside the question of whether international negotiations should be used for states to negotiate the predictions of 'recent outlook studies.' The point is that this paragraph is based on some fundamental premises of neoclassical economics: that nature exists separately from, and is a resource input to, the economic system; that over the long-term there is a limitless supply of natural resources so that supply will meet demand; and that prices reflect only short- to medium-term scarcity rather than long-term depletion. It is this economic framework that leads to stable timber prices during a period of history when global forest cover is in constant decline. Similar examples can be found elsewhere in the global economy, such as short-term falls in oil prices while oil reserves are gradually depleted.[53] Given the short-term orientation of markets, it seems unlikely that price signals can prompt the seismic shift in production and consumption patterns necessary to safeguard natural resources, at least until depletion is considerably more serious than it is now.

Like the IPF, the IFF reproduced a neoliberal discourse that legitimized the expansion of markets and private actors into forests. The attention that the IFF devoted to alternative policies, such as strengthened regulation by publicly accountable authorities, was nugatory in comparison. The ideas of green national accounting and a code of conduct to regulate business were not even raised at the IFF following their dismissal at the IPF (see Chapter 2). Overall, the range of possible policy responses considered by the IFF was considerably narrower than those examined by the Panel.

Rehabilitation of environmentally critical areas

Environmentally critical areas include mountain, coastal, mangrove and small island forests, as well as forests threatened by drought and desertification. The most significant IFF-related achievement in this area came from an intersessional initiative that originated during the IPF process. The Iranian delegation had formed the view that the Panel was overly centred on the needs of major tropical forest countries, with the primary focus being on slowing deforestation. The IPF paid relatively little attention to low forest cover countries, which have some of the most fragile ecosystems in the world. The main problem that these countries face is not slowing deforestation, but rehabilitating and expanding their sparse tree cover. Two Iranian delegates, Mohamed Reza Djabbari, who has represented Iran at the Convention to Combat Desertification, and Bagher Asadi, who served as IFF co-chair, lobbied within the G77 for an intersessional initiative to address the needs of low forest cover countries. The G77 agreed and formally proposed to the IPF that Iran take the lead on such an initiative.[54]

The initiative, which was enabled by the re-emergence of Iran from a period of diplomatic isolation following the Iranian revolution of 1979, was named the Low Forest Cover Countries (LFCC) Process, or Tehran Process. It was sponsored by Iran with assistance from the FAO and UNEP. LFCCs have been defined as countries with a forest cover that is less than 10 per cent of the land area. The causes of low forest cover include a low original endowment of forests, climatic and physical conditions, and land-use pressures.[55] During 2000, 71 countries qualified as LFCCs, although not all of them have taken part in the Tehran Process. Most LFCCs have dryland forests that are threatened by drought, although the definition also admits other forest types. The two countries with the highest forest cover that qualify as LFCCs are the UK (with forest cover of 9.9 per cent of land area) and the Netherlands (9.8 per cent). The countries with the lowest forest cover are Iceland and Saudi Arabia (0.1 per cent each), followed by Iraq, Lesotho and Libya (0.2 per cent). The 71 LFCCs have an average forest cover of just 3.5 per cent.[56] The initiative is primarily aimed at Northern Africa and Asia. Countries from Africa to have taken part include Egypt, Ethiopia, Ghana, Libya, Mali, Nigeria, South Africa and Sudan, while countries from Asia include Armenia, Bangladesh, Iran, Iraq, Kazakhstan, Lebanon, Oman, Tajikstan and Yemen. In economic terms, the forest economies of these countries are insignificant. They have a negligible role in timber production; however, their forests are important ecologically.

The LFCC process was launched in 1999 with the Tehran Declaration, in which countries pledged to formulate strategies to combat desertification and rehabilitate forest cover.[57] The first follow-up meeting was held at the IFF's fourth session. The LFCC initiative is one of the few permanent side products to emerge from the IPF and IFF. Its secretariat is based at the Iranian Agriculture Ministry. Over the long term, some collaboration between the LFCC process and the Convention to Combat Desertification seems essential, given the overlapping state memberships and objectives of these two processes, neither of which are particularly well funded.

In the view of Taghi Shamekhi, who chaired the process in 1998, the LFCC initiative has opened up a new 'front' against deforestation. While the most threatened forest ecosystems are in the tropics, the most fragile forest ecosystems are in the low forest cover countries. As a result of the initiative, Iran and other dryland countries that had previously been marginalized in global forest politics have been able to assume a leading role in an issue of genuine global significance.[58]

The IFF's Category II issues were those on which the IPF had made little or no progress. For many of these issues, the IFF also made no inroads. Genuine debate was constrained due to a variety of factors, including extraneous institutional turf wars (as with TFRK), assertions of state sovereignty (as with NGO proposals on the underlying causes of deforestation) and orthodox economic discourse (as with forest valuation). Of the IFF's 120 proposals, 105 relate to Category II issues (see Table 4.4).

The forests convention debate revisited

The fourth session of the IFF was dominated by negotiations on institutional arrangements. Two post-IFF options were discussed: a forests convention and a new forests body within the UN system.

The position of the EU was dominated by internal coordination problems. In 1996, the EU had formally agreed to support a forests convention.[59] However, this position was not revisited during the IFF process, and by the IFF's fourth session some EU states no longer supported a convention, principally the UK and Sweden. The countries that pushed most strongly for a convention within the EU caucus were Spain and Finland, with support from Germany and the Netherlands.[60] There was also tension between the European Commission and EU member states. A commission representative at the IFF's final session showed some impatience with the anti-convention states and urged all EU members to respect the pro-convention policy of 1996.[61] The result of all this manoeuvring was that the EU, led by Portugal, which held the EU presidency, could offer no clear position on this issue. Despite spending much of the

Table 4.4 *The proposals for action of the Intergovernmental Forum on Forests*

Intergovernmental Forum on Forests category		***Number of proposals for action***
I	*Promoting, facilitating and monitoring the implementation of the proposals for action of the IPF*	
I.a	Promoting and facilitating implementation of IPF proposals for action	7
I.b	Monitoring implementation of the IPF proposals for action	8
II	*Consider matters left pending and issues that need clarification from the work of the IPF*	
II.a	Financial resources	5
II.b	Trade and the environment	8
II.c	Transfer of environmentally sound technologies to support sustainable forest management	14
II.d.i	Underlying causes of deforestation	13
I.d.ii	Traditional forest-related knowledge	5
II.d.iii	Forest conservation and protected areas	13
II.d.iv	Forest research	11
II.d.v	Valuation of forest goods and services	4
II.d.vi	Economic instruments, tax policies and land tenure	7
II.d.vii	Future supply and demand for wood and non-wood forest products and services	7
II.d.viii	Assessment, monitoring and rehabilitation of forest cover in environmentally critical areas	5
II.e	Forest-related work of international and regional organizations and existing instruments	13
Total number of IFF proposals for action		**120**

Notes: Table 4.4 is summarized from UN documentation. The source is not therefore directly comparable with Table 2.6 in Chapter 2, where the original UN source has been screened for areas of overlap. There were no proposals for action for IFF Category III. However, the IFF report contained an annex that recommended the establishment of the United Nations Forum on Forests.

Source: UN document E/CN.17/2000/14, 'Report of the Intergovernmental Forum on Forests at its Fourth Session, New York, 31 January–11 February 2000,' 20 March 2000

IFF's final session in caucus, the EU neither supported nor opposed a forests convention when the plenary reconvened on the final day of negotiations (see Box 4.4).

Unlike the EU, the G77 agreed a synthesized common position, despite a considerably larger membership and a wider diversity of forest types and socioeconomic conditions compared to the EU. With some of its members

Box 4.4 The European Union in international environmental negotiations

Who negotiates on behalf of the EU in international negotiations depends on the issue and whether the European Commission or member states have competence for the issue. The different competences of the EU can be seen as a spectrum. At one end are issues for which the Commission has exclusive competence, while at the other end competence belongs exclusively to member states. In between are varying degrees of shared competence:

- Areas of *Community competence* are provided for in EU treaties, such as the Treaty of Maastricht. For these areas, such as trade, the European Commission speaks on behalf of the EU in international negotiations.
- Areas of *member state competence* are those where policy is decided solely by EU member states with no input from the Commission. Examples are foreign and security policy.
- Areas of *shared competence* are those where both member states and the Commission have a legitimate interest.

The environment has evolved as an area of shared competence, and this has been formally recognized in the draft European Constitution (Article 13). The Single European Act of 1987, which amended the Treaty of Rome of 1957, provides that 'environmental protection requirements shall be a component of the Community's other policies' (Article 130R). The directorate-general with responsibility for environmental negotiations, including forests, is DG Environment (formerly DG XI). But while the European Commission has a role, member states have the right to lead in negotiations. The country holding the EU presidency, which rotates amongst member states every six months, speaks for the EU at international negotiations, except on those issues where the Commission has exclusive competence. So, for example, during the IPF and IFF processes, the European Commission spoke for the EU during negotiations on the trade in forest products since trade is an area of exclusive Community competence.

The EU coordinates its position on forests through a Council Working Group on Forests, which comprises all member states, as well as DG Environment. The working group meets in Brussels before negotiations to agree strategy. When negotiations are in progress, it will usually meet in caucus at least once a day to coordinate positions. The working group is chaired by the presidency. The presidency relies on the European Commission for advice and institutional memory. Commission representatives monitor the negotiations in case they stray into areas of exclusive EU competence, and they check for possible inconsistencies between what the presidency is saying in the negotiations and broader EU policy. They will advise the presidency on what language to use during the negotiations. In everly respect, therefore, the dynamics between the presidency and the European Commission are central to the effectiveness and coherence of the EU as a negotiating entity.[62]

in favour of a convention and others against, the G77 position was that it was premature to launch negotiations for a convention, but that this question should be reconsidered at a later date.

Canada, which has been the most persistent forests convention advocate since 1990, initially tried to entice the G77 countries to support a convention by appealing to their concerns on finance. Canada suggested that it was unlikely that new provisions for finance would be made outside a convention, whereas a convention could include a global forests fund. Canada also noted that the Global Environment Facility would be more likely to fund forest projects if a convention was agreed. Other countries that supported a convention were Chile, Costa Rica, Guatemala, Malaysia, Panama, Poland, Russia and Switzerland. Some former Soviet republics supported Russia by arguing for a convention, namely Belarus, Georgia, Kyrgyzstan, Tajikstan and Ukraine.[63] The countries that most strongly opposed a convention were Brazil and the US, supported by Ghana, India, New Zealand and most South American countries.

The final week of the Intergovernmental Forum on Forests

As the negotiations progressed it was tentatively accepted that any new organ would be called the United Nations Forum on Forests (UNFF), and that it should have a higher status than the IPF and IFF, reporting to either the UN General Assembly or the Economic and Social Council (ECOSOC). Brazil wanted the new organ to have universal membership. Since there was no precedent for an organ reporting to the General Assembly or the ECOSOC to have a larger membership than the parent body,[64] the IFF coordinator, Jag Maini, contacted the UN legal services department for advice during the negotiations. A representative from UN legal services then appeared before the IFF and advised delegates that while it would be unusual for an organ with universal membership to be created that reported to the General Assembly ECOSOC, there was nothing to prevent it providing that the parent body itself accepted this.[65]

Late in the negotiations Canada hardened its position. Canada argued that if there was no consensus for a convention, then it could not support the creation of a UNFF. The text at this stage was relatively clean; but this represented an illusion of progress as Canada proceeded to insert square brackets around all references to the UNFF, noting that Canada would remove these only when other states agreed to a convention. Only then, it was added, would Canada discuss finance. Canada had upped the stakes. This calculated act of brinkmanship led to an all-night negotiating session on the final day.

Almost all developing countries spoke up in favour of increased aid transfers. Algeria, Chile, China, Colombia, Ecuador, Egypt, India, Nigeria for the G77, South Africa, Venezuela and Zimbabwe pushed for a global forests fund. The EU and the US responded that they could not support a fund, although as major donors they were committed to forest-related aid. The head of the US delegation, Stephanie Caswell, sought to deflect attention from the proposal for a fund by stating that the US president had proposed US$150 million in new money for tropical forest conservation, of which 75 per cent would be allocated for bilateral aid and 25 per cent for debt-for-nature swaps.[66] To a brief round of applause from developing country delegates, the chair, Alison Drayton of Guyana, responded 'Thank you, United States, and we look forward to seeing how that proposal emerges from Congress.'[67]

For the final day of negotiations the convention question and finance were the only issues outstanding, and the chair, Bagher Asadi, initiated a series of informal consultations. The plenary negotiations eventually reconvened on the evening of the final day but soon stalled. At 2.20 am the chair suspended the plenary and asked delegates to sort out their differences in a private session. There followed three hours of informal discussions, principally between Brazil, Canada, the EU, the G77 and the US, in between which impromptu EU and G77 coordination meetings took place on the floor. Gradually, the delegates made incremental concessions on language, while giving away nothing of substance. Finally, at 5.55 am, agreement was reached after the exchange of several handwritten drafts.

With respect to a convention, it was agreed that the ESOSOC and the General Assembly would, within five years, 'consider with a view to recommending the parameters of a mandate for developing a legal framework on all types of forests.'[68] This opaque formulation was sufficiently equivocal to allow for several interpretations. With respect to finance, the agreed text committed the ECOSOC and General Assembly to 'take steps to devise approaches towards appropriate financial and technology transfer support to enable the implementation of sustainable forest management, as recommended under the IPF and IFF processes.'[69]

This compromise language did not really satisfy either those who wanted a convention or those who demanded increased forest-related aid. However, Canada agreed to the wording and removed the square brackets that it had earlier inserted around references to the UNFF, despite the text failing to contain a firm commitment to a convention. In fact, Canada had little real option but to agree to the creation of the UNFF. What had appeared a high risk gamble that could push the negotiations towards collapse was merely a carefully controlled negotiating tactic. Had the UNFF not been created, global forest politics would have regressed to the wilderness years of 1992–1995 when there was no international forests dialogue, clearly not a desirable situation for a government striving for the heightened level of international cooperation that would ensue from a convention.

By the time the closing statements were made, the sun was rising over New York. The chair, Bagher Asadi, quipped that the sun was also rising over the UNFF. On that note, the gavel came down to close the IFF.

Informal consultations on the new institutional arrangement

One detail remained outstanding, namely whether the UNFF should report to the General Assembly or to the ECOSOC. The president of the ECOSOC

(Indonesia) asked IFF co-chair Bagher Asadi to resolve this question. Asadi held a series of informal consultations throughout June to September 2000.

The respective advantages and disadvantages of a forum reporting directly to the General Assembly or to the ECOSOC can be summarized thus. The General Assembly has higher status than the ECOSOC. However, it is not as open and inclusive towards NGOs and major groups as the ECOSOC and is a relatively slow and cumbersome body. Unlike the General Assembly, the ECOSOC allows for broad participation, has well-developed relationships with UN specialized agencies, and specializes in coordinating social and economic policy. It tends to make decisions faster than the General Assembly.

The recommendation made as a result of Asadi's consultations was that the UNFF should have universal membership and report to the ECOSOC. This unprecedented proposal was put before the ECOSOC session of 2000, where it generated some controversy. The G77 affirmed the commitment of the developing countries to universal membership. France, speaking for the EU, gave the EU's agreement to the proposal, but noted that the creation of an ECOSOC subsidiary body with universal membership was 'a regrettable precedent, and for that reason the EU had only reluctantly agreed to its adoption.'[70] At the EU's insistence, ECOSOC resolution 2000/35, which formally established the UNFF, stipulated that this 'should not be construed as constituting a precedent.'[71]

The resolution was adopted without a vote in October 2000.

5

United Nations Forum on Forests

Created in 2001 for an initial period of five years, the United Nations Forum on Forests (UNFF) is the only subsidiary body of the ECOSOC with universal membership. It is unique in the UN system as a forum that reports to a parent body of smaller membership.[1] Its creation represented an enhanced international profile for forests. Despite this, its work has been unimpressive. During its first four years, it negotiated just 12 resolutions, most of them weaker than the proposals for action of the IPF and the IFF. On some issues negotiations broke down without agreement. This chapter surveys the work of the UNFF up to and including its sixth session in 2006, which agreed that states should negotiate a non-legally binding instrument on forests.

The working modalities of the United Nations Forum on Forests

Following the agreement of the ECOSOC to establish the UNFF, Germany hosted a workshop attended by 33 countries in November 2000 to discuss the UNFF's work programme.[2] NGOs attending emphasized that the UNFF should concentrate primarily on implementing the IPF/IFF proposals.[3] A synthesis report of the discussions was forwarded to the UNFF's organizational session (February 2001).[4] The first session of the UNFF (June 2001) subsequently agreed a multi-year programme of work structured around 16 elements (see Table 5.1).[5] As with the IPF and IFF, several intersessional initiatives have been held in support of the UNFF (see Table 5.2).

The UNFF's agenda has proved highly inflexible. For example, NGOs lobbied at the fourth session of the UNFF for a global ban on genetically modified (GM) trees, arguing that GM trees can contaminate wild trees and that the risks 'extend over national borders and across generations and are irreversible.' [6] However, the UNFF has allocated no time for discussions on GM trees, which remains one of the major non-issues of international forest politics. There has been no debate in any international institution on the environmental and social consequences of GM trees.[7]

There are five important differences between the modalities of the UNFF and those of its predecessors. First, through an interagency body called the

Table 5.1 *United Nations Forum on Forests: The 16 elements of the multi-year programme of work*

UNFF element	***Lead agency***
1 National forest programmes*	FAO
2 Promoting public participation*	
International	UNFF Secretariat
National	UNDP
3 Combating deforestation and forest degradation	UNEP
4 Traditional forest-related knowledge	CBD
5 Forest-related scientific knowledge	CIFOR/ICRA/IUFRO
6 Forest health and productivity	FAO
7 Criteria and indicators of sustainable forest management	FAO/ITTO
8 Economic, social and cultural aspects of forests	
Economic	World Bank
Social	CIFOR
9 Forest conservation and protection of unique types of forests and fragile ecosystems	UNEP
10 Monitoring, assessment and reporting, and concepts, terminology and definitions*	FAO
11 Rehabilitation and conservation strategies for countries with low forest cover	UNEP
12 Rehabilitation and restoration of degraded lands, and promotion of natural and planted forests	FAO/ICRAF/CCD
13 Maintaining forest cover to meet present and future needs	UNFF Secretariat
14 Financial resources*	World Bank/GEF
15 International trade and sustainable forest management*	ITTO
16 Capacity-building and transfer of environmentally sound technologies*	FAO

Note: * Indicates elements addressed at the second, third, fourth and fifth sessions (other elements were considered at just one of the first five UNFF sessions).
Sources: UN document E/2001/42 (Part II)-E/CN.18/2001/3(Part II), 'Report of the United Nations Forum on Forest on its first session; New York, 11–22 June 2001,' 28 June 2001, pp.16–17; UNFF secretariat information leaflet (undated) 'The Collaborative Partnership on Forests'

Collaborative Partnership on Forests, the UNFF has placed a stronger emphasis on collaboration with forest-related international institutions. Second, some UNFF sessions have included a ministerial segment. Third, there have been regular consultations with forest stakeholders in multi-stakeholder dialogues. Fourth, the UNFF has hosted panel discussions to highlight sectoral and regional issues. Fifth, the UNFF created three expert groups to assist with its work programme. We now consider these five dimensions of the UNFF's work.

Collaborative Partnership on Forests

Created in 2001, the Collaborative Partnership on Forests (CPF) is the successor mechanism to the ITFF. Chaired by the FAO, the CPF comprises 14 forest-related international organizations (compared with eight for the ITFF).[8] It is a high-level informal mechanism. CPF members are responsible to their governing bodies, not to the UNFF. The CPF has no formal status within the UN system, no operational or project management role, and no independent budget. The UNFF cannot direct the CPF or oversee its activities, only provide guidance. At the UNFF's fifth session, indigenous peoples proposed that the Permanent Forum on Indigenous Issues become a CPF member.[9] The proposal was opposed by the US[10] and was not adopted.

CPF members have assumed responsibility for the 16 elements of the UNFF's work programme (see Table 5.1). The CPF has worked to harmonize forest reporting in order to reduce the burden on states that report to several forest-related institutions. CPF members, particularly the FAO, have developed a sourcebook on funding for sustainable forest management. This is a directory of the main domestic, bilateral and multilateral donors that provide funding on forests.[11] Those who hold that the funding problem is due to inadequate information on the accessibility of finance, rather than insufficient funds, argue that the sourcebook will make finance more readily available for fund seekers. Those who hold that the problem is the absence of secure and predictable forest funding argue that the sourcebook will merely increase the number of applicants for already scarce resources.

The CPF's activities have been widely appreciated. At the fourth and fifth sessions of the UNFF, the expressions of thanks that several delegates made to Hosny el-Lakany of the FAO, the first chair and coordinator of the CPF, went well beyond the ritualistic and formulaic statements of appreciation that are often made in intergovernmental fora. It is generally agreed that the CPF has proved an effective mechanism.

Ministerial segments

If the intention behind UNFF ministerial segments was to provide high-level political momentum to the UNFF then they have so far failed. Two ministerial segments have been held. At the second session considerable time was spent negotiating a ministerial declaration to the 2002 World Summit on Sustainable Development in Johannesburg.[12] The declaration concluded by inviting the summit to 'Call on countries and the Collaborative Partnership on Forests to accelerate implementation of the IPF/IFF proposals for action and intensify efforts on reporting to the Forum.' [13] The value of this declaration must be questioned. By issuing it, ministers at the UNFF, a forum open to all UN members, called on other government ministers at the world summit, also

Table 5.2 *Intersessional initiatives in support of the United Nations Forum on Forests*

Name of initiative	***Venue and date***	***Main sponsors***
Shaping the Programme of Work for the United Nations Forum on Forests	Bonn, Germany (November–December 2000)	Australia, Brazil, Canada, France, Germany, Iran, Malaysia, Nigeria
Financing Sustainable Forest Management	Oslo, Norway (January 2001)	Brazil, Denmark, Malaysia, Norway, South Africa, UK, CIFOR
Monitoring, Assessment and Reporting on the Progress Towards Sustainable Forest Management	Yokohama, Japan (November 2001)	Australia, Brazil, Ghana, Indonesia, Japan, Malaysia, Norway, US
Forests and Biological Diversity	Accra, Ghana (January 2002)	Ghana, Netherlands, CBD, UNFF
Transfer of Environmentally Sound Technologies for Mangrove Forests	Managua, Nicaragua (March 2003)	Nicaragua, FAO, ITTO, and the Antigua, Cartagena and Ramsar Conventions
Lessons Learned in Monitoring, Assessment and Reporting on Implementation of the IPF/IFF Proposals for Action (Viterbo Report)	Viterbo, Italy (March 2003)	Brazil, China, Italy, Japan, South Africa, Sweden, Turkey, UK, US, FAO, UNFF
The Role of Planted Forests in Sustainable Forest Management	Wellington, New Zealand (March 2003)	Argentina, Australia, Canada, Chile, Malaysia, New Zealand, South Africa, Switzerland, UK, US, CIFOR, FAO, International Centre for Research in Agroforestry, ITTO
Lessons Learned on Sustainable Forest Management in Africa	Nairobi, Kenya (February 2004), Uppsala, Sweden (October 2004)	Sweden, African Academy of Science, African Forestry Research Network, FAO
Transfer of Environmentally Sound Technologies and Capacity-building for Sustainable Forest Management	Brazzaville, Congo (February 2004)	Brazil, Congo, France, Italy, Norway, Senegal, Switzerland, UK, US
Decentralization, Federal Systems in Forestry and National Forest Programmes	Interlaken, Switzerland (April 2004)	Brazil, Canada, Ghana, Indonesia, Japan, Russia, Switzerland, Uganda, UK, US

Table 5.2 *Intersessional initiatives in support of the United Nations Forum on Forests (Continued)*

Name of initiative	*Venue and date*	*Main sponsors*
Gender and Forestry: Challenges to Sustainable Livelihoods and Forestry Management, Second Worldwide Symposium on Gender and Forestry	Kilimanjaro, Tanzania (August 2004)	Tanzania, CIFOR, IUFRO
Traditional Forest-Related Knowledge and the Implementation of Related International Commitments	San José, Costa Rica (December 2004)	International Alliance of Indigenous and Tribal Peoples of the Tropical Forests
Future of the International Arrangement on Forests	Guadalajara, Mexico (January 2005)	Mexico, US
Practical Solutions to Combat Illegal Logging: Dialogue on Best Practices for Business and Civil Society	Hong Kong, China (March 2005)	The Forests Dialogue
Innovative Financial Mechanisms: Searching for Viable Alternatives to Secure the Financial Sustainability of Forests	San José, Costa Rica (March–April 2005)	Costa Rica
Global Initiative on Forest Landscape Restoration	Petropolis, Brazil (April 2005)	UK, World Conservation Union (IUCN), WWF
Scoping for a Future Agreement on Forests	Berlin, Germany (November 2005)	Germany

Source: UN forests website, www.un.org/esa/forests/gov-unff.html (accessed 15 March 2006)

open to all UN members, to do what the former had already committed to doing.[14]

The fifth session failed to agree a ministerial declaration. While the ministers read pre-prepared statements in one of the UN's showpiece negotiation chambers in New York, the Trusteeship Council, in the basement delegates were trying to agree a ministerial declaration in parallel with a new international arrangement on forests. When agreement on the latter proved impossible, the negotiations for the former also broke down. We return to these negotiations later.

Multi-stakeholder dialogue

The idea of stakeholder consultations has evolved since the mid 1980s when the World Commission on Environment and Development held civil society hearings. The contemporary model of multi-stakeholder dialogue emphasizes open-minded interaction between governments and other stakeholders. It has its origins at the second UN Conference on Human Settlements (Habitat II) in 1996 when a series of dialogues were held between different stakeholders.[15] In 1997, the United Nations General Assembly Special Session review of the Earth Summit commitments agreed to introduce multi-stakeholder dialogues to the CSD.

Although a CSD subsidiary body, the IFF held no multi-stakeholder dialogues, despite NGO pressure. However, in 2000, when recommending the creation of the UNFF, the IFF recommended that the UNFF include multi-stakeholder dialogue segments.[16] This was endorsed by the ECOSOC.[17] Eight 'major groups' have participated in UNFF multi-stakeholder dialogues. They are business and industry; children and youth; farmers and small landowners; indigenous peoples; NGOs; scientific and technological communities; women; and workers and trade unions. The dialogues have consisted of opening statements from representatives of the major groups, after which the floor is opened. Any accredited delegate or observer may make a verbal intervention. The extent to which the dialogues have been genuinely interactive is not clear, as often the spokespeople of major groups have concentrated on making statements that will satisfy their constituencies at the expense of engaging with points made by other groups.

Forest privatization featured prominently in the multi-stakeholder dialogue at the UNFF's fourth session. Lambert Okrah, a campaigner from the Institute of Cultural Affairs in Ghana representing the NGOs, delivered an incisive critique of forest privatization. He noted that local commoners oppose forest privatization, which 'has brought untold hardship to indigenous peoples. The current strategy is to take from the poor and give to the rich.' Okrah went on to note that privatization, which involves government in partnership with business, impoverishes local communities and indigenous peoples, who are deprived of access to common lands.[18] Okrah's intervention was followed by one from Mary Coulombe representing business and industry, and speaking on behalf of the International Council of Forest and Paper Associations (ICFPA). Coulombe stated that the private sector has much to offer forest management since it is often at the 'cutting edge' of forest management technology, and that business needs a secure and predictable environment for investment.[19]

With respect to privatization, no new, shared understandings between NGOs and business have emerged from UNFF multi-stakeholder dialogues. To NGOs and indigenous peoples, there is often little difference between legal and illegal logging: both can displace local forest communities, undermining their livelihoods. To timber and paper businesses, the forest is an investment opportunity. It is difficult to imagine multi-stakeholder dialogues generating a new political discourse that all major groups can subscribe to on a subject as contentious as privatization. The UNFF multi-stakeholder dialogues have brought into sharper focus the differences between those who view the forests as a source of livelihood and those who view it as a source of profit; but they have made no impact on the UNFF's work. No resolutions have been agreed as a result of multi-stakeholder dialogues, which have provided a veneer of participation while leaving unchanged the essentially intergovernmental nature of decision-making. NGOs, indigenous peoples and other major groups have been systematically excluded from drafting groups and from making plenary statements. In many cases, the results of multi-stakeholder dialogues have not been accurately reported to the plenary, leading to increasing disillusionment among major groups.[20] NGOs have argued that 'Unless there are radical changes to ensure the effective consideration of the proposals and views of major groups, the organization of these events should be discouraged.'[21]

At the UNFF's fifth session, Cuba argued against including a multi-stakeholder dialogue segment at the sixth session. To do so, Cuba stated, would detract from the main purpose of the sixth session, namely to conclude the unfinished negotiations of the fifth session. The clear inference of the Cuban intervention was that other stakeholders should have no input to this process. Although most delegates have publicly acknowledged a commitment to multi-stakeholder dialogue, no delegation was prepared to challenge Cuba and take the matter to a vote. (To date, no vote has been taken at the UNFF and none were taken at the IPF and IFF; all decisions have been taken by consensus.) Eventually a compromise was agreed: multi-stakeholder dialogues would be held as informal side events at the UNFF's sixth session, rather than as part of the formal proceedings. Not surprisingly, the result of this decision was significantly lower major group involvement at the sixth session.

Panel discussions

A further difference between the UNFF and its predecessors is that the UNFF includes panel discussions. Up to ten experts sit at the podium usually occupied by officials and make presentations before taking questions from delegates. Subjects on which UNFF panels have been held include the economic aspects of forests, implementation in Africa and forestry in small island developing states. As part of the ministerial segment at the fifth session 'high-level roundtables' were held on forest restoration and forest governance.

The panels, like the multi-stakeholder dialogues, have provided an opportunity for marginalized voices to be heard and questioned. For example, at the UNFF's fourth session, Ole Henrik Magga, the chair of the UN Permanent Forum on Indigenous Issues, stressed that indigenous peoples do not consider themselves poor simply because they do not have access to the cash economy. Poverty has a cultural context: indigenous peoples are poor when access to their land is restricted and when their human rights are violated. Magga emphasized that indigenous peoples should be included in the management of protected areas that cover their traditional land.[22]

There is a notable change of atmosphere during the panel discussions. Powerpoint presentations and question-and-answer sessions provide the ambience of an academic seminar, rather than of international diplomacy. However, few of the panels contain information that well-briefed delegations and their expert advisers do not already know. And as with the multi-stakeholder dialogues, none of the panels has resulted in a resolution or even impacted on the text of a resolution. Many delegates have treated the panels as a sideshow and an opportunity to meet with colleagues in the delegates' lounge. The panels have no clear political role in the formal UNFF process.

Ad hoc expert groups

At its first session, the UNFF agreed to create three ad hoc expert groups that would meet between sessions to consider issues central to the UNFF's programme of work. The expert group model is one in which governments nominate experts in their personal capacities, and not as delegates promoting government positions. In practice, however, many governments appointed senior negotiators to the expert groups, with the unsurprising result that the views expressed by experts often bore a striking resemblance to government policy. The first two groups were tasked with addressing monitoring, assessment and reporting, and finance and the transfer of environmentally sound technologies. The third group was responsible for generating possible options for the international arrangement on forests after the UNFF's initial term of five sessions had expired. We consider the work of this third group later. In the following section, we consider the first two groups.

Lacking in resolutions

Much of the UNFF's second, third and fourth sessions were spent negotiating resolutions. A resolution is a statement of political commitment that has been agreed by a group of states, but which is not legally binding. The UNFF resolutions, like the IPF/IFF proposals for action, form part of the growing body of soft law on forests. Soft law may be defined as those agreements, declarations and statements agreed by intergovernmental organizations that are not ratified through national legislatures and which comprise political as opposed to legal commitments. (The UNFF also negotiates decisions: these should not be considered soft law since they deal only with organizational and procedural matters.) Table 5.3 briefly summarizes the UNFF resolutions negotiated between 2002 and 2004.

There is insufficient space here to deal with all UNFF negotiations. Instead, we confine ourselves to the UNFF's fourth session, which set out to negotiate seven resolutions, but agreed only four (see Table 5.3). First, we consider the negotiations on monitoring assessment and reporting. We then examine the negotiations on finance and environmentally sound technologies, which resulted only in a decision, rather than a resolution. We then consider two series of negotiations that broke down without agreement: traditional forest-related knowledge and enhanced cooperation.

Monitoring, assessment and reporting

The UNFF – or, more accurately, the states that comprise it – has made minimal progress on implementation. The question of compliance has never arisen since the IPF/IFF proposals do not actually *oblige* states to do anything. Following a US proposal that was supported by some developing countries, the UNFF agreed at its first session to voluntary, rather than mandatory, national reporting. The US also argued against collectively agreed implementation targets, insisting that countries should set their own targets and timetables.[23] By its second session, only 16 countries had submitted reports.[24] By the fourth session, the number had risen to 34 countries.[25] The low number of countries submitting national reports to the UNFF indicates a low level of political commitment to the IPF/IFF proposals, which significantly weakens any normative force that they might have.

The UNFF appears to have interpreted part of its function on implementation as generating propositions on the background conditions necessary for implementation. For example, the report of the UNFF's second session notes that 'Sustainable forest management programmes could help maintain the natural resource base and support rural livelihoods by, among other things, protecting soil and water resources and providing employment and income'[26] and 'The development of networks of protected areas and sustainably managed forests representing the full range of forest ecosystems was considered to be important.'[27] These extracts typify the generalized and sometimes banal nature of the UNFF's work on implementation. Existing knowledge and the IPF/IFF proposals have been distilled in a generalized manner, adding little, if anything, that is new.

It was in this context that the UNFF convened its expert group on monitoring, assessment and reporting in December 2003.[28] The report of the expert group was presented to the UNFF's fourth session by Mike Dudley of the UK. It noted that areas of overlap and redundancy exist in national reporting on forests and recommended 'streamlining' of reporting requirements, improved linkages between reporting for national and international purposes, and greater emphasis on lessons learned and emerging issues. It was recommended that 'countries make better use of existing resources' and that CPF members and other organizations strengthen national capacity for national reporting through the provision of financial and technical resources.[29] After the report was presented, delegates commenced the negotiation of a resolution.

The agreed resolution reiterated the voluntary reporting commitments that had previously been agreed. Countries were encouraged 'to include forests and forest-related monitoring and assessment in national development plans and poverty reduction strategy papers where they exist, which could enhance opportunities for international cooperation.' [30] Stronger phrasing was not possible as some states consider commitments that imply an ethos of accountability to other states as an erosion of sovereignty.

Table 5.3 *Resolutions adopted by the United Nations Forum on Forests, 2002–2004*

Number	Title (abbreviated)	Main points (abbreviated)
2/1	Ministerial Declaration to the 2002 World Summit on Sustainable Development	Invites the World Summit on Sustainable Development to advance sustainable forest management as a means of eradicating poverty, reducing land degradation and improving food security, access to safe drinking water and affordable energy
2/2	Implementation of the IPF/IFF proposals and the UNFF Plan of Action	Proposes measures on the implementation of the IPF/IFF proposals with respect to combating deforestation and forest degradation; protection of unique types of forests and fragile ecosystems; low forest cover countries; and the rehabilitation and restoration of degraded lands
2/3	Criteria for Review of the International Arrangement on Forests	Invites countries, the Collaborative Partnership on Forests and major groups to voluntarily provide quantifiable benchmarks to review the effectiveness of the international arrangement on forests
3/1	Economic Aspects of Forests	Invites the Collaborative Partnership on Forests to assist countries to implement full-cost internalization of wood products and non-wood substitutes Invites countries to take action on domestic forest law enforcement
3/2	Forest Health and Productivity	Encourages countries to develop strategies for the transboundary movements of pests and diseases, and to develop forest fire management strategies
3/3	Maintaining Forest Cover to Meet Present and Future Needs	Urges countries to strengthen efforts to combat deforestation and forest degradation and to cooperate on finance, the transfer of environmentally sound technology and capacity-building. Encourages countries to integrate criteria and indicators for sustainable forest management into national forest programmes

Table 5.3 *Resolutions adopted by the United Nations Forum on Forests, 2002–2004 (Continued)*

Number	Title (abbreviated)	Main points (abbreviated)
3/4	Enhanced Cooperation and Programme and Policy Coordination	Encourages partnerships between the Collaborative Partnership on Forests members and governments. Proposes clarification of the relationships between the CBD's ecosystem approach and sustainable forest management, and between the CBD's expanded programme of work on forest biodiversity and the IPF/IFF proposals
3/5	Strengthening the UNFF Secretariat	Urges countries to provide voluntary extra-budgetary contributions to the UNFF trust fund
4/1	Forest-related Scientific Knowledge	Encourages countries to highlight the role of science and research in sustainable forest management and to promote private sector investment in forest research Encourages all stakeholders to promote 'integrated and interdisciplinary research' on forest-related issues
4/2	Social and Cultural Aspects of Forests	Urges countries to integrate the IPF/IFF proposals on the social and cultural aspects of forests into national forest programmes, and to integrate sustainable forest management into poverty eradication and development strategies
4/3	Monitoring, Assessment and Reporting, Criteria and Indicators for Sustainable Forest Management	Encourages forest-related monitoring and assessment in national development plans and poverty reduction strategy papers
4/4	Review of Effectiveness of the International Arrangement on Forests	Invites states to report on the implementation of the IPF/IFF proposals and to respond to a questionnaire on the effectiveness of the international arrangement on forests

Note: In the first column, the first digit indicates the UNFF session, while the second indicates the number of the resolution. So, for example, 4/3 denotes the third resolution of the fourth session.

Sources: UN documents E/2002/42-E/CN.18/2002/14, pp.3–16; E/2003/42-E/CN.18/2003/13, pp.12–22; and E/2004/42-E/CN.18/2004/17, pp.2–9.

Finance and environmentally sound technologies

The negotiations on finance and environmentally sound technologies followed the patterns established in earlier negotiations. Developing countries urged increased resource transfers. Meanwhile the developed states followed the strategy that they had first introduced at the IPF; the range of financial issues under consideration was broadened and it was emphasized that, in addition to official development assistance (ODA), several other routes are available for financial assistance. This strategy served three purposes for the developed countries. First, it deflected attention from the poor ODA records of most developed states (see Chapter 2). Second, it placed developing countries in a defensive position by calling on them to justify why increased forest financing should be met through ODA, rather than through the private sector, national taxation and efficiency savings. Third, by emphasizing that the private sector is a source of forest financing, developed states sought to position their forest corporations in an advantageous position with developing country governments. The issue of financial transfers and the role of private sector investment in tropical forestry have thus become entwined.

The report of the expert group on finance and environmentally sound technologies was presented to the UNFF by Knut Øistad of Norway.[31] Parts of the report have a distinct neoliberal flavour; proposals inserted by experts from developed countries included the use of market mechanisms, private investment, public–private partnerships and 'an enabling investment climate.'[32] The interests of governments from the tropics are also represented, including developing 'an initiative to work with donor countries to mainstream sustainable forest management as one of the important sectors for ODA allocation.'[33]

Qatar, speaking for the G77, made clear the position of developing countries at the start of the fourth session. Qatar noted an 'uneven level of assistance,' a clear reference to the favouring by donors of some countries over others.[34] The increased emphasis of some developed world delegations on private sector finance was a concern for the G77 since private sector commitments are voluntary and do not substitute for commitments made by governments.[35] During the negotiations for a resolution on finance and technology, the G77 noted 'the urgent need for concrete action to meet the 0.7 per cent of GNP [gross national product] target.'[36] The US responded with some irritation: the expert group's report was 'impressive' and 'we hope … that we can avoid the usual cliché statements of "We need more money."'[37] These two interventions set the tone for the negotiations. Developed states, in particular the US, argued that several sources of finance were available and developing states should explore these rather than look to developed states for increased ODA. The G77 countered by asserting that financial resources from other avenues have not materialized and developing countries remain dependent on ODA. The result was zero-sum game bargaining. Eventually, the draft text was abandoned, and 'informal discussions' were held.[38] In other words, negotiations between the main protagonists – the G77, US and EU – took place behind closed doors with no observers admitted.

The outcome was a text of just two short substantive paragraphs: member states, CPF members and other organizations were encouraged to 'take concrete action' on the recommendations of the expert group, and the UNFF decided 'to give further consideration in its programme of work to the issues of finance and transfer of environmentally sound technologies, including the recommendations of the expert group.'[39] This text was so weak that it was downgraded from a resolution to a decision.

Agreeing to disagree

Draft resolutions are living documents that evolve and develop during negotiations. Sometimes a resolution is agreed. It then becomes a piece of soft international law that may later be cited in legal opinions and international law journals. Sometimes the agreed text is so thin that it has no legal or political significance: the decision on finance and environmentally sound technologies is an example. In other cases, negotiations break down with no agreement at all. At the UNFF's fourth session, this happened twice. We now consider these negotiations.

Traditional forest-related knowledge

In February 2004, three months before the UNFF negotiations on traditional forest-related knowledge (TFRK), the seventh session of the Convention on Biological Diversity (CBD) agreed decision VII/19, committing states to negotiations for an international regime on access to biological resources and the sharing of the benefits arising from the utilization of these resources (see Chapter 9). This decision had a direct bearing on the TFRK negotiations, which were chaired by Xolisa Mabhongo (South Africa). The first draft of this resolution was prepared by the UNFF Secretariat following consultations (see Box 5.1). The intention of the preambular paragraphs was to document the lessons learned from implementation of the IPF/IFF proposals on TFRK, while the substantive paragraphs were intended to describe future actions.

The negotiations during the first reading of 12 May 2004 revealed, in addition to the usual disagreements on financial and technical assistance, two main lines of conflict. First, there was contention on the word 'rights' with reference to indigenous people/peoples (see Box 5.2). Canada favoured the replacement of 'rights' by 'interests.' The Canadian delegation explained that rights to traditional knowledge and land have not been fully defined in Canada. The response of the G77 was that the word 'rights' had been agreed at the UNFF's third session and should be retained. The US supported the G77 on this point.

The Canadian position illustrates how domestic affairs may affect international negotiations. Section 35 of the 1982 Constitution Act of Canada recognizes and affirms aboriginal and treaty rights, but does not define them.

Box 5.1 First draft of a United Nations Forum on Forests resolution on traditional forest-related knowledge

11 May 2004

The United Nations Forum on Forests:

Taking note of the views exchanged by countries, as well as major groups, at its fourth session on the status of countries' efforts to implement the IPF/IFF proposals for action related to traditional forest-related knowledge, which identified progress, obstacles and lessons learned;

Highlighting the following lessons learned through the exchange of country experiences;

More effective measures to recognize, protect and maintain traditional forest-related knowledge could enhance sustainable forest management;

Countries should further explore the different ways of protecting traditional forest-related knowledge and the rights of indigenous people and local communities. Thus, examples and lessons learned in this regard should be effectively shared and utilized;

The identification and the further exploration of the possible synergies between scientific and traditional forest-related knowledge could enhance sustainable forest management;

The involvement of holders of traditional forest-related knowledge in management decisions, in accordance with national laws and regulations, could result in improved decision-making for the preservation of traditional forest-related knowledge and contribute to sustainable forest management;

At the international level various organizations are doing work related to the protection of traditional knowledge including the CBD, WIPO and FAO. It is therefore important for the UNFF to take into account the work of these organizations;

1. Urges countries to continue to take necessary action to further safeguard traditional forest-related knowledge, including the further development of national legislation aimed at regulating access to and protection of traditional forest-related knowledge as well as the rights of indigenous and local communities;

2. *Encourages* countries and regional and international organizations to further explore and develop the diverse systems for the protection of traditional forest-related knowledge including *sui generis* systems;

3. *Calls upon* countries to ensure the fair and equitable sharing of benefits arising from the utilization of traditional forest-related knowledge. In this regard notes decision VII/19 of the Convention on Biological Diversity on the negotiation of an international regime for access and benefit sharing from the utilization of genetic resources;

4. *Urges* the international community to take into account the need to respect the national access and benefit sharing regimes and laws of the countries wherein the traditional knowledge custodians reside;

5. *Encourages* countries to integrate traditional forest-related knowledge into national forest programmes and into formal education schemes in order to increase awareness and understanding. In this regard also encourages countries to help preserve and promote further application of this knowledge for forest management purposes;

6. *Invites* the members of the Collaborative Partnership on Forests to support national and regional actions that promote the preservation of traditional forest-related knowledge and its application in sustainable forest management;

7. *Urges* countries and members of the Collaborative Partnership on Forests and other relevant international organizations to continue to identify and develop the linkages between scientific forest-related knowledge and traditional forest-related knowledge taking into account national laws and regulations;

8. *Encourages* countries to continue to develop, in consultation with local communities, methods of compiling and managing their registers and databases in accordance with national laws. In this regard also encourages countries to ensure that documentation and cataloguing do not adversely affect holders of traditional forest-related knowledge through misappropriation or use in ways not anticipated when holders gave the information;

9. *Urges* the international community to provide financial and technical support to developing countries for the protection of traditional forest-related knowledge;

10. *Encourages* cooperation between the UNFF, WIPO, CBD and FAO on issues related to the documentation and protection of traditional forest-related knowledge.

Source: UNFF Working Group I, 'Vice-Chairman's Text, Traditional forest-related knowledge, 11 May 2004' (draft)

Box 5.2 'Peoples,' 'people' and international law

Civil society organizations have long campaigned for the term 'indigenous peoples' to be used in intergovernmentally negotiated outputs since under international law the word 'peoples' carries legal connotations of self-determination, namely the right of peoples to determine the conditions under which they live.[41] In international forest negotiations, states refused until 2002 to accept the term 'indigenous peoples' (plural). In the IPF and IFF proposals for action, only the softer expressions 'indigenous people' (singular) or 'indigenous communities' were accepted.[42] In 2002, the lobbying efforts of civil society organizations were rewarded when, in Article 25 of the Johannesburg Declaration on Sustainable Development, states agreed the following: 'We reaffirm the vital role of the indigenous peoples in sustainable development.'[43] The following year the phrase 'indigenous peoples' was used for the first time in a negotiated output from an international forests institution when the third session of the UNFF recognized that 'Secure land tenure and property rights are vital to the well-being of indigenous peoples and local communities who live in and around forests.'[44] In 2005, the World Summit of heads of state and government pledged commitment, in an outcome that was adopted by the UN General Assembly, to 'the advancement of the human rights of the world's indigenous peoples.'[45] While the recognition of the pluralized expression 'indigenous peoples' is significant, its practical ramifications for indigenous peoples' rights remain to be seen.

Some court cases were in progress during the UNFF negotiations, and the Canadian authorities wished to avoid the word 'rights' lest agreeing to it complicated domestic litigation processes. The Canadian delegation was thus instructed not to agree to the word unless approval was obtained from the Department of Justice.[40]

Second, there were different views on who should have access to biological resources, who should benefit from using them, and which international institutions should handle these issues. The G77 opposed international rules on access, arguing that each sovereign state should set its own rules on the actors that may have access to its biological resources.[46] But while committed to national rules on access, the G77 favoured international rules stipulating how the benefits – in other words, the financial returns – from biological resource use should be shared between the businesses that use biological resources and the stakeholders in countries from which the resources are harvested. These rules, asserted the G77, should be negotiated at the CBD. The G77 was opposed by the US, which wanted the deletion of 'to ensure' before 'the fair and equitable sharing of benefits' (see Box 5.1, paragraph 3). The US proposed a softer formulation – 'Calls upon countries to take appropriate measures for the fair and equitable sharing of benefits' – noting that this is the wording agreed in Agenda 21. The US then proposed deletion of the reference to decision VII/19 of the CBD.[47]

It should be noted here that the US government uses various means, such as bilateral diplomacy and international aid, to try to leverage the access of US corporations to the biological resources of developing countries. The US has not ratified the CBD and opposes multilateral rules on benefit-sharing, which could reduce the profits made by US corporations under current international intellectual property rights law. This explains its opposition to CBD decision VII/19. Note that the US is willing to cite from Agenda 21, adopted in 1992, but is unwilling to accept reference to a CBD decision agreed a few months earlier. The selective citation of language agreed elsewhere is a common negotiating tactic by government delegates when seeking to legitimize their positions.

The first reading of the draft resolution had proceeded in normal UN style. Although there was by now some heavily bracketed text, there was nothing to suggest that the negotiations would not conclude with an agreed resolution.

The second reading was due to resume at 2 pm on 13 May 2004. However, the G77 met in caucus for much of that day. When the negotiations resumed nearly two hours late, there was a major shift in the G77's negotiation strategy. The G77 announced that the UNFF was not the appropriate forum for discussing TFRK and intellectual property rights, hence it wished a very short resolution. The G77 proposed a single substantive paragraph:

> *[The UNFF] urges countries to continue to safeguard and protect TFRK, including through the development and further development of national and international legislation, ensuring that these activities do not adversely affect the holders of TFRK through misappropriation of use in ways not intended when holders gave information.*

The G77 added that if the paragraph was 'peppered with references to access and other issues [then] we will have a lot of difficulties.'[48]

With the seventh session of the CBD having agreed to negotiate a regime on access and benefit-sharing, the G77 in general, and Brazil in particular, now believed that developing country interests on traditional knowledge were best realized at the CBD. The possible ramifications on G77 interests of a lengthy UNFF resolution on TFRK would be hard to fathom, especially in the heat of negotiations. There were no advantages to the G77 in protracted negotiations for a TFRK resolution. In fact there could have been costs: the developing countries could have inadvertently consented to text that could later be invoked as a precedent against developing countries at the CBD.[49]

Canada, New Zealand, the US and Ireland on behalf of the EU[50] stated that they favoured working with the existing text. At the request of the chair, the G77 agreed to this and a second reading of the draft commenced. When the afternoon session ended without agreement, it was agreed to convene an un-programmed evening session. Many delegates did not return for this.[51] With no interpreters or microphones, delegates, major groups and the secretariat gathered in a more informal arrangement around a table at the front of the negotiation chamber. The US worked to secure language that TFRK was a national-level issue, not an international one.[52] The G77 now opposed any mention of access to TFRK, saying that this was being addressed by the CBD. It became clear that the overriding concern of the G77 was to protect TFRK from unauthorized access. The G77 also proposed deleting draft paragraph 9 on finance (Box 5.1, p.105 above), inserting in its place:

> Calls upon *the international community to provide financial and technical support to developing countries for the protection and preservation of TFRK and its application in sustainable forest management where appropriate.*[53]

The US responded that it could accept this proposal if reference was made to 'indigenous and community-led initiatives.'[54] The G77 replied that this was unacceptable since it implied that finance was being given 'only for these initiatives.' The G77 then added: 'I'll be frank. The international community provides financial assistance for its own interests, namely to gain knowledge in this case; hence, we cannot agree to this.' To the G77, language allowing developed governments and private actors to provide financial assistance direct to local community groups could legitimize the access of these actors to traditional knowledge while bypassing national governments. By opposing the linkage of financial assistance to 'indigenous and community-led initiatives,' the G77 was protecting its position that access to biodiversity was a national concern and that developed countries should respect national-level access laws.

Later, the EU proposed adding after 'national access and benefit-sharing regimes and laws,' a phrase on the need to take into account 'international obligations as appropriate.'[55] This drew opposition from the US and the G77, although for different reasons. For the US, the phrase 'international obligations' could be read as the CBD. The US has not ratified this instrument and opposes the CBD's commitment to an international regime on access and benefit-sharing. The basis for the G77's opposition was twofold: it refuses to recognize that states should have any international obligations with respect to access, only on benefit-sharing; and the EU's amendment could be interpreted as obligations outside the CBD, such as WTO rules on intellecutal property rights.

After further proposals and counterproposals, most of which were opposed by the G77, the US stated that 'The resolution is now so weak we cannot agree to it. We would rather have no resolution at all than a bad resolution.'[56] After a brief adjournment, the G77 reiterated that it wanted a short resolution. The EU, Canada and New Zealand then agreed with the US that no resolution was possible. After some diplomatic statements that the outcome was regrettable and that the exchange of views had been positive, the negotiations ended without agreement.[57]

Enhanced cooperation and policy and programme coordination

Negotiations also broke down for a resolution on the seemingly apolitical subject of enhanced cooperation. The central point of contention was the conceptual relationship between sustainable forest management (SFM) and the 'ecosystem approach' that has been adopted by the CBD (see Chapter 9).[58] Deliberations on what appeared to be technical differences became politically charged. As with the TFRK negotiations, the main difficulty was the role of the CBD in international forest policy. Some developed states wanted a resolution to agree that SFM was a means of implementing the ecosystem approach.[59] The G77 objected to this since it wished to keep mention of any concept that was synonymous with the CBD absent from UNFF resolutions. The G77's aspirations for an access and benefit-sharing regime and its desire to avoid any blurring of jurisdiction between the UNFF and CBD thus contributed to the demise of a second UNFF resolution.

Negotiating the new international arrangement on forests

The UNFF's third expert group was tasked with generating options for the international arrangement on forests after the UNFF's fifth session. It was given a laborious title, the Ad Hoc Expert Group on Consideration with a View to Recommending the Parameters of a Mandate for Developing a Legal Framework on All Types of Forests,[60] a name that replicated the deliberately equivocal consensus language agreed at the IFF's final session in 2000 (see Chapter 4). Held in September 2004, the group was chaired by Tim Rollinson (UK) and Andrea Alban Duran (Colombia). Experts spoke in their personal capacities. Rollinson emphasized that he wanted a process that was as inclusive as possible. In addition to nominated experts, major groups and delegates attending as observers were permitted to make interventions.[61] The group prepared a list of options for a new international arrangement on forests. However, there was insufficient time for interactive dialogue between experts, and to enable further discussion a country-led intersessional meeting, also chaired by Rollinson and Duran, was held in Mexico in January 2005.

During these meetings the options crystallized into two main categories. The first was a legally binding instrument, either a forests convention or a forests protocol to another convention.[62] It was noted that a protocol must fall within the objectives of the parent convention. A CBD forests protocol could, therefore, address forest biodiversity, but not other forest-related issues. At present, the option of a CBD forests protocol is unlikely, with the CBD preoccupied with access and benefit-sharing. The second category of options centred on strengthening the UNFF. Two proposals made at the expert group and Mexico meetings featured prominently in later UNFF negotiations. The first was for quantifiable and time-bound targets: for example, the rate of deforestation should be reduced by *x* per cent by, say, 2015.[63] The second was for the negotiation of a non-legally binding instrument.[64]

By the time the UNFF's fifth session convened, the proposal for a non-legally binding instrument had been informally developed, with various non-papers being circulated advocating a voluntary code for sustainable forest management, a *Codex Sylvus*. (A non-paper is a document that is circulated among delegates to float proposals and suggestions; non-papers are not formal proposals and are non-attributable.) The name *Codex Sylvus* owes its inspiration to the *Codex Alimentarius*, a body of non-legally binding but widely implemented food standards administered by the FAO and World Health Organization (WHO).[65] The idea of a code illustrates that there is no simple distinction between soft law and hard law, and that the two can be seen as poles to a continuum.[66]

The UNFF's consideration of a non-legally binding instrument or voluntary code would prove far more extensive than that of the IPF (see Chapter 2). In principal, such an instrument has the attraction of providing a bridge between pro- and anti-convention states. For states wishing to strengthen international cooperation on forests, including the pro-convention states, a non-legally binding instrument offered the possibility of a stronger regulatory framework than the IPF/IFF proposals and UNFF resolutions. For states opposed to a convention – including the most powerful opponent, the US – a non-legally binding instrument would have the advantage of strengthening the UNFF politically while avoiding legally binding commitments. Against this it can be argued that anti-convention states might oppose a non-legally binding instrument in case it were eventually to lead to legal codification. A precedent for this is the non-legally-binding International Undertaking on Plant Genetic Resources of 1983, which was later renegotiated as the International Treaty on Plant Genetic Resources of 2001 (see Chapter 4).

The formal position of the forest industry, voiced by the International Council of Forest and Paper Associations, was that business neither supported nor opposed any particular international arrangement on forests.[67] However, the idea of a code was endorsed by some North American forest industry leaders,[68] which explains its support by the US and Canada. There was a feeling among business that with many industrial sectors adopting codes, the forest industry should move towards this voluntarily, rather than risk being pushed towards it later.

Language that the IFF had agreed in 2000, and after which the UNFF's third expert group had been named, returned to haunt the UNFF. Delegates were debating when a new international arrangement on forests should be reviewed. Many developing country delegates were not present, as the G77 was meeting separately in caucus. The US proposed that states would 'In 2015 consider, with a view to recommending, the parameters of a mandate for developing a legal framework on all types of forests.'[69] Within 30 minutes, the G77 returned, having agreed a position. Speaking on behalf of the G77, Jamaica then proposed the same formulation as the US; states would 'In 2015 consider, with a view to recommending ...,' etc.[70]

In making the proposal the US was using the IFF's language as a precedent, while aware that agreeing to it would not commit the US to a forests convention any more than agreeing to the same language at the IFF had bound the US. The US proposal can thus be seen as the cynical invocation of ambiguous previously agreed language to push consideration of a convention further into the future. The anti-convention states in the G77 favoured the IFF wording for the same reason. The pro-convention G77 states appeared to have agreed to it for the same reason that they had accepted it five years previously: it was as close to an agreement for a convention as it was possible to get.

But G77 unity was paper thin. By the second week of the UNFF's fifth session, divisions between the developing countries proved so deep that the G77 fractured as a negotiating caucus. Developing countries now negotiated individually. The main axis of conflict was between the Latin American countries, with the Central American countries, supported by Argentina and, to a lesser degree, by Chile and Mexico, arguing in favour of a convention, opposed by the Amazonian Pact countries (led by Brazil).[71] The two main timber-producing states of Southeast Asia were also divided, with Indonesia inclining against a convention, while Malaysia inclined in favour. Other developing countries that indicated support for a convention were Cuba, China, Cambodia and Iran. Outside the G77, a convention was supported by Canada, Switzerland, South Korea and the EU.[72]

The fracture of the G77 significantly slowed the speed of the negotiations. There were two other key areas of disagreement. First, and as expected, was financial assistance. Developed countries resisted calls from developing countries for a global forests fund that mirrored those made at the IPF and the IFF. However, developed countries did advocate that the FAO's National Forest Programme (NFP) Facility and the World Bank's Programme on Forests (PROFOR) should create trust funds made up of voluntary donations (see Box 5.3). The negotiations also saw the return of a principle that has a chequered history in international forest politics, namely 'common but differentiated responsibilities.' In 1992, the concept was written into the Framework Convention on Climate Change to signify that while all states share responsibilities for tackling global warming, those states that historically have emitted the most greenhouse gases are the most responsible and should thus carry the burden of adjustment.[73] The concept was proposed by the G77 during the UNCED forest negotiations, but it was opposed by developed states and does not appear in the UNCED Forest Principles or Chapter 11, 'Combating deforestation,' of Agenda 21.[74] Three years later, developed states agreed to its mention in the IPF proposals for action.[75] Thereafter, the concept waned; it does not feature in the IFF proposals or the resolutions agreed by the UNFF up to and including 2004. At the UNFF's fifth session, it was proposed by Ecuador, India, Iran and Syria.[76] The practical application of the principle for forest policy is unclear. It is used by some developing countries to claim aid on the basis that developed countries bear most historical responsibility for deforestation through high demand for timber and other forest products. However, this line of argument tends to negate the repeated assertions of

Box 5.3 The NFP Facility and PROFOR

The FAO's NFP Facility and the World Bank's PROFOR support the implementation of national forest programmes and other national forest policy initiatives. Operational since 2002, the NFP Facility is the offspring of the TFAP Coordination Unit created during the 1980s. Its mandate extends to all forest regions. PROFOR was initially created by the UNDP in 1997. It relocated to the World Bank in 2002.

The NFP Facility provides practical donor support to countries developing national forest programmes. It has expertise in field projects, stakeholder participation, policy implementation and governance. PROFOR's speciality is in analytical work, such as knowledge generation and problem-solving, in four thematic areas: livelihoods, governance, innovative finance and cross-sectoral cooperation. To ensure the exploitation of synergies between them, the NFP Facility and PROFOR hold their annual meetings back to back, followed by a joint meeting.

Source: NFP Facility and PROFOR (undated) 'The NFP Facility and PROFOR – two initiatives in support of national forest policy: How do the two interrelate?,' information leaflet;

developing countries that they have sovereignty over their forests. It also downplays the contributions that political and economic elites in developing countries have made to tropical deforestation.

The second key disagreement was the issue of global targets that had first been proposed at the UNFF's third expert group.[77] Quantifiable and time-bound targets were favoured by Canada, Costa Rica, the EU, Mexico, Norway, South Korea and Switzerland, but opposed by Brazil, India, Indonesia, Iran, Peru and the US. Brazil and the US, in particular, negotiated aggressively on this issue. Late in the negotiations, the EU and Canada dropped their insistence on quantifiable targets, asking for a *quid pro quo* commitment to strong time-bound commitments from other countries. When this concession was not reciprocated, the negotiations collapsed.[78] A contributory factor to the failure of the fifth session to reach agreement was that the developed countries had different visions of a code, with the US favouring a statement of general political commitments, while the EU and Canada advocated a more detailed code of practice.[79]

After the negotiations ended it was agreed to hold two further UNFF sessions. The draft resolution from the UNFF's fifth session, bracketed in its entirety, was forwarded to the sixth session, which convened in February 2006. Negotiations were complicated by the continuing fragmentation of the G77, which had failed to agree a common forests strategy since it broke up during the fifth session. The main developing country caucuses were the Amazonian Pact,

Central American countries, African Group and the Association of Southeast Asian Nations (ASEAN). The EU argued that the sixth session should both initiate and conclude the text of a non-legally binding instrument. This proved overambitious, although states eventually agreed a draft ECOSOC resolution containing a commitment to negotiate for a non-legally binding instrument. This instrument should pursue four global objectives:

1 Reverse the loss of forest cover worldwide through sustainable forest management, including protection, restoration, afforestation and reforestation, and increase efforts to prevent forest degradation.
2 Enhance forest-based economic, social and environmental benefits, including by improving the livelihoods of forest dependent people.
3 Increase significantly the area of protected forests worldwide and other areas of sustainably managed forests, and increase the proportion of forest products from sustainably managed forests.
4 Reverse the decline in official development assistance for sustainable forest management and mobilize significantly increased new and additional financial resources from all sources for the implementation of sustainable forest management.[80]

The time-bound dimension to these objectives, on which the EU and Canada had insisted at the fifth session, was lost in negotiation, with states agreeing only to make progress towards their achievement by 2015. 2015 was also agreed as the year when states would review the effectiveness of the international arrangement on forests. In agreeing this, the UNFF extended its life for an additional nine years, although it will now meet only every second year.

Argentina, the EU, Canada and the Central American states pressed successfully for inclusion in the draft resolution of a 'legally binding instrument' as a future option.[81] But Canada also expressed disillusionment with the UNFF and said that it was prepared to consider options outside the UN. Canada arranged an invitation-only event during the second week of the sixth session, inviting only states that had previously expressed interest in a forests convention. Countries that attended included Argentina, Canada, Chile, China, Costa Rica, Finland, Germany, Guatemala, Ghana, Japan, Kenya, Mexico, the Netherlands, South Africa and Spain.[82] Only Argentina, Costa Rica and Mexico had publicly intimated that they might be interested in pursuing a forests convention outside the UN. At the time of writing (June 2006), Canada has yet to garner a sufficient critical mass of countries to negotiate a forests convention outside the UN. The agreement to negotiate a non-legally binding instrument is likely to marginalize the Canadian initiative for the time being.

The sixth session also saw a reiteration of the familiar positions on financing. Significantly, however, the draft ECOSOC resolution included a commitment to review funding mechanisms, including 'the possibility of setting up a voluntary global funding mechanism as a contribution towards achieving the global objectives and implementing sustainable forest management.'[83] This is the first reference to a possible global forests fund in a textual output agreed by a UN forest institution. Reference is also made to the need to strengthen as funding sources the NFP Facility, PROFOR and the Bali Partnership Fund of the ITTO.[84] The text also contains reference to the principle of common but differentiated responsibilities. Following a proposal from Croatia, the UNFF agreed to recommend that 2010 be designated the International Year of the Forests.

It is likely that the UNFF will shift towards a two-tiered approach, with the UNFF meeting every two years and regional meetings concentrating on implementation being held every other year. Regional processes may be structured around either the ECOSOC regional commissions or the FAO regional forestry commissions.[85] The view that regional processes are necessary can be seen as part of an historical cycle in which the political locus of global forest policy moves over time between the international and regional levels. During the mid 1980s, the FAO created an international mechanism, the Tropical Forestry Action Plan (TFAP). By 1990, the TFAP was seen as overly centralized and removed from the political realities on the ground. The solution was a restructured TFAP, with responsibility devolved from FAO headquarters in Rome to the FAO regional offices.[86] By the mid 1990s, there was consensus that an international forest policy was necessary, hence the creation of the IPF. There is now a widespread recognition that the UNFF as a purely international process has proved ineffective and that a shift in focus back to the regional level is needed.

Concluding thoughts

The UNFF has incorporated a broader range of activities than the IPF and IFF. Unlike its predecessors, it has established expert groups and held high-level ministerial segments, multi-stakeholder dialogues and panels. But these different activities usually operate in isolation from each other. The panels, like the multi-stakeholder dialogues, have exposed delegates to critical voices that would not otherwise be heard; but neither has had a visible impact on the intergovernmental negotiations. Even major groups are now questioning the value of multi-stakeholder dialogues. The country-led and NGO-led intersessional initiatives continue because they are perceived to have value in their own right through promoting information-sharing and network-building; but there is no longer any pretence that they affect formal UNFF decision-making. At the UNFF's fifth session, ministers read statements to each other, while in separate rooms the substantive negotiations gradually foundered. The three expert groups had no tangible impact on the negotiations. Overall, the UNFF has developed a peculiar type of *disconnected* politics. The various pieces do not connect to yield a coherent whole.

If and when a non-legally binding instrument on forests is agreed, it will add to the body of soft law on forests that has emerged over the last 15 years. This comprises the 1992 Forest Principles; Chapter 11, 'Combating deforestation' of Agenda 21; the IPF proposals for action of 1997; the IFF proposals for action of 2000; and the UNFF resolutions. Given this existing body of soft law, what will a non-legally binding instrument on forests achieve?

One view is that such an instrument will prove significant. The UNFF, with active support from the Collaborative Partnership on Forests, is now confronting its own weaknesses, with the result that a revitalized international arrangement on forests is being created. The proposed non-legally binding instrument can build on and strengthen the commitments made in existing soft law, and will usher in a new era of international cooperation on forests built around the four global objectives and a vibrant regional structure.

Against this it can be argued that the UNFF, despite its high profile in the UN system, has reached the law of diminishing marginal returns, and there is little to be gained from trying to agree further political commitments.[87] While the IPF and IFF agreed the proposals for action and catalysed work on national forest programmes, these sorts of benefits cannot be continually reaped. Most of the UNFF resolutions that have been agreed are weaker than the IPF/IFF proposals. The UNFF has completely failed as a guiding body that provides leadership and direction to other forest-related institutions. There is nothing intrinsic to a non-legally binding instrument on forests that will make it necessarily stronger than the soft law on forests that has previously been agreed. According to this view, if such an instrument is agreed, it will at best yield only incremental gains. And if it merely reiterates existing commitments, then it will prove an irrelevancy.

6

The Certification Wars

Forest certification is the process by which an independent third party certifies that a forest management process or forest product conforms to agreed standards and requirements.[1] Two types of standards may be promoted in certification schemes: systems-based standards and performance-based standards (see Box 6.1). The certification wars referred to here are the value-based disagreements and conflicts between the proponents of the different non-state, market-based forest certification schemes that have emerged since the mid 1990s, particularly between the Forest Stewardship Council (FSC) and various business-promoted schemes that have challenged the FSC. The struggle between the FSC and these competitor schemes has assumed the form of a struggle for global hegemony and, as with all hegemonic struggles, those involved, namely the supporters of the different schemes, have sought to gain legitimacy and authority through action across a broad range of sites. This chapter provides an analytical overview of the certification wars in, more or less, chronological order. We argue that the certification wars symbolize a deeper conflict about who makes the rules of global environmental governance, and in whose interests.

Opening salvo: The Forest Stewardship Council

By the late 1980s, NGOs were becoming disillusioned with the failure of the International Tropical Timber Organization (ITTO) to address the sustainable management of tropical forests. In 1988, the WWF stated that 'if the ITTO fails to actively promote tropical forest conservation … then conservation organizations will have to seek other mechanisms to achieve this.'[3] One year later the UK delegation to the ITTO proposed a labelling system for sustainably produced tropical timber.[4] The proposal was blocked by tropical timber-producing countries, with the ITTO considering the proposal 'a veiled attempt to … encourage the current campaign of boycott against the import of tropical timber products.'[5]

The rejection of the proposal led to the WWF making good on its threat to seek other mechanisms.[6] In 1991, WWF formed a certification working group with some other NGOs, including Greenpeace and the Rainforest Alliance, which in 1990 had created Smart Wood, the world's first independent forest

Box 6.1 Systems-based standards and performance-based standards

Systems-based standards (sometimes called management or process standards) focus on the means of forest management; that is, the management systems by which forest owners and managers review their objectives. When systems-based standards are certified, it is not the forest that is assessed, but the management system, such as a forest management plan or monitoring system. Performance-based standards focus on the ends of forest management; that is, the goals and results that forest owners or managers must attain. When performance-based standards are certified, the forest itself is assessed. To gain certification using performance-based standards, a forestry organization may have to manage its forests with an agreed buffer zone size or with clearcuts that do not exceed a stipulated size.[2]

Systems-based standards have been criticized as they can be certified without a visit to the forest. They can certify different forestry organizations carrying out similar activities to very different performance standards. Purely performance-based standards focus on ends, but can neglect the means by which these ends are achieved. A good forest certification scheme thus requires a mix of systems-based and performance-based standards. Environmental NGOs tend to prefer performance-based standards that focus on the maintenance or improvement of environmental quality. Social NGOs insist that auditors should talk to communities in order to assess performance and not rely solely on documentary evidence. The FSC, which was created largely by NGOs, has a strong performance-based element. Forest businesses tend to favour the flexibility of systems-based standards, particularly those that the business sector itself has developed. Industry-promoted schemes all have a strong systems-based component, although none entirely neglects performance-based standards.

certification scheme.[7] The timber trade was represented mainly by small-scale producers. The working group agreed to form the FSC, an independent certification scheme for well-managed forests. The term sustainably managed forests was eschewed due to the competing definitions and controversy that surround this term. It was agreed that the FSC should deal with all forests, and not just tropical forests as the ill-fated ITTO proposal had intended. At this stage support from the private sector consisted of a handful of environmentally concerned individuals in business, principally from the UK. Hubert Kwisthout, who had formed the Ecological Trading Company to promote the import of wood from well-managed sources, was a key figure,[8] as was the environmental director of British retailer B&Q, Alan Knight, who played an important role both in delivering B&Q's support and in the creation of the FSC. It was originally planned to establish the FSC as a foundation with a board of trustees, but no membership structure. The World Rainforest Movement (WRM) reacted to this by mobilizing a large body of support from NGOs, which collectively persuaded the groups establishing the FSC to adopt an open membership structure with voting rights, accountability mechanisms and complaints procedures. The FSC has survived as a viable and robust mechanism throughout the certification wars as its broad constituency base and participatory approach has provided it with a legitimacy that it would have lacked if established as a foundation.[9]

The FSC founding assembly, held in Toronto in 1993, subsequently agreed an original institutional format, with decision-making authority vested in a bicameral system of two chambers: a social and environmental chamber holding 75 per cent of votes, and an economic chamber, with representatives from forest owners and the retail sector, holding 25 per cent of votes. This arrangement was later changed to a tripartite structure with social, environmental and economic chambers, each holding one third of voting rights. This new arrangement separated social stakeholders (such as forest workers, trade unionists and indigenous peoples) from conservation groups, and increased the share of votes of economic stakeholders from one quarter to one third. This was a pragmatic shift to attract greater business support. The three chambers have voting parity between developed and developing country stakeholders. The revised constitutional arrangements ensure that no single chamber, group or region dominates.

In 1994 the FSC agreed nine principles for well-managed forests. These principles were revised in 1996 and 2000, during which time a tenth principle on plantations was added (see Box 6.2). The plantations principle has proved divisive as many tropical plantations occupy former primary forestland. Some NGOs, including Greenpeace and Friends of the Earth, argued that the FSC would lose credibility by certifying plantations, which are not ecologically representative, cannot support the same level of biodiversity as natural forests, cannot provide the same returns of non-timber forest products, and do not provide the cultural and recreational services of natural forests.[10] The admission of plantations has allowed the FSC to certify more forests, with approximately one third of FSC-certified tropical timber coming from plantations.[11] In Brazil, approximately three-quarters of FSC-certified timber is harvested from plantations, while in South Africa plantations account for almost all FSC-certified timber.[13] During 2003, some NGOs, including the WRM, Friends of the Earth and NOVIB (Nederlandse Organisatie voor Internationale Bijstand), urged the FSC to suspend certification of large-scale plantations pending a policy review.[14] In September 2004 the FSC announced a plantations policy review, although it did not suspend the certification of plantations.[15]

Embedded within the FSC principles are potentially competing ownership claims. FSC principle 1 stipulates respect for national law, while principle 3 requires respect for the customary rights of indigenous peoples. There is no necessary contradiction between the two, as long as national law recognizes traditional land claims (see Chapter 1). But where this is not so there is a clear potential for conflict. In such circumstances, indigenous peoples have favoured the FSC. Other certification schemes have no equivalent to principle 3.

Rather than certify forests itself, the FSC accredits independent third-party certifying organizations to do this.[15] FSC-accredited certifiers include Smart Wood and Scientific Certification Systems in the US. A major UK certifier is the Soil Association. Over half of FSC forests worldwide are certified by Société Générale de Surveillance (SGS).[16] Outright refusal is rare, and where a forest area does not meet FSC standards the certifier usually issues a list of corrective action requests to be completed before certification is given. The most common conditions stipulated in corrective action requests are an improved management plan, improved monitoring, written environmental impact guidelines, and the protection of a representative sample of existing forest ecosystems.[17] A central feature of FSC certification is the chain of custody. This is the route along which timber travels from the forest to the retail outlet, including all intermediate stages when timber passes from one custodian to another, such as from warehouse to ship to railway, and so on. The FSC scheme thus involves both forest certification and supply chain certification.

The FSC principles and criteria are not to be confused with the various regional criteria and indicators (C+I) for sustainable forest management that have been produced by intergovernmental bodies since the mid 1990s. A criterion is an element of sustainable forest management. For each criterion there are several indicators. An indicator gauges an aspect of a criterion. It is a quantitative or qualitative variable that can be measured or assessed to detect changes over time. Nine regional C+I processes have been developed that between them cover 150 countries and 85 per cent of the world's forest area (see Table 6.1). It is generally accepted that the diversity of the world's forest types rules out a set of global criteria and indicators. However, a comparative analysis of the nine C+I processes reveals seven criteria of sustainable forest management common across all processes. They are:

1 extent of forest resources;
2 forest health and vitality;
3 productive functions of forests;
4 biological diversity;
5 protective functions of forests;
6 socioeconomic benefits and needs;
7 legal, policy and institutional framework.

While these seven criteria are embedded within the FSC's principles, there are some important conceptual and practical differences between the FSC and the C+I processes (see Table 6.2). C+I are tools for determining the status of forests at a given time and for measuring trends over time. They cannot be used to make claims that a forest management regime has attained a certain standard. In fact, a C+I scheme could be used to show that an area of forest scores 'low' on all criteria. As none of the C+I schemes have normative benchmarks, they provide evidence neither of sustainability nor of unsustainability.

Box 6.2 Forest Stewardship Council principles for forest stewardship

1 *Compliance with laws and FSC Principles:* Forest management shall respect all applicable laws of the country in which they occur, and international treaties and agreements to which the country is a signatory, and comply with all FSC Principles and Criteria.
2 *Tenure and use rights and responsibilities:* Long-term tenure and use rights to the land and forest resources shall be clearly defined, documented and legally established.
3 *Indigenous peoples' rights:* The legal and customary rights of indigenous peoples to own, use and manage their lands, territories and resources shall be recognized and respected.
4 *Community relations and workers' rights:* Forest management operations shall maintain or enhance the long-term social and economic well-being of forest workers and local communities.
5 *Benefits from the forest:* Forest management operations shall encourage the efficient use of the forest's multiple products and services to ensure economic viability and a wide range of environmental and social benefits.
6 *Environmental impact:* Forest management shall conserve biological diversity and its associated values, water resources, soils, and unique and fragile ecosystems and landscapes, and, by so doing, maintain the ecological functions and the integrity of the forest.
7 *Management plan:* A management plan – appropriate to the scale and plan of the operations – shall be written, implemented, and kept up to date. The long-term objectives of management, and the means of achieving them, shall be clearly stated.
8 *Monitoring and assessment:* Monitoring shall be conducted – appropriate to the scale and intensity of forest management – to assess the condition of the forest, yields of forest products, chain of custody, management activities and their social and environmental impacts.
9 *Maintenance of high conservation value forests:* Management activities in high conservation value forests shall maintain or enhance the attributes which define such forests. Decisions regarding high conservation value forests shall always be considered in the context of a precautionary approach.
10 *Plantations:* Plantations shall be planned and managed in accordance with Principles and Criteria 1 to 9, and Principle 10 and its Criteria. While plantations can provide an array of social and economic benefits, and can contribute to satisfying the world's needs for forest products, they should complement the management of, reduce pressures on, and promote the restoration and conservation of natural forests.

Source: FSC (2002) *FSC Principles and Criteria for Forest Stewardship, Document 1.2, Revised February 2000*, Oaxaca: FSC

Table 6.1 *Criteria and indicators for sustainable forest management: Nine processes*

Name	**Details**	**Date and place adopted**	**Adopted by**
ITTO	7 criteria and 66 indicators at the national and forest management unit levels for humid tropical forests	March 1992, Yokohama, Japan	28 tropical timber producing countries (also endorsed by 25 tropical timber consuming countries)
Dry-Zone Africa Process	7 criteria and 47 indicators at the national level for dry-zone forests	November 1995, Nairobi, Kenya	28 countries
Ministerial Conference on the Protection of Forests in Europe (MCPFE, or the Pan-European process)	27 quantitative indicators and 101 descriptive indicators at the regional and national levels for European forests	June 1993, Helsinki, Finland, and June 1998, Lisbon, Portugal	36 countries
Montreal Process	7 criteria and 67 indicators at the national level for non-European temperate and boreal forests	February 1995, Santiago, Chile	12 countries
Tarapoto Process (or the Amazonian Process)	1 criterion and 7 indicators at the global level, 7 criteria and 47 indicators at the national level, and 4 criteria and 22 indicators at the forest management unit level for Amazonian forests	February 1995, Tarapoto, Peru	8 countries

Table 6.1 *Criteria and indicators for sustainable forest management: Nine processes (Continued)*

Name	***Details***	***Date and place adopted***	***Adopted by***
Near East Process (sponsored by the FAO and the UNEP)	7 criteria and 65 indicators at the regional and national levels for dry forests in Asia, the Arabian Peninsula and Northern Africa	October 1996, Cairo, Egypt	30 countries
Lepaterique Process (or the Central American Process)	4 criteria and 40 indicators at the regional level, 8 criteria and 42 indicators at the national level, with additional criteria and indicators at the forest management unit level, for Central American tropical forests	January 1997, Tegucigalpa, Honduras	7 countries
African Timber Organization	5 principles, 2 sub-principles, 26 criteria and 60 indicators at the national and regional levels for tropical forests in Africa	January 1993, Libreville, Gabon	13 countries
Dry Forests in Asia	8 criteria and 49 indicators at the national level for dry forests in Asia	December 1999, Bhopal, India	9 countries

Sources: FAO (2004) *Report: FAO/ITTO Expert Consultation on Criteria and Indicators for Sustainable Forest Management*, Rome: FAO, Appendix 2, pp.89–93; Ministerial Conference on the Protection of Forests in Europe (2000) 'Brief description and number of countries participating in the major international on-going processes on criteria and indicators for sustainable forest management,' mimeo dated 3 March 2000

Despite their differences, the near-simultaneous arrival in international forest politics of C+I and forest certification caused some confusion among policy-makers. Much NGO activity during the mid 1990s concentrated on lobbying policy-makers to adopt the FSC and not to use C+I schemes, which many NGOs concede are useful tools, to make claims on sustainable forest management.[18] However, despite the differences between C+I and forest certification, some FSC competitor schemes have been based, in part, on C+I processes.

As a market-based scheme, the FSC relies for its success on demand from environmentally discerning consumers and retailers that cater to this demand. However, the proponents of the FSC have not left the success of the scheme entirely to free market forces, but have engaged in a form of demand manipulation through NGO-sponsored forest and trade networks (originally called buyers' groups). The first such network was the 1995 Group, which WWF-UK established in 1991 following a report to the ITTO that less than 1 per cent of the world's tropical forests were sustainably managed.[19] The aim of the 1995 group – now the WWF-UK Forest and Trade Network – is to encourage companies to use only timber from well-managed sources. Not all buyers' groups are WWF sponsored. Some NGOs that initially disagreed with the FSC's plantations principle have organized buyers' groups, including Friends of the Earth in Brazil and the Netherlands. These groups form part of the Global Forest and Trade Network, which in 2004 spanned 30 countries.[20] According to one estimate, more than half of global demand for FSC-certified timber is created by the Global Forest and Trade Network.[21]

Table 6.2 *Differences between criteria and indicator processes and Forest Stewardship Council certification*

Criteria and indicators for sustainable forest management	***FSC certification***
Regional-level processes	Global-level process
To be applied mainly at the national level	To be applied at the forest management unit level
Descriptive: aims to measure and depict trends in forest management over time	Prescriptive: aims to stipulate normative standards and requirements, and to enable assessment on whether these standards have been met
Used mainly by governments and forest policy-makers	Used mainly by market players: forest owners, retailers and NGOs

Source: adapted from Rametsteiner, Ewald and Simula, Markku (2002) 'Forest certification: An instrument to promote sustainable forest management?,' *Journal of Environmental Management*, Vol. 67, No. 1, pp.87–98

The North American counterattack

Some forest owners would rather have avoided forest certification given the extra costs involved; but when customers and retailers began demanding certified timber, many found that they had little alternative but to engage with the FSC. The first recourse of business was to contest authority within the FSC. As we have seen, business argued for and won a larger share of voting rights at the FSC. However, it was clear that the FSC would make only marginal and occasional concessions to business. While the FSC aims to be a flexible scheme that is responsive to different ecological and socioeconomic conditions, 'excessive' flexibility to business demands would erode the currency of the FSC, namely its high standards, and alienate environmental and social groups.

Hence many businesses have sought to undermine the FSC through the creation of competitor certification schemes. Forest certification has become another rule-making arena that business has set out to capture, just as it has tried to colonize other regulatory spaces, such as domestic environmental governance and multilateral environmental regimes. The conflict between the FSC and the competitor schemes is the central axis on which the certification wars have been fought. In 1993, the year that the FSC founding assembly took place, preparations began in North America to create two competitor schemes: the sustainable forest management standard of the Canadian Standards Association (CSA), and the US Sustainable Forestry Initiative (SFI).

Canadian Standards Association

The CSA scheme was created at the request of the Canadian Sustainable Forestry Certification Coalition, which had been formed by the Canadian Pulp and Paper Association, now the Forest Products Association of Canada. The scheme is based on the environmental management system standards of the International Organization for Standardization (ISO) (see Box 6.3). There has been controversy on whether the ISO should be involved in environmental management, given that the ISO 14000 series does not mention international environmental agreements.[22] Harris Gleckman and Riva Krut argue that the ISO 14000 series emphasizes 'environmental *conformance* (to an internal set of standards), not environmental *performance*'[23] and that it 'reverses the direction of global environmental performance standard-setting, whether public or private.'[24] To Jennifer Clapp, the ISO 14000 series of standards for environmental management systems will, 'at best maintain the status quo on ... [environmental] problems and at worst actually exacerbate them.'[25] David Downes notes that ISO 14000 will only verify whether a forest management system 'is likely to meet the environmental goals set by the management company itself.'[26] Two leading figures in the creation of the FSC, Chris Elliott and Matthew Wenban-Smith, have criticized ISO 14000 for allowing companies to define their own objectives and for permitting companies engaged in similar activities to have different performance standards.[27]

Most CSA certifications have been carried out using ISO 14000 standards. The criteria and indicators agreed by the Canadian Council of Forest Ministers and the Montreal Process for non-European temperate and boreal forests have been used as a framework for standard-setting. The CSA is principally, although not exclusively, a systems-based standard. Fred Gale has criticized the criteria and indictors of the Canadian Council of Forest Ministers as 'firmly embedded in an industrial approach to forestry and ... thus not enough, in themselves, to establish sustainable forestry.'[28] Under the CSA scheme each company draws up a sustainable forest management plan. However, there are no common performance targets or minimum thresholds across the scheme, hence standards vary substantially from case to case.[29] The scheme does not prohibit the use of GM trees and has weak procedures for involving indigenous peoples. It places no restrictions on the establishment of new plantations.

Sustainable Forestry Initiative

There are some similarities between the Sustainable Forestry Initiative (SFI) and the CSA. Both are based on ISO 14000 standards and both were created by a nationwide forest industry group in 1993. In the case of the SFI this was the powerful American Forest and Paper Association (AFPA). During 2002, the SFI became institutionally independent from the AFPA, thus enabling the SFI to escape the 'foxes guarding the hen house' argument. The SFI now elects an independent board of 15 members, of whom two-thirds are from non-industry interests.[36] It remains a condition of AFPA membership that a forest business participates in the SFI.[37] The size of the AFPA gives the SFI a readymade constituency in the US. The SFI provides a list of indicators for companies to address, although each company can adapt these indicators when producing its own standards. The scheme does not address social issues, such as the rights of forest workers and indigenous peoples. It permits the use of GM trees, herbicides and pesticides.[38] In March 2005, over 90 scientists submitted a letter to the SFI board claiming that the scheme:

> *... does not discourage logging and buying of wood from the most biologically diverse and sensitive areas ... allows for the conversion of native and natural forests to single species pine plantations [and] ... allows for logging practices that can be harmful to habitat and water quality, including large-scale clear-cutting.*[39]

According to FERN, the SFI is 'one of the least credible of all schemes' and it certifies 'near status quo' practices.[40] The SFI has prevailed over the FSC in the US, in part because most US wood products are sold domestically rather than exported. Hence, and as Benjamin Cashore and colleagues have previously argued, supporters of the FSC in the US cannot rely for support from markets at the lower end of the supply chain where there is high demand for FSC timber, as in Western Europe. NGO efforts have thus had to take place almost exclusively within the US.[41]

Box 6.3 The ISO 14000 series on Environmental Management Systems

Created in 1946, the International Organization for Standardization (commonly referred to as the ISO) is a worldwide federation of national standards bodies. During the 1990s, it agreed a set of standards for environmental management systems (EMS): the ISO 14000 series. ISO 14000 can be applied to environmental management in any sector. It stipulates the elements of an EMS, including analysis of environmental impacts, a programme of environmental objectives and a commitment to continuing improvement.[30] ISO 14000 does not prescribe environmental performance objectives. Instead, each participating organization sets its own objectives. ISO 14000 standards are thus systems-based rather than performance-based. As the introduction to ISO 14001 notes:

> *It should be noted that this standard does not establish absolute requirements for environmental performance beyond commitment, in the policy, to compliance with applicable legislation and regulations to continual improvement. Thus, two organizations carrying out similar activities but having different environmental performance may both comply with its requirements.*[31]

Any organization can apply for its EMS to be ISO registered. An independent third-party 'registrar' will audit an organization's systems and assess whether they comply with ISO 14000. Since the ISO is not a product certification body, forest products cannot be labelled as 'ISO certified,' and ISO 14000 cannot be used to make claims on the sustainability of forest products. However, a company's forest management systems can be registered as conforming with ISO 14000 standards. There is no mechanism within ISO for chain of custody certification.

ISO EMS standards represent a 'blurring' of public and private international law.[32] The standards are established largely by the private sector. Despite the dominance of business and industry in ISO standard-setting, ISO standards have been recognized within the World Trade Organization (WTO)[33] and thus have some status as public standards in international trade law.[34] ISO 14000 standards can be challenged at the WTO if a WTO member believes that they constitute a barrier to trade under the provisions of the WTO's Technical Barriers to Trade (TBT) Agreement.[35] However, there is no WTO dispute mechanism for challenging the standards on environmental grounds.

The opening of the European front: World War!

The Pan-European Forest Certification (PEFC) scheme was launched in 1999 by forest owners in six European countries; Austria, France, Finland, Germany, Norway and Sweden.[42] Throughout the late 1990s, many private forest owners in Europe had opposed the FSC. They lobbied the European Commission to develop a European Union framework for forest certification that would be suitable for small forest owners.[43] However, the Commission has issued no directive on forest certification.[44] The creation of the PEFC was principally a forest owner reaction against the FSC, although it was also a reaction to the European Commission's reluctance to intervene in favour of European forest owners.

The PEFC is a mutual recognition framework through which national certification schemes can recognize each other as having equivalent standards. The PEFC uses the criteria and indicators adopted by the Ministerial Conference on the Protection of Forests in Europe (MCPFE) and the MCPFE's Pan-European Operational Level Guidelines as the framework for national-level standard-setting.[45] Since the MCPFE is an intergovernmental forum and PEFC is principally an association of forest owners, PEFC's use of MCPFE documents has been controversial. As we have seen, there are important conceptual differences between C+I and certification schemes. Furthermore, the MCPFE's Pan-European Operational Level Guidelines explicitly state that the guidelines 'cannot be used in isolation to determine sustainability in management. Their purpose is to identify complementary actions at the operational level which will further contribute to sustainability of forest management.'[46] Those European forest owners who established the PEFC did not approach the MCPFE secretariat for advice. Had they done so, the MCPFE secretariat would have advised, first, that as an intergovernmental forum the MCPFE does not endorse or oppose any certification scheme, and, second, the MCPFE C+I and Operational Level Guidelines were not designed to be used as a standard-setting system.[47] The creation of the PEFC thus led to some confusion on forest certification, especially since the PEFC adopted the term 'pan-European,' which is the informal name for the MCPFE process and the prefix for its Operational Level Guidelines.

The PEFC has attracted criticism from NGOs, some claiming that the scheme is less rigorous than the FSC. The PEFC does not prohibit or limit future conversions of forest to plantations, and it does not prohibit GM trees.[48] The national PEFC scheme in Finland, the Finnish Forest Certification Scheme, has been criticized by WWF for allowing the logging of old growth forests in the Kainuu region and for threatening the habitat of the Siberian jay bird in the Virat region.[49] The PEFC has only weak provisions on the rights of indigenous peoples. As a result, the scheme favoured by indigenous peoples is the FSC. For example, FSC principle 3 provides a stronger recognition than Swedish national law of the rights of the Sami people to graze their reindeer on private forest land in northern Sweden.[50] The PEFC provides no such recognition, so unsurprisingly the Sami people prefer the FSC.

Table 6.3 *Comparison between the Forest Stewardship Council and FSC competitor schemes*

	FSC	**FSC competitor schemes**
Created by	Environmental NGOs and some socially concerned retailers	Forest owners and/or forest industry
Type of scheme	Primarily performance-based, with a large systems-based component	Primarily systems-based, with some performance-based components
Rule-making authority	Rules set by a tripartite arrangement of economic, social and environmental stakeholders	Rules set principally by business and forest owners, although other actors may have advisory roles
GM trees	Prohibits the use of GM trees	Permits GM trees
Plantations	Establishment of new plantations should be limited	No policy to limit new plantations
Indigenous peoples	The customary and legal rights of indigenous peoples should be respected	Weak or non-existent safeguards for the rights of indigenous peoples

Sources: Cashore, Benjamin (2002) 'Legitimacy and the privatization of environmental governance: How non-state market-driven (NSMD) governance systems gain rule-making authority,' *Governance*, Vol. 15, No. 4, pp.503–529; Cashore, Benjamin; Auld, Graeme and Newsom, Deanna (2004) *Governing Through Markets: Forest Certification and the Emergence of Non-state Authority*, New Haven: Yale University Press

During 2003 PEFC was relaunched as a worldwide framework, the Programme for the Endorsement of Forest Certification schemes (retaining the PEFC acronym). There are some important differences between, on the one hand, the FSC and, on the other, the PEFC and the other competitor schemes (see Table 6.3). PEFC allows for purely systems-based national schemes to be endorsed, although most national schemes also have a performance-based component. Although the PEFC has established itself as a mutual recognition system, it has no mechanism for ensuring that the different national schemes offer similar standards.[51] PEFC is ISO 14000 compatible, with some European national forest certification schemes applying for ISO 14000 registration in addition to PEFC certification. Norway is one example.

Whereas the PEFC endorses national certification schemes, the FSC certifies individual forests, some of them very small. This explains the different total areas of forest that the two schemes have certified. By the start of 2005, the two schemes had certified comparable areas of forest; between 50 million and 55 million hectares.[52] During 2005, the PEFC, which had earlier recognized the national forest certification schemes of Australia and Chile, recognized the Brazilian national scheme, as well as the CSA and SFI. This led to a huge increase in PEFC-certified forests, so that by the end of December 2005 the PEFC had certified over 186 million hectares of forests in 19 countries.[53] This is a crude average of more than 9.8 million hectares per country. By this time the FSC had certified over 68 million hectares in 66 countries,[54] a crude per country average of just over 1.03 million hectares per country, less than one ninth that of the PEFC. The trend is clear: when the PEFC becomes established in a country, it certifies far more forests than the FSC.

Assuming that there is a high level of demand for certified timber, businesses that produce or sell certified timber will gain market share at the expense of those that do not. While many factors may determine changes in market share between businesses, those companies and forest owners supporting the competitor schemes clearly wish to ensure that a certification scheme created by conservation interests will not be one of them. The creation of the FSC competitor schemes can be seen as a defensive move to protect relative market shares. But not all businesses have elected to support the competitor schemes. Had they done so the FSC could not survive. Some companies have declared exclusive support for the FSC, such as the British retail outlet B&Q and the publishers of this book, Earthscan. Other businesses support the FSC, but without offering exclusive support. Examples include the Swedish firm IKEA and the UK-based publishing group Random House, which in 2006 became the first commercial book publisher to receive FSC chain-of-custody certification.[55] Within an individual company, policy can shift over time. The Home Depot in the US initially opposed the FSC. A few years later, it declared exclusive support for the FSC.

While the FSC enjoys widespread support from NGOs, such support is by no means unconditional. Friends of the Earth and Greenpeace initially declined to play a formal role in the FSC after the founding assembly's decision to grant voting rights to those with an economic stake in the timber industry.[56] And as we have seen, the decision to include FSC principles on plantations is opposed by many NGOs. NGO criticism of the FSC is now principally focused on implementation. In a study carried out for the Indonesian Environment Forum (WALHI) and the Indigenous Peoples' Alliance of the Archipelago (AMAN), the Forest Peoples Programme recommended in 2003 that the FSC suspend certification in Indonesia after finding that the Indonesian state 'lacks effective measures for securing customary rights to land and forests.'[57] The study found that only 12 per cent of Indonesia's forests had been gazetted, with even fewer concessions properly demarcated, meaning that most of Indonesia's logging concessions were technically illegal.[58] While the FSC endorsed the main findings of the report, it did not suspend certification. The confusion about the legality of Indonesian logging remains unresolved.[59]

The most thorough critique of the FSC, based on studies of implementation in Brazil, Thailand, Malaysia, Ireland and Canada, was a 2002 report, *Trading in Credibility*, from the Rainforest Foundation. Like the Forest Peoples Programme, the Rainforest Foundation found that there had been difficulties in implementing FSC principles 2 and 3. It also found that many timber companies were directly implicated in human rights abuses, and that the FSC permits logging in primary rainforest. Because there were 'direct economic relations' between certifiers and forest managers, there was a vested corporate interest in granting certification to applicants that were 'in breach' of FSC principles.[60] The FSC responded that certifiers are paid irrespective of whether they approve certification, hence they have no direct financial interest in certifying forests; that the FSC 'has never and will never certify (or maintain certification of) any company, community or private forest owner' involved in human rights abuses; and that to protect the livelihoods of indigenous and local peoples, the FSC allows the possibility of logging in high conservation value forests 'subject to extreme precautions' with the biological value of such forests protected under FSC principle 9.[61] *Trading in Credibility* was presented to the FSC's board of directors, and the Rainforest Foundation were invited to present their findings.[62]

NGO support for the FSC is thus qualified and conditional. But although many NGOs have criticized the FSC, they have rarely gone so far as to oppose the FSC,[63] and never to advocate support for a competitor scheme in lieu of the FSC. On the contrary, NGO critiques of the competitor schemes tend to be far more stringent. For example, in a report entitled *Certifying Extinction?* Greenpeace and two Finnish NGOs criticized the PEFC-endorsed Finnish Forest Certification Scheme (FFCS) for promoting logging in old growth forests and rewarding poor forest management practices. The NGOs urge Finland's timber companies to move towards FSC certification.[64] The FFCS responded by claiming that endangered species are constantly monitored and calling the title of the report 'emotional, political and misleading.'[65] Accusation and counteraccusation are a central element of the certification wars.

Mutual recognition: Ceasefire proposal or flanking manoeuvre?

It is sometimes suggested that the existence of several schemes is desirable since it offers choices to retailers and consumers, who can decide which scheme best meets their needs. However, this is a supporting argument based on neoliberal 'market knows best' logic, as the existence of so many different schemes has little intrinsic merit. FSC competitor schemes were not created to offer more choice but to weaken the FSC, and one way in which they have done this is by confusing customers. Having successfully generated confusion, the main competitors – the SFI, CSA and PEFC – then endorsed a proposal for mutual recognition between schemes which, it was claimed, would eliminate this confusion. Mutual recognition is a concept with different applications. As we have seen, the PEFC is a mutual recognition framework through which national schemes recognize each other as having equivalent standards. Our interest here is on mutual recognition between international schemes.

The proposal for mutual recognition came from the International Forest Industry Roundtable (IFIR), a group created with the help of the WBCSD. The IFIR was a forest industry group with members that included the Finnish Forest Industries Federation, the Brazilian Pulp and Paper Association and the Confederation of European Paper Industries (CEPI). No certification schemes participated in the IFIR discussions, although the bodies that had created the North American schemes – the AFPA and the Canadian Pulp and Paper Association – took part. The IFIR proposed a mutual recognition framework that would embrace all 'credible' certification schemes, which would be considered 'equivalent.' Criteria and indicators would define the elements of a credible scheme. (The use of the terms 'criteria and indicators' was surprising, given that the proposal was intended to eliminate confusion; the IFIR's proposed criteria and indicators for certification schemes should not, of course, be confused with the regional C+I schemes for sustainable forest management.) It was proposed that the criteria would include conformity of a certification scheme with sustainable forest management principles, participation and a commitment to continual improvement. All schemes passing an agreed threshold would be considered equivalent and credible. Those schemes that did not meet the threshold would not qualify.[66] The SFI, CSA and PEFC promptly supported the proposal.[67] The proposal was not an attempt to create a unified global scheme, although it was suggested that an international mutual recognition system could have a single global trademark.

There are two ways of viewing the IFIR's proposal. The first is as an honest endeavour to reduce customer confusion and to provide a more stable policy environment by eliminating the uncertainties caused by competing schemes. The second view is that as mutual recognition would only be as strong as its weakest scheme, the proposal was a forest industry attempt to outmanoeuvre the FSC. Under mutual recognition a cynical retailer could claim that its policy was to sell timber consistent with the highest scheme, then stock only timber certified by the weakest scheme that qualified as 'credible' under the mutual recognition framework. In short, under mutual recognition there would be no incentive for retailers to sell timber produced according to the higher standards.

The FSC's reaction was summed up by its former head, Timothy Synott, in 2000. Synott noted the IFIR's definition of a mutual framework as 'reciprocal arrangements under which one standards body or system recognizes and accepts other standards and certification systems as being substantively equivalent in intent, outcomes and process.' Noting the IFIR's suggestion that a single global trademark would be desirable, Synott responded: 'Absolutely! These elements all provide a full description of the FSC system.'[68] The FSC opposed the IFIR's proposal, but declared its willingness to work with any certification scheme that met FSC standards. The PEFC responded that it 'has a functioning mutual recognition program in place' and 'is now the dominant forest certification recognition scheme in the world.'[69]

The FSC's opposition to mutual recognition was supported by the NGO community. WWF stated that 'only certification under the FSC system can be considered to reach satisfactory performance levels and thus provide an adequate incentive for improving forest management worldwide.'[70] FERN, Friends of the Earth, the German group Robin Wood and the Forest Peoples Programme argued that the weakest scheme in a mutual recognition framework would constitute a liability that would extend to all other schemes, and that no certification scheme 'is likely to intentionally sacrifice its credibility by accepting, as its own, the serious weaknesses of other programs.'[71] Greenpeace argued that 'Mutual recognition must not become a process for weakening standards. We reject the IFIR ... proposal as fundamentally flawed and a significant step backwards for forests, forest certification and consumers.'[72]

One reason for business support for mutual recognition was that it would have entitled transnational corporations that currently use different schemes in different countries to claim that they operate a consistent worldwide policy on forest certification. But undoubtedly another reason for business support was a wish to further weaken the normative pull of FSC standards. While the IFIR sought to frame mutual recognition as a technical process to be agreed using criteria and indicators, it is best seen as a thinly disguised political move to rout the FSC. The idea retains support among the competitor schemes, although it foundered when it became clear that the FSC would not support it. Without the involvement of the FSC, a mutual recognition framework would not eliminate customer confusion; indeed, it would bring the differences between the FSC and other schemes into sharper focus. The PEFC was relaunched as an international scheme shortly after the FSC declared that it would not enter an international mutual recognition framework.

The search for coalition allies

One reason why the certification wars are unresolved is that there is no commonly accepted authoritative definition of an international standard-setting organization. So the FSC and the PEFC have sought to garner legitimacy as standard-setting organizations through recognition from other actors. Each has sought recognition from governments through adoption as recognized certification schemes for state-owned forests. FSC standards have been used to certify state-owned forests, including in the UK and Latvia.[73] In some other countries, PEFC national schemes have been used. Some German state-owned forests have been certified using PEFC, and others using the FSC.[74] The legitimacy of some certification schemes has also been enhanced through recognition in government timber procurement policies. France is developing a procurement policy favouring the FSC and equivalent schemes.[75] The FSC and the CSA have been accepted in the UK as schemes that provide evidence of legality and sustainability.[76]

A target for both the FSC and PEFC has been the International Accreditation Forum (IAF). The IAF is a world association of accreditation bodies. Created in

1986, IAF members are national-level organizations that accredit other bodies with the authority to certify whether an organization's management processes and products conform with an agreed standard. (The IAF refers to this process as conformity assessment, rather than certification.) Whereas the ISO is a worldwide federation of national standards bodies, the IAF is a worldwide federation of national accreditation bodies. ISO and IAF work closely together, and the IAF has endorsed the ISO 14000 series. The IAF is developing a single international programme of conformity assessment for its members, the intention being to provide a predictable international working environment for businesses.[77] It aims for mutual recognition between national accreditation bodies, the eventual objective being to 'cover all accreditation bodies in all countries in the world, thus eliminating the need for suppliers of products or services to be certified in each country where they sell their products.'[78] The authority of the IAF is recognized by the European Commission, which usually requires a national accreditation body to receive IAF membership before it is considered competent in the EU. The IAF's slogan of 'Certified once, accepted everywhere' was attractive to the PEFC, as was its endorsement of the idea of mutual recognition between national bodies.

In 1999, the FSC helped to create the International Social and Environmental Accreditation and Labelling (ISEAL) Alliance, a group of international voluntary standard-setting, accreditation and certification organizations that aim to promote ecological sustainability and social justice in trade. There were eight founding member organizations of the ISEAL Alliance (see Box 6.4). ISEAL aims to ensure greater compatibility between accreditation bodies promoting trade in products produced according to strong environmental and social criteria. ISEAL's Code of Good Practice for Setting Social and Environmental Standards aims to promote standards that are least trade restrictive.[79] All ISEAL members are non-profit making.

The IAF has been unwilling to recognize ISEAL or its member organizations. In 1998, the International Organic Accreditation Service (IOAS) applied for IAF membership.[80] Its application was rejected. In 2000, ISEAL announced that it had approached the IAF and suggested a second IAF chamber be formed, made up of international accreditation bodies, that would complement the existing chamber of national bodies. This proposal was rejected. In 2002, ISEAL applied for associate status at the IAF. It was turned down.[81] Significantly, however, the IAF has granted associate status to other international bodies, including, in 2004, the PEFC. The PEFC secretary general subsequently announced that 'We will now actively participate in IAF's accreditation harmonization processes to ensure that accredited forest ... certificates issued in one part of the world are recognized elsewhere.'[82]

At the heart of the IAF–ISEAL disagreement are two very different visions on who makes the rules of global governance, and whether the rules should include ecological and social criteria. IAF recognition of PEFC, which came shortly after PEFC was relaunched as an international scheme, has aided the PEFC's efforts to become the world's dominant forest certification scheme.[83] Both PEFC and IAF are organizations comprising national-level bodies.

Box 6.4 Founding members of the International Social and Environmental Accreditation and Labelling (ISEAL) Alliance

- *Forest Stewardship Council (FSC):* promotes the trade of timber from forests that are well-managed according to ten principles and associated criteria.
- *Marine Stewardship Council (MSC):* set up along the lines of the FSC model, the MSC has developed a label and environmental standard for well-managed fisheries.
- *Fairtrade Labelling Organizations International (FLO):* promotes international fairtrade standards that aim to contribute positively to 'disadvantaged producers.'
- *International Federation of Organic Agriculture Movements (IFOAM):* promotes the worldwide adoption of agricultural systems based on the principles of organic agriculture.
- *International Organic Accreditation Service (IOAS):* accredits certification bodies that certify organic products and runs the accreditation programme of IFOAM.
- *Marine Aquarium Council (MAC):* includes in its membership marine acquarium animal collectors and aims 'to conserve coral reefs and other marine ecosystems by creating standards and certification for those engaged in the collection and care of ornamental marine life from reef to aquarium.'
- *Rainforest Alliance:* as well as establishing the Smart Wood programme and being a founding member of the FSC, the Rainforest Alliance serves as the international secretariat for the Sustainable Agriculture Network and was a founding member of the Sustainable Tourism Certification Network.
- *Social Accountability International (SAI):* aims to combat discrimination in the workplace through the implementation and monitoring of voluntary and verifiable social standards. It aims to improve working conditions and fight sweatshops through its international workplace standards.

Source: ISEAL Alliance website: www.isealalliance.org/membership/founding.htm (accessed 24 January 2005)

They thus share a similar structure and organizational culture. But this is only part of the reason why PEFC has succeeded in penetrating the IAF, whereas ISEAL and its member organizations have not. Business interests are prominently represented both in the national forest certification organizations that comprise the PEFC and the national-level accreditation bodies that are members of the IAF. And, as we have seen, business is profoundly reluctant to accept environmental and social standards drawn up by civil society groups. The IAF and the PEFC each represent economic interests that are cautious of the ecological and social justice standards promoted by ISEAL member organizations. IAF members may well have looked at the FSC's unwillingness to enter into mutual recognition with other forest certification schemes and concluded that the FSC and its ISEAL partners would, if admitted to the IAF, seek to impose their values on IAF member organizations.

Since ISEAL's application for IAF associate status was rejected in 2002, ISEAL has remained interested in working with the IAF to pursue the common goal of strengthening accredited certification around the world. It is slowly establishing a dialogue with the IAF, and in 2005 an ISEAL representative attended an IAF meeting as an observer.[84] The ISEAL–IAF relationship could thaw further in the future.

The question of legitimacy can also be approached from a more scholarly perspective. Fred Gale has formulated five criteria of legitimacy: scientificity (in other words the extent to which a certification scheme considers broader ecological values); representativity (the range of interests that participate in a scheme); accountability (the interests to which a scheme is accountable, including indigenous and local communities); transparency (the public availability of information); and equality (the extent to which timber producers are treated equally within and between countries). He applied these criteria to the SFI, CSA and FSC, rating each criterion on a scale of low, medium and high. He concluded that, overall, the FSC has a 'high' legitimacy ranking, while the CSA and SFI each rank 'low to medium.'[85]

A second ceasefire proposal: Legitimacy thresholds

The concept of legitimacy is also central to a proposal from the WBCSD to reconcile the various forest certification schemes. Called the legitimacy thresholds model, the idea is, at present, largely at the conceptual stage. It has been promoted since 2002 through a voluntary global partnership, The Forests Dialogue (TFD) (see Box 6.5).

The legitimacy thresholds model acknowledges that different users of forest certification schemes have different ideas as to what constitutes legitimacy. Given the many different users of certification schemes, and given further the different criteria of sustainable forest management, a particular user may consider certain elements of, say, scheme A, to be legitimate, with other, different, elements of scheme B also seen as legitimate. So, for some users no scheme may be accepted in its entirety. Different schemes may have different merits for different users. For example, government procurement bodies need schemes that pass a high threshold with respect to legally sourced timber, while indigenous peoples' groups require schemes that pass a high threshold on the rights of forest peoples.

So the legitimacy threshold model holds that for any given attribute of sustainable forest management there may be different thresholds of legitimacy; for example, low, medium and high. The intention is for these different thresholds to correspond to the needs of different user groups. The model

Box 6.5 The Forests Dialogue

Organizations involved in the creation of The Forests Dialogue in 1999 included the World Bank, the World Resources Institute (Washington), the International Institute for Environment and Development (London) and the WWF. The WBCSD, which in 1996 established a sub-group, the Sustainable Forest Products Industry working group, also helped to establish The Forests Dialogue.

The Forests Dialogue aims to admit only member organizations with international convening power and the authority and resources to initiate and implement new processes. It is intended as a non-confrontational process to address the constraints to sustainable forest management, build trust, share learning and promote collaborative action. It is structured around five global forest issues:

1 forest certification;
2 illegal logging;
3 forests and biodiversity conservation;
4 intensive forest management;
5 forests for the alleviation of poverty.

aims to agree criteria for these thresholds to enable users to assess which schemes pass which thresholds for particular attributes of sustainable forest management. The model, it is intended, will allow users to differentiate according to their needs. It will also allow users from developing countries to adopt a phased approach to certification, starting at low thresholds and moving to stronger thresholds over time.[86]

The legitimacy thresholds model aims to move beyond the mutual recognition proposal, which, the WBCSD has acknowledged, 'is perceived by NGOs to equate to "lowest common denominator" standards.'[87] It differs from mutual recognition in some important respects. The idea behind mutual recognition was for criteria to be used to judge individual schemes; there would be just one threshold with all schemes surpassing that threshold being considered 'equivalent,' even though above the threshold there could be significant differences between schemes. The legitimacy threshold model is more complicated and allows for broader differentiation (see Table 6.4).

Like mutual recognition, the legitimacy thresholds model has shifted the focus of policy-makers away from certification schemes towards frameworks that will assess the schemes.[88] For the model to be implemented, it needs to define legitimacy and its different thresholds. The model is at the design stage, and it is unclear whether it will be successful. The large number of variables in the model could prove its undoing since each variable is a potential source

Table 6.4 *Differences between mutual recognition and the legitimacy thresholds model*

	Mutual recognition	***Legitimacy thresholds model***
What is assessed?	Forest certification schemes in their entirety	The different attributes of sustainable forest management within different schemes
Thresholds	One threshold, above which a scheme is considered credible	Several thresholds, corresponding to the needs of different user groups
Objectives	To provide a global 'umbrella' framework encompassing all credible schemes	To allow differentiation between schemes according to the attributes of sustainable forest management and user needs, and to enable a phased approach to certification

of disagreement. And the more disagreements there are, the greater will be the uncertainty over the model as a whole. Even if a model is agreed that has the confidence of the proponents of the main certification schemes, it would then need to affect the decisions of timber buyers, suppliers and consumers if it were to be more than a paper exercise.

At a meeting of TFD in October 2004, the WWF and WBCSD issued a joint statement in which they noted they had 'divergent views' on mutual recognition but were committed to developing the legitimacy thresholds model.[89] There are two possible reasons why the WWF has chosen to align itself with the WBCSD in support of the model. First, the forest industry considered WWF and other NGOs uncooperative during the mutual recognition debate. By aligning itself with the WBCSD to support the legitimacy thresholds model, the WWF demonstrates a commitment to working with other actors to resolve forest certification conflicts. Second, WWF may confidently expect that any agreed model will show that the FSC passes the highest legitimacy threshold (however defined) on most, if not all, attributes of sustainable forest management.

Forest certification and international trade law

The FSC was created out of the unwillingness of the ITTO to approve a timber labelling scheme. However, it is doubtful that an ITTO scheme would have been permissible under international trade rules. There is no provision in the international trade system that allows states to discriminate in favour of timber harvested from sustainable sources and against timber produced from unsustainable sources. Discrimination between like products according to the production and processing methods (PPMs) used in their manufacture was prohibited under the GATT of 1947 and remains so since the creation of the WTO in 1995.

Forest certification schemes can be seen as voluntary PPM schemes. It is sometimes claimed that because forest certification schemes involve the private sector, exclude governmental membership and are voluntary, they are not covered by the WTO's Technical Barriers to Trade (TBT) Agreement. In fact, Annex 2 to the TBT Agreement includes voluntary standards.[90] However, the status of forest certification schemes within this agreement has yet to be tested through a WTO ruling.

The status of certification schemes at the WTO was raised at the IPF, which considered whether certification should be considered a non-tariff barrier. The EU stated that it should not;[91] but the lack of clarity on the meaning of the TBT Agreement prevented delegates from agreeing to include in the IPF's final report the phrase 'Voluntary certification and eco-labelling are not considered to be non-tariff barriers.'[92] At the IFF, Brazil stressed that certification schemes should confirm with the TBT Agreement. The EU adhered to its IPF position that certification was not a technical barrier to trade and argued for references to the TBT Agreement to be deleted.[93] The final IFF text merely notes that the 'IFF took note of the work of the World Trade Organization (WTO) with regard to voluntary eco-labelling schemes.'[94]

A 2003 status report from the WTO's Committee on Trade and Environment neither endorsed nor condemned labelling schemes, noting that 'voluntary, participatory, market-based and transparent environmental labelling schemes are potentially efficient economic instruments in order to inform consumers about environmentally friendly products.'[95] It is possible that the provisions of the TBT Agreement could prohibit FSC labelling on the grounds that it is not in line with ISO standards, was not developed by national standards-setting bodies and is a technical barrier to trade.[96] It is difficult to predict with certainty whether or not a WTO panel would rule voluntary certification illegal if a challenge were to be brought. Even if no case is brought, a WTO ruling on another matter could have ramifications for the legality of forest certification schemes. A question mark will continue to hang over the status of certification schemes in international trade law until such a ruling is made.

Concluding thoughts

Sooner or later all wars end. How will the certification wars end? There are two dimensions to this question. First, will there be a *consolidation of schemes*? Consolidation could take place through mergers, through some schemes going out of business or through an overarching framework such as mutual recognition. Second, will there be a *convergence of standards*?

At present, both consolidation and convergence are taking place. Following scheme proliferation during the mid 1990s, some consolidation of schemes has occurred. Schemes from outside Europe, including those from Brazil, Canada,

Chile and the US, are, in effect, merging with the PEFC. The medium-term future will see two global schemes: the FSC and the PEFC. Convergence of standards is also happening through the competitor schemes raising their standards and the FSC being forced into some tactical concessions in order to keep industry engaged.[97] Some national competitor schemes have attracted criticism for being weak, and to retain credibility they have raised their standards. A future downward drift of standards is extremely unlikely.

Although several FSC national schemes, under pressure from the competitors, have been forced into some concessions, large-scale compromises from the FSC will not happen. The currency of the FSC is its high standards. However, rigorous standards are worthless if no one abides by them. The FSC will not win the certification wars if its standards are so far removed from the realities forest owners face that they become merely aspirational, with no independent normative pull. However, there is an equal and opposite problem. Certification standards cannot merely be what the forest industry want them to be, so that they merely reflect existing forest management practices and require no behavioural change. Certification schemes have to change forest management, or they are meaningless.

In this respect, forest certification schemes face the same problems as other forms of governance, such as international law. International law, including multilateral environmental agreements, reflects, on the one hand, shared values and, on the other, the power of states. International law cannot merely consist of abstract and laudable values, but neither can it just reflect existing power configurations: in both scenarios there would be no shift in state behaviour. To be effective, international law must mediate between state power and shared values.[98] Similarly, certification standards must mediate, between, first the market power of forest owners and the timber industry, and, second, the values of sustainable forest management.

The futures of the FSC and the PEFC are, to a large extent, interdependent. Like two planetary bodies, these schemes exert a gravitational pull on each other. We can continue the astronomy analogy by asking: which body has the greater 'mass'? The answer again lies in the interplay between power and values. In terms of power, namely support from industry and hectares certified, the PEFC undoubtedly is the strongest. As long as the PEFC exists, the FSC can never achieve the success it would have achieved had it been the only global scheme. However, in terms of values, namely the stringency of standards, the FSC is more authoritative. As long as the FSC exists, it will continue to exert an upwards pull on the standards of other schemes, which, after all, would not even exist were it not for the FSC.

But while convergence of standards is likely to continue, this does not necessarily mean complete convergence. The more popular the competitor schemes, the more difficult it will be for the FSC to raise its standards lest more forest owners elect for the less rigorous competitors. A likely future scenario, therefore, is one with FSC standards stuck at their present ceiling, with the competitor schemes gradually raising their standards. The speed of convergence will depend on the relative support that the two schemes enjoy. If there is a shift in favour of the FSC, the result will be a faster rate of standard-raising from the PEFC. Similarly, a relative shift in favour of the PEFC will mean that any future raising of PEFC standards will happen at a slower pace, possibly coming to a complete halt. As long as the competitor schemes maintain a broad base of forest owner support, complete convergence of standards is unlikely.

It is in the boardroom that the certification wars will be won and lost. Boardroom decisions are based on cost–benefit analysis and corporate risk management. With respect to cost–benefit analysis, it needs to be remembered that forest certification as a form of governance is grounded not in the norms of sustainability or conservation, but in the market and the assumptions of neoclassical economics. The inner logic of forest certification is not sustainable forest management *per se*, but market share and profit and loss accounts. Few businesses enter a forest certification scheme out of ethical concerns. The future of the FSC depends on creating a structure of market incentives and disincentives that will lead forest owners and timber retailers to view their interests in a new light. For example, a retailer will calculate the benefits of certification, such as attracting new customers, increased market share, marketing an environmentally sustainable image and avoiding negative NGO campaigns. And it will calculate the costs, such as the extra expense of buying certified timber and the inconvenience of switching suppliers. Where the anticipated benefits exceed the anticipated costs, the rational utility-maximising business will adopt a certification policy. Where the costs exceed the benefits, the business will see no advantage to certification. Similarly, the rational business will evaluate the costs and benefits of different schemes, and will adopt a weaker scheme if the cost–benefit calculus indicates this to be the most financially advantageous policy.

Of course, not all future costs and benefits can be accurately gauged. With the certification wars having created an uncertain policy environment, there are risks whatever policy a company chooses. Supporting the FSC involves some financial costs, with no guarantee that these costs will be recouped through increased market share. Supporting an FSC competitor carries the risk that a company will invest in a weak scheme that may not retain credibility over the long term. So, whatever scheme a business adopts depends on its internal corporate risk strategy. No business wishes to be on the losing side when the certification wars finally end. Since it is not clear which scheme or schemes will win, for some businesses the most sensible risk management strategy is to adopt standards that are consistent with the most rigorous scheme available, thus ensuring compliance with the 'winner,' whoever that may be. Such a strategy avoids the risk of incurring additional future costs to adjust to the standards of the winner should the wrong scheme be backed. At present, this strategy means adopting the FSC, or at least standards that can be made FSC compliant relatively easily. But other businesses have opted for certification by competitor schemes on the basis that the FSC will not triumph over the long term. Given that forest certification is a form of market-based governance, for the FSC to prevail it must somehow find a way of increasing both the market costs for those businesses that elect for the competitor schemes and the market benefits for those that opt for the FSC.

But sustainability cannot be achieved solely by changing the cost-benefit calculations of market players. The FSC has failed to build a stronger base in the tropics because many logging companies are not prepared to forward concessions for certification that will fail due to unresolved land tenure conflicts with local communities. This recognition has led to a shift in international forest politics towards legality and governance reform. Voluntary certification schemes are failing to reform the forest sector in the tropics since they do not result in changes to forest law and governance, which is often where the main problems are. It is to this subject that we now turn.

7

New Policies to Counter Illegal Logging

Until the mid 1990s, illegal logging was a non-issue in international politics. When the problem was recognized, it was considered a national-level matter rather than a legitimate foreign policy issue. But within a decade of the first mention of illegal logging in an intergovernmentally negotiated textual output, four linked regional processes that, between them, covered the important forested countries of Asia, Africa and Europe had been created to tackle illegal logging through forest governance and law enforcement reforms.

We begin by considering illegal logging as a worldwide problem. We then examine how the problem was recognized as an international issue. The bulk of the chapter explores the regional processes initiated to address illegal logging in the period from 2001 to 2005. These processes are developing some innovative supply- and demand-side measures to tackle illegal logging. It is argued, however, that to greater or lesser degrees they have been constrained by the neoliberal international trade system, which does not permit countries to impose unilateral bans against the import of illegally logged timber. Indeed, trading in illegally logged timber is not even recognized as a crime under international law.

Anatomy of a global problem

At present there is no internationally agreed definition of illegal logging, which may be defined as logging practices that violate national law. Domestic forest law varies from country to country, and within countries the law may change over time. In some Latin American countries people may still stake a legal claim to land through forest clearance, although in others such practices have been made illegal. So what constitutes illegal logging varies according to time and space.

The World Bank has estimated that illegal logging costs the legal forest industry more than US$10 billion per year and deprives governments of about US$5 billion in revenue.[1] Illegal logging includes encroachment on forestlands by the rural poor clearing land for shelter, subsistence and fuelwood. However, far more serious is illegal logging by unscrupulous timber companies. Illegal practices include logging outside concession boundaries, cutting more timber than stipulated in concession contracts, logging in protected areas and felling protected tree species.[2] Furthermore, illegal logging is part of a broader problem of malpractice and crime associated with the timber trade. As Mark Taylor has argued, control of the natural capital that tropical forests represent is a form of political power. In many countries politicians use the allocation of timber concessions as a mechanism to reward supporters.[3] Public officials may engage in corrupt practices when awarding logging concessions, such as stipulating conditions that only favoured businesses can satisfy, restricting public information on the availability of a concession to restrict competition, leaking confidential information and bribe-taking.[4] Forests are spaces that conceal other illegal activities, such as illicit drug cultivation, illegal mining and guerrilla armies. The poor transport infrastructure in many forested regions often makes law enforcement difficult.

Similar patterns of forest destruction caused by illegal logging can be observed on a worldwide scale. The problem is not confined solely to the tropics. In the US, illegal logging on public lands is estimated to cost more than US$1 billion per annum. The problem affects old growth cedar in Washington, cherry in New York and the koa tree in Hawaii.[5] In Canada, the Saskatchewan Environmental Society and Rainforest Action Network (RAN) reported in 2005 that a major corporation, Weyerhaeuser, had been illegally logging in Canadian boreal forests.[6] This case touches on some broader forest policy issues since the area in question is certified as sustainably managed by the Canadian Standards Association. RAN has called on Weyerhaeuser customers to boycott the company's products until all of the company's operations are certified by the FSC.[7]

But while illegal logging affects developed countries, the most heavily afflicted regions are Asia and the Pacific, Africa, the former Soviet Union and Latin America. We now provide some brief snapshots of the illegal logging problems in these regions.

Asia and the Pacific

TRAFFIC International[8] has documented extensive illegal forest activity in the Asia–Pacific region, with hundreds of thousands of hectares of tropical forests being logged illegally.[9] Illegal activities include timber smuggling, undergrading, misclassification of species and illegal processing.[10] Countries with an illegal logging problem in the region include Cambodia, Indonesia, Malaysia and the Solomon Islands.[11]

In Papua New Guinea during the 1980s, the Barnett Commission on the timber industry documented widespread illegal logging, with some companies 'roaming the countryside with the self-assurance of robber barons; bribing politicians and leaders, creating social disharmony, and ignoring laws in order to gain access to, rip out and export the last remnants of ... valuable timber.'[12] The Barnett Commission found evidence of widespread transfer pricing.

Transfer pricing occurs when a company exporting a product, in this case timber, declares a sale price that is lower than the actual sale price. This reduces the tax liability of the transfer in the country where the timber was logged. The Barnett Commission found that transfer pricing was a major activity in most of the companies it studied, many of them Japanese.[13] Illegal logging remains a major problem in Papua New Guinea. In 2004, Greenpeace reported that the Malaysian timber corporation Rimbunan Hijau was a major player in illegal logging in Papua New Guinea.[14] Rimbunan Hijau responded by threatening to sue Greenpeace International, which replied by claiming the company was 'hoping that the threat of litigation will silence its critics... We're confident our report will hold up in court.'[15]

The Philippines has an illegal logging problem that dates back at least to the 1970s. Official figures for Filipino timber exports to Japan have frequently been less than Japanese figures for imports of Filipino timber, the difference being attributable to the trade in illegally sourced timber that has evaded, or been ignored by, Filipino officials but has been recorded by the Japanese authorities.[16] In 1992, the Philippines banned tree felling in virgin forests; but by 2001 seizures of illegally logged timber were increasing.[17] Claims have been made in the national parliament that local government officials are involved with illegal loggers, who are among the major campaign contributors to politicians.[18]

In Suharto's Indonesia, timber barons, particularly the ethnic Chinese, enjoyed a favoured status. By the time Suharto was ousted in 1998, the 3 per cent Chinese of the population controlled 70 per cent of the Indonesian economy.[19] Prominent among the Chinese timber barons was Mohamad 'Bob' Hasan, who at various times was trade and industry minister and head of the Indonesian delegation to the International Tropical Timber Organization (ITTO). Hasan wielded so much influence that it was claimed he wrote legislation favouring his rattan and plywood companies.[20] The fall of Suharto in 1998 led to democratic elections, and in 2001 Hasan was sentenced to six years' imprisonment for defrauding the state and misusing US$75 million of public funds.[21] (It is notable that Hasan was sentenced for financial crime, not for the widespread forest destruction that his companies precipitated.) The change of regime has led to a repositioning of Indonesia's role in international forest politics, as we see below. However, corruption remains deeply entrenched with widespread illegal logging.[22] The WWF has found that illegal logging in Sumatra has depleted the remaining habitat of the Sumatran tiger, which is now on the brink of extinction.[23] In 2005, the Environmental Investigation Agency (EIA), an NGO based in London and Washington, and its Indonesian partner, Telapak, found that crime syndicates routinely ship to China illegally logged timber, including the luxurious hardwood merbau (*Intsia* spp). Much of this timber is shipped onwards to Europe.[24] Malaysian businesses regularly import illegally logged ramin (*Gonystylus* spp) from Indonesia across the Kalimantan–Sarawak border on Borneo and by sea from Sumatra to peninsular Malaysia.[25]

Africa

Illegal logging is widespread across the Congo Basin. In 2002, the World Bank urged the Cameroon government to take action against the offending companies.[26] The Central Africa Republic has formed an armed unit to protect the country's forests from illegal logging and fraud.[27] Other Congo Basin countries with an illegal logging problem include Gabon and Congo.

However, the most seriously affected country is the Democratic Republic of the Congo (DRC, formerly Zaire). In 2000, the UN Security Council established a panel of experts on the illegal exploitation of natural resources in the DRC. During 2002, the panel named 85 corporations from Europe, North America and South Africa that appeared to have violated the Organisation for Economic Co-operation and Development (OECD) Guidelines for Multinational Enterprises, including Anglo-American PLC, a conglomerate with interests in forestry, and Dara Forest, a Thai logging company.[28] The panel found evidence that illegally logged timber had been exported to Uganda with the active involvement of an 'elite network' supported by the Uganda People's Defence Forces and allied rebel militias.[29] In 2003, the Security Council noted continuing illegal exploitation of the DRC's natural resources.[30] The following year the Rainforest Foundation criticized plans to expand the DRC's logging industry.[31]

The Security Council has also taken an interest in Liberia, where rebel leaders have used illegally sourced timber and minerals to finance war. Sanctions against the export of timber from Liberia were imposed in 2003 and renewed in 2004 after the Security Council noted that the transitional government had made 'only limited progress' towards ensuring that revenues from the timber industry are not used to fund conflict.[32] Elsewhere in West Africa illegal logging of teak is a serious problem in the Asubima Forest of Ghana, with illegal loggers felling and transporting trees at night to evade detection.[33]

The former Soviet Union

The forests of Siberia have been coming under increasing threat from legal and illegal traders since the collapse of communism. One of the main routes by which illegally logged timber leaves Russia is by sea to Japan.[34] Illegally logged timber is also transported to the Black Sea coasts of Russia and Georgia for export by sea to Europe. Georgia itself has an illegal logging problem, with chestnut (*Castanea sativa*) particularly vulnerable. A major destination for Russian timber is China, which, following severe flooding caused, in part, by widespread logging, introduced a national logging ban in 1998.[35] Sawmills have been built on the Chinese side of the Russian border to process this timber. Some Russian timber entering China is re-exported to Japan and the US. (China also imports illegally felled timber from Burma.[36]) Finally, some illegally felled timber from north-western Russia crosses the border into Finland, where it is laundered into the extensive Scandinavian timber industry.[37] False customs declarations are often used. For example, an exporter may ship a high value species such as Siberian pine (*Pinus siberica*),[38] but declare a low value species.[39] The illegal trade in forest products to Europe has depressed roundwood prices in Russia and the EU.[40] WWF Latvia has documented evidence of illegal logging in the three former Soviet Baltic states that are now EU members: Estonia, Latvia and Lithuania.[41]

Latin America

Reports from the ITTO reveal significant illegal logging in Peru and Brazil. In Peru, between 70 to 90 per cent of all timber logged is illegally harvested, with the species most affected being cedar and mahogany.[42] The Brazilian Institute of Environment and Renewable Natural Resources (IBAMA), whose staff often operate under dangerous conditions confronting illegal loggers in remote forest regions, makes regular seizures of illegally felled timber, particularly mahogany.[43] In 2003, illegal logging in the Amazon prompted the listing of bigleaf mahogany (*Swietenia macrophylla*) on Appendix II of the Convention on International Trade of Endangered Species of Wild Fauna and Flora (CITES) (see Chapter 9). In 2005, IBAMA launched major operations against illegal loggers in the Amazon, including one that broke a major timber crime syndicate that involved some IBAMA staff. The syndicate had used forged permits to transport timber to over 400 ghost companies.[44] Illegal logging also affects most of the countries of Central America.[45]

Illegal logging is thus a significant problem of global scope, with illegal loggers profiting from the globalization of the legal timber industry. The problem has economic, environmental and social dimensions. Environmentally, it destroys wildlife habitat and species. Governments and public authorities suffer lost tax revenues. Legal businesses lose financially through the flooding of markets with illegally sourced timber. Illegal logging degrades the resource base of communities, causes population displacement and degrades forest public goods. It involves the systematic capture of resources from public authorities, legal businesses and the general public. Given the severity and scale of illegal logging, it was only a matter of time before states started to cooperate to address the problem.

Recognizing the issue

During the early 1990s, the political climate of mutual suspicion between tropical forest countries and developed countries blocked international cooperation on forests. Many developed country governments questioned the commitment to halting deforestation of tropical countries, while the latter often viewed forest conservation proposals as disguised trade barriers. Meanwhile, environmental activists persistently highlighted the global illegal logging problem by conducting extensive research in the afflicted forests, often at considerable personal risk.

The new spirit of international cooperation on forests during the mid 1990s created the political space for illegal logging to be recognized as an international issue. When the second International Tropical Timber Agreement was negotiated between 1992 and 1994, environmental NGOs pressed for the illegal trade to be mentioned.[46] They were partially successful. The International Tropical Timber Agreement of 1994 became the first international legal agreement to allude to illegal logging, although it did so using a euphemism: states agreed to 'keep under continuous review the international timber market ... including information related to *undocumented trade*' [emphasis added].[47] In 1994, the illegal timber trade was still a truth that dare not speak its name.

It would be a further two years before the phrase 'illegal logging' was mentioned in an intergovernmentally negotiated textual output. This was during the second session of the IPF in 1996 when two NGOs, the Global Forest Policy Project and Global Witness, asked the US delegation, headed by the US State Department, to support mention of illegal logging in the IPF's outputs. US negotiator Jan McAlpine raised the issue but faced some resistance, with some developing governments insisting that illegal logging was a national-level issue.[48] Eventually, two mentions of illegal logging were agreed. First, the IPF agreed that reducing illegal logging was one means by which countries could help to mobilize additional financial resources.[49] Second, the IPF noted that market transparency 'would also help focus attention on adverse forest practices, such as illegal logging.'[50] These brief references paved the way for agreement of an IPF proposal for action that invited 'countries to provide an assessment and share relevant information on the nature and extent of illegal trade in forest products, and to consider measures to counter such illegal trade.'[51] In 1997, the IPF was replaced by the IFF, which like the Panel agreed a proposal for action that mentioned illegal logging: countries were called on 'to consider appropriate national level actions and promote international cooperation to reduce the illegal trade in wood and non-wood forest products.'[52]

Despite their weak wording, the IPF and IFF proposals had finally established illegal logging as an international issue. However, this represents virtually the sum total of action taken by UN institutions on the issue during the 1990s. After the International Tropical Timber Agreement of 1994 entered into legal effect, the ITTO made no effort to tackle the 'undocumented trade' throughout the rest of the decade. However, environmental NGOs maintained pressure on the issue, and their efforts bore fruit when the Group of Eight Developed Countries (G8) declared an interest in illegal logging.

The G8 Action Programme on Forests

As an entity, the G8 allows the major developed countries to highlight their priorities without having to reach a consensus with developing countries.[53] The first mention of forests in a G7 or G8 communiqué was at the 1987 G7 summit in Venice, which noted the need to halt tropical deforestation. In 1989, the G7 summit in Paris lent its support to the Tropical Forestry Action Plan.[54] The communiqué issued at the 1990 G7 summit in Houston stated that G7 leaders 'are ready to begin negotiations in the appropriate fora as expeditiously as possible on a global forests convention or agreement.'[55] After Houston, the G7/G8 was silent on forests until 1997, when the G8 summit in Denver called on countries to implement the IPF proposals.

The following year, the G8 summit in Birmingham formally adopted an Action Programme on Forests to run for four years.[56] The US State Department, with support from the UK, had pushed strongly for the action programme, which had five themes: monitoring and assessment; national forest programmes; privatization; protected areas; and illegal logging.[57] The word programme was a misnomer, as the G8 is not an implementation or project management body, and the commitments made at Birmingham merely required G8 governments to report on three types of action carried out in support of the five themes: domestic actions, bilateral assistance programmes and support for intergovernmental processes. The final reports[58] presented at the G8 summit of 2002 in Kananaskis, Canada provided no evidence that G8 countries had taken part in a collective programme of work. To Alexander Horst, the action programme 'had nothing new to offer, either contentwise or financially, especially for developing countries.' It was 'mere rhetoric.'[59] Why then did the G8 adopt the action programme?

First, it provided a stocktaking of G8 government policies to address illegal logging following pressure from the G8's domestic timber industries, which had expressed concerns at the illegal trade. Second, the action programme can be seen as a public relations exercise to demonstrate resolve on forest issues following the huge multidimensional crisis in Indonesia throughout 1997 and 1998 of severe forest fires, economic collapse, capital flight and political upheaval. Third, the launch of the action programme indicated that G8 governments did not consider the IFF or the World Commission on Forests and Sustainable Development to be effective mechanisms for tackling illegal logging. Finally, the G8 used the action programme to promote key neoliberal themes to which most G8 members, particularly the US and the UK, are committed. The action programme stressed the involvement of the private sector in forests: 'It is the responsibility of each government to involve all private sector stakeholders in achieving sustainable forest management.' Emphasis was placed on 'voluntary codes of conduct' and 'private voluntary market-based mechanisms.'[60] As we have seen, outside the G8 most forests are under state management, while the forest corporations that are most likely to benefit from increased private sector involvement in tropical forests are from G8 countries (see Chapter 1).

The action programme ended in 2002. It had served as a notice of intent that some G8 countries were serious about addressing illegal logging. Foremost among them was the US. At the 2000 G8 summit in Okinawa, the US announced that it was planning a more ambitious initiative to address illegal logging in Asia.[61] This was the Forest Law Enforcement and Governance (FLEG) Ministerial Conference held in Bali in 2001.

The launch of the Forest Law Enforcement and Governance process

The cheapness of illegally logged timber undermines legitimate businesses, which, as a result of pressure from consumer groups and certifying companies, have to meet stricter sustainability standards than was the case a decade ago. The American Forest and Paper Association (AFPA) has long been concerned about how the illegal trade depresses prices and reduces the demand for exports of US roundwood, sawnwood and wood panels.[62] The AFPA and environmental NGOs thus share an opposition to illegal logging, although for different reasons. Faced with increasing concern from the AFPA and NGOs, the US State Department became committed to pursuing the issue beyond the G8 action programme.

State Department official Jan McAlpine, who, it will be recalled, first raised the issue of illegal logging at the IPF, did not consider it worthwhile pursuing illegal logging in UN negotiating institutions, which are time consuming, tend to have a culture of defensiveness and are poorly equipped for dealing with problems on the ground. McAlpine thought it best to focus on regions. There would need to be several regional initiatives, each focusing on the dimensions of the problem peculiar to that region.[63] She approached an international organization with huge convening power and influence – the World Bank – which agreed to work with the US on the issue.[64] The State Department was thus able to harness the Bank's resources behind what had now become a US foreign policy issue. However, it was not simply a case of the State Department using the World Bank as a proxy (although the location of the Bank in Washington, DC, clearly serves the interests of the US government, the World Bank's major shareholder, more than any other). The Bank had already developed considerable expertise on illegal logging, sponsoring workshops on the dimensions of the problem in the Mekong Basin (Phnom Penh, June 1999) and in East Asia (Jakarta, August 2000).

McAlpine worked with John Hudson of the UK's Department for International Development in proposing to the World Bank a ministerial conference on illegal logging that the Bank would co-host. Whereas the Bank's previous involvements in illegal logging had been at the technical level, McAlpine and Hudson insisted that an intergovernmental meeting was necessary at which governments would commit politically. They settled on East Asia as the region in which to launch the first regional process. The World Bank's prior involvement in this region made it a natural choice. The Indonesian government agreed to host a ministerial conference on illegal logging in Bali.[65]

This would have been unthinkable during Suharto's rule. With influential timber traders having colonized the inner recesses of the Indonesian state, there was no possibility of the Suharto regime moving against illegal activities in the forest sector, many of which involved Hasan's companies. However, the government of President Megawati Sukarnoputri initiated an abrupt change of Indonesian forest policy that included seeking international support to address illegal logging. In 2000, Indonesia reported to the ITTO that 'illegal logging was a serious threat to Indonesian forests.'[66] Indonesia invited an ITTO technical mission to visit the country to report on the country's forest sector. The mission reported in September 2001. It identified several factors that were prominent in the spread of 'rampant illegal logging,' in particular a breakdown in law enforcement, the unregulated expansion of unsustainable wood processing industries and the neglect of the rights of local communities.[67] The mission concluded that illegal logging 'has to be tackled on a war footing.'[68] The crimes documented by the ITTO mission are similar to those reported by the Barnett Commission in Papua New Guinea and include the underdeclaration of harvesting volume, transfer pricing and tax avoidance. Fire was used to clear forests illegally to free land for other uses, particularly palm oil plantations.[69]

Following a preparatory meeting in Jakarta in April 2001, the FLEG Ministerial Conference, co-hosted by the Indonesian government and the World Bank, opened in Bali on 11 September 2001. The conference was a success, despite being overshadowed by the terrorist attacks in New York and Washington. Twenty countries were represented, 11 of them at ministerial level.[70] Indonesian NGOs represented included Telapak and WALHI, while international NGOs included Greenpeace and the Environmental Investigation Agency.[71] Global Witness, which two years earlier had been appointed as an independent forest monitor in Cambodia, also attended (see Box 7.1). There were two days of technical discussions, followed by a ministerial segment.

The ministerial declaration has historical significance as the first intergovernmental statement to elaborate political measures to address illegal logging (see Box 7.2). The declaration recognizes that 'illegal logging and associated illegal trade directly threaten ecosystems and biodiversity in forests,' resulting in 'serious economic and social damage upon our nations, particularly on local communities, the poor and the disadvantaged.'[74] Attached to the declaration was an 'indicative list of actions' to which states are not formally committed, but which they can consider when promoting forest governance reforms. The declaration came just five years after the first intergovernmental reference to illegal logging at the IPF.

Ministers agreed to create a regional task force to advance the objectives of the declaration. Other stakeholders, including industry and civil society, were invited to form an advisory group to the task force. The Washington office of the Environmental Investigation Agency now organizes and chairs the advisory group. The reaction from civil society groups was mixed. NGOs that attended the ministerial conference tended to be satisfied with the outcome. Nigel Sizer of The Nature Conservancy said that the declaration surpassed what NGOs had expected, while Dave Currey of the Environmental Investigation Agency was 'encouraged' by the declaration.[75] However, indigenous peoples' groups argued that a major tension in the FLEG process is the extent to which it ignores forest peoples. In the absence of legal recognition of the customary rights of indigenous peoples, as well as the rights of other forest peoples, an exclusive focus on law *enforcement* would uphold an unjust system of law that would perpetuate the social exclusion of those denied legal rights.[76] People

Box 7.1 Independent forest monitoring: The case of Global Witness and Cambodia

Global Witness first investigated timber exploitation in Cambodia in 1995, focusing on the western areas of the country then dominated by the Khmer Rouge. As it was too dangerous to enter Khmer Rouge-controlled areas, most of Global Witness's investigative work took place on the Thai side of the Thai–Cambodian border. Global Witness found that the Khmer Rouge was active in the transport of illegally felled timber to Thailand, with much of the transport supplied by Thai timber companies. Global Witness later turned its attention to Cambodian timber companies, many of which have engaged in illegal logging outside concession boundaries. During the late 1990s, the donor community, in particular the World Bank, pressed the Cambodian government to introduce independent forest monitoring. In December 1999, Global Witness was appointed as an independent forest monitor in Cambodia, with financial support from the UK and Australian governments. This represented the first case of a government working with a foreign NGO to monitor the implementation of national forest policy.

Global Witness gathered data and compared it with data gathered by two Cambodian government departments: the Department of Inspection in the Cambodian Ministry of Environment; and the Forest Crimes Monitoring Unit in the Department of Forestry and Wildlife. Global Witness investigated the reasons for any discrepancies in the data. In 2001, the Cambodian government reacted negatively to Global Witness after it published reports documenting that regulations had been ignored and indicating that some public officials were involved in illegal logging.[72] In December 2003, the government accused Global Witness of stirring civil unrest in the capital Phnom Penh over illegal logging. At this point the relationship between Global Witness and the Cambodian government broke down. The contract between the two parties ended in July 2004. Although the Cambodian authorities have not banned Global Witness as an organization, Global Witness staff were prevented from entering Cambodia in 2005.[73]

without legal protection would, to law enforcement agencies, be targets for enforcement measures. According to this view, where the rights of forest peoples have no legal status, reform of the law should, logically, precede law enforcement. International illegal logging policy has yet to resolve this major underlying tension.

A few months after the Bali summit, the second session of the UNFF agreed a ministerial declaration to be sent to the World Summit on Sustainable Development. This called 'for immediate action on domestic forest law enforcement and illegal international trade in forest products.'[77] This language

Box 7.2 Forest Law Enforcement and Governance, Ministerial Declaration, Bali, Indonesia, 13 September 2001 – Main Commitments (summarized)

- Intensify national efforts and strengthen bilateral, regional and multilateral collaboration to address forest crime, particularly illegal logging and the associated illegal trade.
- Undertake actions, including among law enforcement authorities within and among countries, to prevent the movement of illegal timber.
- Explore how the export and import of illegally harvested timber can be eliminated, including the possibility of a prior notification system for commercially traded timber.
- Improve forest-related governance within countries in order to enforce forest law, better enforce property rights and promote the independence of the judiciary.
- Involve stakeholders and local communities in forest decision-making.
- Improve economic opportunities for those relying on forest resources to reduce the incentives for illegal logging.
- Institute appropriate forest policy reforms relating to the granting and monitoring of concessions, subsidies and excess processing capacity.
- Give priority to the most vulnerable transboundary areas.
- Monitoring and assessment of forest resources.
- Strengthen capacity to prevent, detect and suppress forest crime.

Source: Forest Law Enforcement and Governance, East Asia Ministerial Conference, Bali, Indonesia, 11–13 September 2001, Ministerial Declaration, paras 11–23

was rubber stamped at the 2002 World Summit on Sustainable Development.[78] 2002 also saw the launch of a new arrangement, the Asia Forest Partnership (AFP).[79] The AFP aims to provide a framework for cooperation in five areas, namely 'good governance and forest law enforcement, developing capacity for effective forest management, control of illegal logging, control of forest fires and rehabilitation of degraded lands.'[80] However, given the recent launch of the Asian FLEG process, the need for another regional partnership has yet to be demonstrated. The AFP can be seen as an attempt by Japan to reclaim control over Asian forest dialogue at the expense of FLEG, which may be perceived as Anglo–American driven. It remains to be seen whether a working relationship between FLEG and the AFP can be established.

Bilateral agreements between Indonesia and other countries

The Indonesian government continued to demonstrate its commitment to tackling illegal logging after the Bali summit. Following the report of the ITTO technical mission, Indonesia announced a temporary ban on exports of round logs.[81] Indonesia then negotiated bilateral agreements to combat illegal logging with four countries: the UK (April 2002), Norway (August 2002), China (December 2002) and Japan (June 2003).[82] These agreements are not part of the regional FLEG process in Asia, although they do form part of the new system of governance to address illegal logging that FLEG has catalysed. The most comprehensive agreements are those with the UK and Japan: a comparative analysis of the two agreements reveals that, in many respects, the former served as a model for the latter, with identical wording for some commitments. Indonesia's agreement with China also draws from the agreement with the UK. The briefest agreement is with Norway.[83] Table 7.1 details the main commitments agreed in the four bilateral agreements.

Bilateral agreements have obvious weaknesses. They have a limited geographical reach and can be easily avoided by trading through a third country. Furthermore, and given the differences between the four agreements negotiated to date, to rely solely on bilateralism would result in an uneven system of governance in which different commitments would apply across different bilateral relationships. The UK's strategy in concluding an agreement with Indonesia was not to promote a global network of similar bilateral agreements, but to demonstrate what could later be developed at the EU level.[84] In this respect, the strategy was successful: as we see below, the EU has since adopted the idea of bilateral voluntary partnership agreements between the EU and timber producing states.

The Africa Forest Law Enforcement and Governance process

The Asian FLEG conference of 2001 provided the momentum and the model for a similar process in Africa. Like its Asian predecessor, the Africa FLEG process involved a preparatory meeting (Brazzaville, Congo, in April 2002). This was followed by a ministerial conference attended by timber exporting countries and key donors in Yaoundé, Cameroon (October 2003). During the pre-ministerial negotiations some delegates stressed the need to remain within WTO rules, a theme that had also informed the Asian FLEG negotiations. The InterAfrican Forest Industries Association called for an international response to tackle the forests conflicts in Liberia. Global Witness announced that it had signed an independent forest monitoring agreement with the government of Cameroon.[85] As at the Bali meeting, the Yaoundé meeting produced a ministerial declaration with an appended list of indicative actions. The

Table 7.1 *Main commitments between parties to the bilateral agreements on illegal logging agreed in 2002 and 2003 between the government of Indonesia and the governments of the UK, Norway, China and Japan*

	Paragraph number in the agreement between the government of Indonesia and the following countries			
Main commitments	***UK***	***Norway***	***China***	***Japan***
Identify legislative reform required to prevent illegal logging	1.a			
Develop, test and implement systems for the verification of legal compliance based on independently verified chain-of-custody tracking	1.b			2(1)
Provide technical and financial capacity-building to Indonesia	1.c			
Provide possible support for the development and enforcement in Indonesia of policy reforms, laws and regulations, and for the capacity-building of judicial, legislative and administrative institutions		3		
Encourage the involvement of civil society in monitoring implementation	1.d		2.2	2(2)
Involve civil society in increasing public awareness on rainforest timber trade		6		
Improve economic opportunities for local communities				1(3)
Collect and exchange data and information	1.e	5	2.3, 2.4	2(3)
Develop effective collaboration between enforcement agencies	1.f		2.4	2(4)
Encourage action by industry	1.g			
Establish an action plan	2			4
Review implementation	3			5
Settle disputes amicably by negotiation	4			7
Identify illegally harvested products			2.1	
Enhance economic cooperation in the forest sector and facilitate normal forest trade			2.5	
Encourage cooperation on the criteria of sustainable forest development			2.6	
Initiate human resources development				2(5)

Source: adapted from the full texts of the bilateral agreements on illegal logging between Indonesia and the UK, Norway, China and Japan, as published in *International Forestry Review*, Vol. 5, No. 3, pp.223–229

Box 7.3 Africa Forest Law Enforcement and Governance, Ministerial Declaration, Yaoundé, Cameroon, 16 October 2003 – Main Commitments (summarized)

- Strengthen institutional reforms in the forest sector.
- Facilitate the mobilization of financial resources.
- Review the effect of structural adjustment programmes on forest law enforcement.
- Promote better economic opportunities for forest dependent communities.
- Invite cooperation between law enforcement agencies within countries and internationally.
- Strengthen the capacity of all relevant institutions and groups.
- Involve stakeholders, including local communities and rural populations, in forestry sector decision-making.
- Re-establish good governance in post-conflict situations.
- Respect property and usufruct rights, including traditional forest-related knowledge.
- Strengthen laws and regulations for hunting and the bushmeat trade.
- Integrate law enforcement within national forest programmes.
- Invite representatives from the private sector and NGOs to form advisory groups for sub-regional task forces.

Source: Africa Forest Law Enforcement and Governance, Ministerial Conference, Yaoundé, Cameroon, 13–16 October 2003, Ministerial Declaration, paras 1–30

ministerial declaration stresses the need to strengthen political commitment and capacity, mobilize financial resources and promote cooperation between law enforcement agencies (see Box 7.3).

The Africa FLEG process aims to work through and strengthen existing mechanisms, of which three are likely to prove central. First, the Congo Basin Forest Partnership, launched with US sponsorship at the World Summit on Sustainable Development, aims to encourage donors to engage in the forests of the region. It has no independent implementation role.[86] The six founding members of the partnership are Cameroon, the Central African Republic, the Democratic Republic of the Congo, Equatorial Guinea, Gabon and the Republic of the Congo.

Second, in February 2005 at a forest summit in Brazzaville, ten Central African countries recognized the Conference of Ministers in Charge of Forests in Central Africa (COMIFAC) as the sole decision-making body on forests for the region. The ten countries were the six founding members of the Congo Basin Forest Partnership, plus Burundi, Chad, Rwanda, and São Tomé and Principe.

This meeting also agreed a trilateral accord permitting free movement of park staff between Cameroon, the Central African Republic and the Republic of Congo in the Sangha Tri-National Conservation Area, thus allowing staff to work across national borders to counter illegal logging and poaching.[87]

Third, in September 2005, the first Intergovernmental Meeting on Great Apes agreed the Kinshasha Declaration on Great Apes. Signed by 16 great ape range states and 6 donor countries, the declaration commits states to protecting the habitats of the great apes. It applies principally to the gorilla and chimpanzee range states of the Congo Basin, as well as to orang-utan range states in Southeast Asia, notably Indonesia. The declaration is an interesting example of how an initiative by UN agencies eventually resulted in an intergovernmental initiative. In 2002, the United Nations Environment Programme (UNEP) and the United Nations Educational, Scientific and Cultural Organization (UNESCO) launched the Great Apes Survival Project (GRASP), a multisectoral partnership involving public and private actors. The Kinshasha Declaration endorsed GRASP.[88] The meeting made several references to the role of illegal logging in destroying the habitat of great apes.[89]

The EU's Forest Law Enforcement, Governance and Trade (FLEGT) Action Plan

The Asian and African FLEG processes are primarily supply-side approaches to reduce illegal logging at source in tropical timber producing countries. To complement and support these processes the EU, as a major timber importer, has developed the FLEGT Action Plan to combat illegal logging.[90] This promotes both supply-side measures, by providing assistance to developing and former communist countries, and demand-side measures to curtail the trade of illegally logged timber to the EU. The action plan was approved in Council Conclusions during the same month that the Africa FLEG process was launched (October 2003).[91]

The development of the action plan provides an illustration of how actors from outside the European Commission can make a decisive impact on EU policy. During the preparation of the EU action plan, the UK government engaged the Royal Institute of International Affairs (Chatham House) to prepare possible measures that the action plan could endorse. Chatham House consulted with a broad range of stakeholders. In collaboration with the NGO FERN, Chatham House produced a study – *Controlling Imports of Illegal Timber: Options for Europe* – that contained several recommendations that have become key elements in the FLEGT action plan. They include voluntary partnership agreements between producer countries and the EU on timber licensing;[92] the adoption by member states of procurement policies stipulating the purchase of timber from legal sources;[93] promoting private sector initiatives, including codes of conduct;[94] and the exercise of due diligence by export credit agencies and financial institutions when funding logging projects.[95] We now examine these four elements.

Timber licensing and voluntary partnership agreements

The EU may conclude voluntary partnership agreements (VPAs) with producer countries that commit to exporting to the EU only legally logged timber. Although there is some illegal logging within the EU, the licensing scheme focuses on producer countries outside the EU where the problem is more serious.[96] With no internationally agreed definition of illegal logging, each VPA will contain an agreed definition. Definitions will vary according to national and local conditions in the producer country. Timber shipped between the EU and producer countries with which the EU has concluded a VPA must be accompanied by a licence. Licences should be 'forgery-resistant, tamper-proof and verifiable.'[97] African countries are likely to be most affected by the licensing scheme, as the supply of legal timber in Africa is low in relation to demand from the EU. Asian and Latin American countries will be affected to lesser degrees.[98]

It is likely that most VPAs will include both issues of European Community competence and of member state competence (see Box 4.4 in Chapter 4). They will thus have a dual legal base.[99] For issues of Community competence, notably trade, the European Commission will lead on negotiations; the legal base of any agreed provisions will rest on European Council decisions, which will be binding on all EU states. Most VPAs are also likely to include issues of member state competence, such as development assistance. Member states will not be bound by these provisions unless they ratify them individually. The legal adoption of VPAs within the EU will thus be complex and time consuming. (As this book went to press, deliberations on the legal base of VPAs within the EU were continuing within the European Commission and between member states.)

The licensing scheme has been designed to be compatible with the WTO. The FLEGT approach of voluntary licensing was adopted as a stronger approach, such as a mandatory licensing scheme applying to all countries, would probably have encountered a challenge at the WTO. Current legal opinion is that the FLEGT licensing scheme is compatible with the GATT and that it does not constitute a prohibition of, or a barrier to, trade under the WTO's Technical Barriers to Trade (TBT) Agreement.[100] However, this has not yet been tested through a WTO challenge. By respecting WTO law, the licensing scheme has an inbuilt weakness. It will be illegal to try to import to the EU timber without a licence from a VPA country; but it will not be illegal to import illegally logged timber from non-VPA countries. Since no country is obliged to adopt the scheme, illegal loggers who successfully evade the authorities in the producer country can then circumvent the licensing scheme by exporting timber to a country with no VPA for onward shipment to the EU. To close this loophole, several NGOs urged the European Commission to present legislation designating it a crime to import into the EU any illegally sourced timber or timber products.[101] However, this would almost certainly violate WTO law unless the EU agreed a prohibition multilaterally with other states.[102] This would have to be agreed either at the WTO or in a multilaterally environmental agreement with trade restriction measures that do not fall foul of the WTO. There are precedents for this; previous multilateral environmental agreements prohibit or restrict the international trade of ozone-depleting chemicals, hazardous wastes and endangered species.[103] However, the negotiation of a multilateral prohibition would require political will, and at present key states are opposed, notably the US.

So far no VPAs have been concluded with producer countries. Different EU states are conversing with individual producer states on possible VPAs. Germany has entered into a dialogue with Cameroon, France with Gabon, the UK with Ghana, and the Netherlands with Malaysia.[104] Over time, VPA countries can expect to gain increased access to EU markets and to capture additional revenues. While traders of illegally logged timber will seek to divert timber through non-VPA countries, this option will gradually be closed off as more VPAs are concluded. Countries that do not conclude VPAs can continue to trade with the EU as before. However, as more VPAs are concluded, there is likely to be reduced demand from the EU for timber from non-VPA countries.[105] The licensing scheme thus relies, in part, on demand for the scheme from producer countries that wish to protect and increase their market share.

Public procurement

At the 2000 G8 summit in Okinawa, Greenpeace lobbied for G8 governments to adopt green procurement policies, including buying FSC-certified timber products.[106] In 2001, the EU and Japan agreed to examine 'ways to combat illegal logging, including export and procurement practices.'[107] In 2002, the WWF, noting the significant imports to G8 countries and China of illegally logged timber, recommended that the governments of these countries commit to purchasing timber only from legal and well-managed sources.[108] Because of the enormous purchasing power of these countries, such a policy would help to shift timber production patterns to a more sustainable and legal basis.

In the EU, procurement, unlike trade, is a member state competence. The FLEGT action plan recommends that states make use of their competence with respect to procurement,[109] although the final decision is one for individual governments. If all EU governments were to agree to purchase only legally sourced timber, a strong market signal would be sent to producing countries. At present, only two EU governments – Denmark and the UK – have introduced timber procurement policies that require evidence of legal sourcing. France, Germany and the Netherlands are developing such policies.[110]

Private sector initiatives

The action plan provides for the European Commission to 'promote private-sector initiatives, including support for ... the adoption of high standards in codes of conduct, transparency in private sector activities, and independent monitoring.'[111] In the two EU countries that have adopted government procurement policies – Denmark and the UK – the private sector has responded by developing codes of conduct.[112] The UK Timber Trade Federation (TTF) has developed a Responsible Purchasing Policy that is intended to 'act as a "fast track" for TTF members wanting access to central governments.'[113] This illustrates that a shift in government procurement policy can help prompt new private sector initiatives as business reacts to new market conditions.

At the Pan-European level, the Confederation of European Paper Industries (CEPI) announced in 2005 a code of conduct that committed member organizations to purchasing only legally logged timber, to respecting national laws and to ensuring that the legality of wood purchased was documented.[114] So far, CEPI has not announced any procedures for implementing and verifying the code of conduct.

Due diligence

Export credit agencies provide government-guaranteed loans to corporations engaged in financially risky investments. They are major public financial actors and the value of their financial transactions collectively exceeds that of the World Bank.[115] Due diligence is the exercise of caution to ensure that legitimate finance is not used for illegal activities.[116] During the preparation of the action plan, Chatham House and FERN recommended that EU governments adopt 'binding environmental and social rules' to ensure that export credit agencies which finance timber logging fund only legal operations.[117] Instead, the action plan adopted a softer emphasis; the European Commission should 'foster the development of specific procedures for environmental and social due diligence for Export Credit Agencies.'[118]

One option for promoting the exercise of due diligence is the Equator Principles adopted by the International Finance Corporation and major investment banks in 2003 (see Chapter 8). The export credit agency of Denmark has adopted the Equator Principles. A major difficulty in promoting due diligence is tracking how finance is used in the forest industry.[119] The FLEGT action plan has not addressed this problem.

The action plan also includes supply-side measures by providing for improved development assistance to promote governance reforms in timber producing countries, including independent monitoring, auditing and strengthening civil society.[120] Much of this assistance will be directed at countries with which the EU has concluded a VPA. There is likely to be considerable political bargaining between the EU and producer countries over development assistance packages. The EU will seek to link development assistance with VPAs in order to make the licensing scheme feasible, while individual producer countries will seek to extract the best terms from the EU on financial and technical assistance in exchange for agreeing to a VPA.

The G8 Summit at Gleneagles, 2005

In 2005, the UK government used its presidency of the G8 to press for progress on international climate policy and poverty alleviation in Africa. A secondary objective was strengthened demand-side measures on illegal logging.

An internal US State Department memorandum outlined the US strategy for the G8 negotiations. The memorandum stated that 'Demand-side actions involving new import or procurement regulations/restrictions are unacceptable.' The US would 'work with Canada to hold back procurement and other unacceptable demand-side actions, and with Russia and Japan, to dissuade them from supporting UK.'[121] The US, it was noted, should seek commitments consistent with the President's Initiative Against Illegal Logging. Announced in 2003, this initiative contains no demand-side measures and outlines only supply-side responses, such as country capacity building, community-based actions and technology transfer for monitoring systems.[122]

It was clear that the US wanted to go beyond merely abstaining from demand-side measures itself, and wished to use its agency to stop other governments from taking such measures. The memorandum attracted press coverage in the UK after it was leaked to the BBC's *Newsnight* programme.[123] A spokesperson for the Competitive Enterprise Institute in Washington commented to *Newsnight* that 'green trade' is not necessarily a step in the right direction: 'We think that trade should be as free as possible and these other issues as to involving [sic] environment and so on should be secondary to free trade, secondary to the major considerations of the WTO.'[124]

Shortly after the leak the first G8 Environment and Development Ministerial Conference took place in England. With the US opposed to joint demand-side measures, ministers agreed only to 'take steps to halt the import and marketing of illegally logged timber, for example by giving appropriate powers to our border control authorities through voluntary bilateral trade agreements or other arrangements, consistent with WTO rules.'[125] The word 'appropriate' deprived this phrase of all substantive content. In another paragraph that lacked hard commitment, ministers agreed to 'encourage, adopt or extend public timber procurement policies that favour legal timber.'[126]

The G8 summit of heads of state and government at Gleneagles in July 2005 took place following the Live8 concerts intended to pressure G8 leaders to agree measures to reduce poverty in Africa. Like the Asia FLEG summit, the first day of the summit was overshadowed by a terrorist attack, with bombings on London's transport infrastructure. The G8 heads of state and government agreed only that countries would act alone on illegal logging: 'We endorse the outcome of the G8 Environment and Development Ministerial conference on illegal logging. To help further our objectives in this area we will take forward the conclusions endorsed at that meeting, with each country acting where it can to contribute most effectively.'[127]

Of the non-EU members of the G8, Japan is the only government to have developed a policy of public procurement of timber from legal sources.[128]

The Europe and North Asia Forest Law Enforcement and Governance process

Four months after the Gleaneagles summit the third FLEG ministerial meeting was held in St Petersburg.[129] This aimed to initiate a cooperative process between European and North Asian countries that would, in part, address the illegal trade between Russia and China. China occupies an important intermediary position in the international timber trade, with an estimated two-thirds of China's timber imports being re-exported.[130] The participation of China was thus important. Environmental NGOs, including Forest Trends and the Beijing-based Global Environmental Institute, helped to persuade China to participate in the St Petersburg ministerial, which the Chinese delegation approached in a constructive manner.[131] Eighteen EU member states attended, a significant increase on EU representation at the Asian and African FLEG ministerials and a reflection of the impact of the FLEGT action plan. As with the Asian and African meetings, the St Petersburg conference agreed a ministerial declaration supported by an indicative list of actions.

The business view at St Petersburg was somewhat divided. The Finnish corporation Stora Enso sought to protect itself from cost increases by arguing that law enforcement costs should fall on government, rather than legal operators, a point reiterated by the International Council of Forest and Paper Associations.[132] Stora Enso also challenged commitments made in the FLEGT action plan: timber licensing and procurement policies should be avoided as the 'main measures.'[133] However, the Ilim Pulp Enterprise of Russia saw certification, labelling and licensing as 'key instruments to combat illegal logging.' The Swedish firm IKEA also stressed the importance of certification.[134] Civil society groups pressed unsuccessfully for a timebound follow-up process.[135] Despite some differences between them, there was open and constructive cooperation between the private sector and civil society groups, resulting in a set of joint civil society/private sector positions that was fed into the ministerial-level negotiations. The IUCN facilitated civil society participation, while The Forests Dialogue facilitated the joint civil society/private sector dialogues.[136]

The importance of transboundary cooperation, particularly along the porous Russian–Chinese border, was emphasized, with Friends of the Siberian Forests stating that illegal logging in Russia would persist unless China took action.[137] The Chinese authorities are taking a stronger line against illegal loggers operating within China, although the willingness of the authorities to deal with the illegal trade that transits through China has yet to be demonstrated. However, China did agree to mention in the St Petersburg declaration of the need to 'give priority to and strengthen transboundary cooperation between countries with border areas which require coordinated actions.'[138]

Forty-three countries endorsed the declaration (see Box 7.4). This brought the number of countries involved in a FLEG process to 90 (see Table 7.2).

Box 7.4 Europe and North Asia Forest Law Enforcement and Governance, Ministerial Declaration, St Petersburg, Russia, 16 October 2003 – Main Commitments (summarized)

- Strengthen interagency cooperation among law enforcement and judicial authorities.
- Formulate actions under clearly defined targets, including monitoring of implementation.
- Recognize the rights of forest dependent communities, taking into consideration traditional laws and practices and traditional knowledge.
- Engage stakeholders in the formulation of forest laws and policies.
- Develop and implement anti-corruption tools, including codes of conduct.
- Promote the establishment of third-party audited traceability systems.
- Strengthen international cooperation using existing structures.
- Strengthen transboundary cooperation between countries with border areas which require coordinated actions.
- Facilitate technology transfer and information-sharing.

Source: Europe and North Asia Forest Law Enforcement and Governance, Ministerial Conference, St Petersburg, Russia, 16 October 2005, St Petersburg Declaration, paras 1–29

The issue of illegal logging has spilled over into other international institutions. The expanded programme of work on forest biological diversity, agreed by parties to the Convention on Biological Diversity in 2002, promotes forest law enforcement, including legislation to address illegal activities and capacity-building for effective law enforcement.[139] One of the objectives of the International Tropical Timber Agreement of 2006 is 'Strengthening the capacity of members to improve forest law enforcement and governance, and address illegal logging and related trade in tropical timber.'[140] When it enters into legal effect this agreement will replace the International Tropical Timber Agreement of 1994, which refers only to the 'undocumented trade.' However, just one month after the 2006 agreement was concluded, the fragility of the international consensus on this issue was exposed when India, China and the Amazonian Pact countries, led by Brazil, opposed the phrase 'illegal logging' in the draft ECOSOC resolution negotiated by the UNFF's sixth session (see Chapter 6).[141] The compromise solution was 'illegal practices according to national legislation and illegal international trade in forest products.'[142]

The refusal of the Amazonian Pact countries to agree to 'illegal logging' at the UNFF is significant given the absence of a FLEG process in the Americas. One obstacle to a South American FLEG is the suspicion of most countries on the continent to multilateral cooperation on forests; the Amazonian Pact

Table 7.2 *Countries involved in the four regional Forest Law Enforcement and Governance processes, as of November 2005*

Country	East Asia and Pacific FLEG (launched in Bali, September 2001)*	Africa FLEG (launched in Yaoundé, October 2003)**	EU FLEGT action plan (adopted October 2003)***	Europe and North Asia FLEG (launched in St Petersburg, November 2005)****
Albania				X
Angola		X		
Armenia				X
Austria			X	X
Azerbaijan				X
Belarus				X
Belgium		X	X	
Benin		X		
Bosnia-Herzegovina				X
Botswana		X		
Burkina Faso		X		
Bulgaria			X	X
Burundi		X		
Cambodia	X			
Cameroon		X		
Canada	X	X		X
Central African Rep		X		
China	X			X
Congo	X	X		
Croatia			X	X
Cyprus			X	
Côte d'Ivoire		X		
Czech Republic			X	
D R of Congo		X		
Denmark			X	X
Estonia			X	X
Ethiopia		X		
Finland			X	X
France		X	X	
Gabon		X		
Gambia		X		
Georgia				X
Germany		X	X	X
Ghana	X	X		
Greece			X	X
Guinea		X		
Guinea-Bissau		X		
Hungary			X	
Indonesia	X			

Table 7.2 *Countries involved in the four regional Forest Law Enforcement and Governance processes, as of November 2005 (Continued)*

Country	East Asia and Pacific FLEG (launched in Bali, September 2001)*	Africa FLEG (launched in Yaoundé, October 2003)**	EU FLEGT action plan (adopted October 2003)***	Europe and North Asia FLEG (launched in St Petersburg, November 2005)****
Ireland			X	
Italy		X	X	X
Japan	X			X
Kazakhstan				X
Kenya		X		
Kyrgyzstan				X
Laos	X			
Latvia			X	X
Lesotho		X		
Lithuania			X	X
Luxembourg			X	
FYR Macedonia				X
Madagascar		X		
Malta			X	
Mauritius		X		
Moldova				X
Mongolia				X
Mozambique		X		
Namibia		X		
Netherlands			X	X
New Zealand	X			
Niger		X		
Nigeria		X		
Norway				X
Papua New Guinea	X			
Philippines	X			
Poland			X	X
Portugal			X	X
Romania			X	X
Russia				X
Senegal		X		
Serbia & Montenegro				X
Seychelles		X		
Slovakia			X	X
Slovenia			X	X
South Africa		X		
Spain			X	X
Sweden			X	X

Table 7.2 *Countries involved in the four regional Forest Law Enforcement and Governance processes, as of November 2005 (Continued)*

Country	East Asia and Pacific FLEG (launched in Bali, September 2001)*	Africa FLEG (launched in Yaoundé, October 2003)**	EU FLEGT action plan (adopted October 2003)***	Europe and North Asia FLEG (launched in St Petersburg, November 2005)****
Switzerland		X		X
Tajikistan				X
Thailand	X			
Togo		X		
Turkey			X	X
Uganda		X		
Ukraine				X
United Kingdom	X	X	X	X
United States	X	X		X
Uzbekistan				X
Vietnam	X			
Zambia		X		
Zimbabwe		X		
Total At least one FLEG/FLEGT process	**Total East Asia and Pacific FLEG**	**Total Africa FLEG**	**Total FLEGT Action Plan**	**Total Europe and North Asia FLEG**
90	**15**	**39**	**29**	**43**

Notes:
* With the exception of Canada, New Zealand and Papua New Guinea, all countries in this column sent delegates to the inaugural FLEG ministerial meeting in Bali in September 2001. Canada, New Zealand and Papua New Guinea did not attend the 2001 ministerial, but have since sent representation to the East Asia–FLEG regional task force.
** Countries in this column sent delegates to the Africa FLEG ministerial meeting in Yaoundé, October 2003.
*** The EU FLEGT action plan is legally binding on its 25 member states. Also associated with the action plan, although not formally bound by it at the time that it was adopted, were Bulgaria and Romania as accession states due to join the EU in January 2007, and Turkey and Croatia as candidate states.
**** Countries in this column adopted the St Petersburg declaration of November 2005.

Sources: Lists of countries that endorsed the ministerial declarations of the first FLEG ministerial conference (September 2001), the Africa FLEG ministerial conference (October 2003) and the Europe and North Asia FLEG ministerial conference (November 2005). The names of countries participating in the Asia regional task force but which did not attend the September 2001 ministerial conference have been obtained from 'East Asia Forest Law Enforcement and Governance: Follow-up to the September 2001 Bali Ministerial Declaration, Minutes, Meeting on Task Force, Sunday, 19 May 2002, Bali Convention Center' (mimeo), p.1; and World Bank (2003), Press release, 'Regional Task Force Aims to Combat Illegal Logging and Forest Crimes,' 5 February 2003.

countries have a longstanding opposition to a forests convention. A Latin American FLEG process is more likely in Central America, where there is a stronger tradition of transboundary cooperation on forests, evinced by the Central American Forests Convention of 1993.

Concluding thoughts

The creation of regional processes to combat illegal logging is taking place because UN institutions such as the UNFF are, with justification, considered too slow to deal with the complex issues involved. The US State Department did most to catalyse international cooperation on illegal logging. Without the economic power that the US brought to bear on this issue, it is unlikely that sufficient political momentum would have developed to enable the first regional meeting in Bali to take place. But international political processes usually assume a life of their own. They rarely remain under the control of their creators, even when the creator is the global hegemon. As the FLEG processes have evolved, the logic of halting illegal logging has suggested trade controls, such as import restrictions and licensing.

While previous US administrations have, on a selective basis, supported trade restrictions in pursuit of environmental goals, for the Bush administration no such measures are tolerated. While the US administration exercised its power to promote the FLEG processes, it was also pursuing neoliberal policy objectives in parallel negotiations on trade liberalization with Latin American countries on the Central American Free Trade Agreement (CAFTA) and the Free Trade Area of the Americas (FTAA) agreement. These proposed agreements contain no measures to address illegal logging. Environmental NGOs have warned that due to the poor environmental governance of many Latin American countries, further trade liberalization will exacerbate illegal logging and result in increased US imports of cheap illegally sourced timber.[143] But for the US political and business elite trade liberalization is a more salient issue than illegal logging. The economic benefits that the US can realize from expanding international trade vastly exceed those from tackling illegal logging.[144] This explains the 'unacceptability' to the US of demand-side measures at the G8 summit of 2005.

Neoliberal principles have also informed the EU's illegal logging policy. The constitutional neoliberalism of the WTO has formed the limits to the FLEGT action plan and licensing scheme. While the WTO does not *per se* rule out demand-side measures, such as import controls, it does rule out such measures that are unilaterally imposed. Import controls are permitted only if they are agreed multilaterally or in voluntary bilateral agreements. With no multilateral support for a global trade ban on illegally logged timber, the EU thus opted for a voluntary scheme. In distinction, it is emphasized, the opposition of the US to demand-side measures goes beyond respect for the WTO agreements; it is a neoliberal ideological aversion to interfering in international trade on environmental grounds. Trade barriers are anathema to the Bush administration unless they provide clear economic benefits to the US economy, as with the steel tariffs imposed temporarily in 2002.

Like voluntary market-based certification schemes, the FLEGT licensing scheme has accepted the WTO agreements as inviolable, recognizing that to clash with the WTO would be counter-productive. Forest certification schemes and the FLEGT licensing scheme thus reflect neoliberal precepts: both are voluntary, and both rely on market-based incentives to change timber production patterns by increasing the demand-side pull from the public and private sectors for timber that has been produced with respect for certain ethical principles. But constructing the FLEGT licensing scheme within an international trading order that prioritizes trade liberalization over all other considerations, so that importing countries cannot discriminate between legally and illegally logged timber, has led to a bizarre consequence. The possibility that a licensing scheme to combat the trade in illegally sourced timber might itself be illegal under international trade law has resulted in a scheme that allows countries that currently export illegally logged timber to the EU to continue to do so. The problem here is not with the FLEGT action plan, but with the WTO. Under WTO law, it is not an offence to trade in illegally logged timber (although, it should be noted, it is an offence to trade in products manufactured in violation of international IPR law, such as fake CDs or imitation brand-name clothing). The WTO agreements represent a stronger normative framework than multilateral environmental law, and, as we see in Chapter 10, they form a significant constraint to an environmentally sustainable global governance by privileging corporate interests.

By championing both trade liberalization and international efforts against illegal logging, the US is now in the ambiguous position of providing development assistance to strengthen law enforcement and governance in countries with an illegal logging problem, while refusing to introduce demand-side measures to block imports of illegally logged timber. Should the US continue to remain opposed to demand-side measures, should the EU's licensing scheme prove successful with a large number of partner countries participating, and should the government timber procurement policies being developed by Japan and some European countries also prove effective, then there will, over the long term, be a significant restructuring of international timber trade flows. More legally logged timber will flow to the EU and Japan. And the illegal trade will increasingly gravitate towards the United States.

8

The World Bank's Forests Strategy

The World Bank Group comprises five institutions: the International Bank for Reconstruction and Development (created in 1945 and the world's largest development assistance lender); the International Development Association (an affiliate of the International Bank for Reconstruction and Development); the International Finance Corporation (which facilitates private sector lending); the Multilateral Investment Guarantee Agency; and the International Centre for the Settlement of Investment Disputes.[1] Two separate but related forest policy tracks have developed within the World Bank Group. First, the International Bank for Reconstruction and Development, along with the International Development Association, has adopted three forest policy initiatives: the 1978 forestry policy, the 1991 forests strategy and the 2002 forests strategy.[2] Second, in 1998 the International Finance Corporation adopted a forestry policy. In this chapter we examine these initiatives before considering the adjustment lending policy of the International Bank for Reconstruction and Development; while adjustment lending policy is not 'forest policy,' it can have a profound effect on forests. We also examine the role of the World Bank Group in promoting private sector forest governance.

The International Bank for Reconstruction and Development is commonly called the World Bank, which is how we refer to it now. Ultimate decision-making authority for the World Bank rests with the board of governors, which meets annually and comprises senior finance and treasury ministers. Routine decisions are taken by the board of executive directors (usually called the executive board), which meets at least once a week. Thinking within the World Bank has long reflected both neoliberal and interventionist elements. Because of this the Bank, perhaps uniquely among international organizations, has attracted criticism from both left and right. From the left, critics have argued that the World Bank neglects its mission to help the world's poor; that it serves as an agent of rich countries by prising open the economies of developing countries to foreign investment; and that it imposes a market-based neoliberal economic model on these countries. From the right, the Bank has been criticized as an ineffective bureaucracy; for preventing the efficient operation of markets by pursuing discredited Keynsian interventionist policies; and for subsidizing inefficient economies through inappropriate lending.[3] Both neoliberal and interventionist elements recur throughout the history of the World Bank's forest policy, although the strength of these two elements in relationship to each other has fluctuated over time.

The World Bank's first loans for forestry were made in 1949. Between then and 1978, there was no Bank-wide forestry strategy. Many investments were made without full knowledge of the different local claims to forests. As the World Bank acknowledged, 'land that seemed likely to be available for forestry development in fact proved not to be, often because it was common property or because it was subject to multiple competing claims.'[4] Approximately 95 per cent of Bank forestry investment during this period went to commercial operations, such as logging, saw mills, pulp mills and plantations.[5] During the mid 1970s, the World Bank began preparing its first policy on forests.

The 1978 forestry policy

The 1978 policy sought to broaden the World Bank's approach to forestry investment to meet the needs of the rural poor while paying stronger attention to environmental considerations. However, the economic benefits of commercial logging featured prominently, with the policy stating:

> *It is estimated that more than two-thirds of the natural closed forest area in developing countries is subject to no regulation or control... The extraction of this resource, however, provides valuable foreign exchange that can provide potential benefits to a much larger population.*[6]

An internal World Bank review of the policy noted that it favoured 'traditional modes of development organization' within sectoral boundaries, such as forestry, agriculture and transport. It argued that intersectoral action beyond the forest sector was needed instead. The review also noted inadequate analysis after projects were completed so that 'post evaluation has become a very difficult exercise.'[7]

Criticism of the 1978 policy outside the World Bank was considerably harsher. Cheryl Payer argued that the policy resulted in continued heavy investment in forest industries and that despite claiming to help the poor, it blamed the poor for deforestation, while avoiding the role of commercial operations in forest degradation. The policy directed resources not at forest conservation, but at replacing logged forests with plantations, and it advocated that the cost of reforesting logged landscapes should be the responsibility of the host government rather than timber corporations.[8] To Payer, the policy was more industry oriented than people oriented. Local people had no real input and were merely employed in the service of industrial forestry, for example in planting tree seedlings.[9]

The policy was reviewed during the late 1980s, by which time the international focus on tropical forests had sharpened. In 1988, there was global media attention on the Brazilian Amazon following the assassination of the rubber tappers' union leader, Chico Mendes. In 1989, the International

Institute for Environment and Development (IIED) found that just one eighth of 1 per cent of the world's tropical forests was sustainably managed.[10] The same year severe deforestation forced Thailand to implement a national logging ban. Meanwhile, environmental NGOs were lobbying against the destructive effects of World Bank-funded mega-development projects in Brazilian Amazonia. The Polonoroeste project involved the colonization of remote Amazonian regions through the construction of a highway through Brazil to Peru. Although the project included the creation of protected areas and Indian reserves, it caused substantial deforestation.[11] The Grande Carajás iron ore and mining project led to a development rush and agricultural colonization, fuelling widespread deforestation. The EU was a co-funder of the Carajás project, and in May 1989 the European Parliament expressed concern at the ecological damage that it had caused.[12] World Bank-financed projects in Congo, Ivory Coast and Malaysia were also criticized for causing deforestation.[13] In 1990, the Tropical Forestry Action Plan, for which the Bank was a co-sponsor and core funder, received extensive criticism.[14] The Bank faced concerted pressure from environmental and human rights NGOs to cease funding logging operations in tropical forests. It was in this changed international policy environment that, in 1990, the World Bank announced a funding moratorium on new forest projects pending a new forests strategy. In 1991, the new strategy was unveiled.

The 1991 forests strategy

The 1991 strategy represented a shift towards a more preservationist approach. Its strongest provision was the commitment that 'the Bank Group will not under any circumstances finance commercial logging in primary tropical moist forests.'[15] Throughout the late 1980s, World Bank investment in tropical forests had become politically untenable, and the commercial logging ban was welcomed by civil society groups. The strategy also stipulated that commercial lending, including for plantations, will be 'conditional on government commitment to sustainable and conservation-oriented forestry.' This entailed 'setting aside adequate compensatory preservation forests to maintain biodiversity and safeguard the interests of forest dwellers, specifically their rights of access to designated forest areas.'[16] The strategy thus adopted a far tougher approach to forest businesses than the 1978 policy.

The strategy also pledged that the World Bank would give 'closer attention to infrastructural and other land-using projects and will work to minimize their potentially negative effects on forests ... the Bank's economic and sector work should address forestry issues within a multisectoral context.'[17] This can be read as a commitment to avoiding the damaging mega-projects that had been criticized during the 1980s. The World Bank also committed itself to involving forest-dwellers in its projects, including 'the participation of local people' and 'the promotion of nonwood products of natural forests to benefit such people.'[18] As in the 1978 policy, poverty alleviation was emphasized.[19]

The strategy reflected neoliberal thinking by envisioning a conservation role for new markets; 'the development of ecotourism and of markets for nontimber forest products such as medicines, berries and fodder could increase the value of the forest in relation to nonforest use of the land.'[20] It was also argued that there was potential for trade between developed and developing countries in carbon sequestration. The World Bank estimated that the sequestered value of carbon in one hectare of the Amazon 'may be in the range of $375–$1625.'[21]

Overall, the 1991 strategy was silent on the mechanisms by which many of its commitments could be realized. Ans Kolk criticized it for giving no indication of how it could be implemented 'in the current context of government deficits and reduction of state functions in economies under adjustment.'[22] Even its most impressive commitment, the commercial logging ban, applied only to the tropics and not to old growth temperate and boreal forests. The tropical only emphasis had an unintended consequence at the 1992 UNCED by hardening the view in the G77 that developing countries were being denied the right to exploit their forest resources. This explains, in part, the strong line of the G77 against international cooperation on forests at the UNCED.

During the same year that the Bank adopted the 1991 strategy, it collaborated with two UN programmes to create the Global Environment Facility (GEF) (see Box 8.1). The GEF followed the Bank's strategy of refusing to fund logging projects in primary tropical forests.[22]

The World Bank's alliance with the World Wide Fund for Nature, 1998–2005

The WWF is one of the more 'insider' NGOs, with a respected reputation for working with government and business when it can, while also providing strong criticism when it believes this is necessary. During the early 1990s, the WWF criticized World Bank forest lending, including recommending that the executive board withhold approval of a forest management and logging project in Guinea, advice that the executive board ignored.[25] However, by the mid 1990s, the two organizations were pursuing complementary policies on certification and protected areas, and in 1998 they announced the formation of the World Bank–WWF Alliance for Forest Conservation and Sustainable Use. The alliance aimed by the end of 2005 to have created 50 million hectares of new protected areas, to have secured under effective management an additional 50 million hectares of threatened protected areas, and to have brought under independently certified sustainable management 200 million hectares of forest (see Table 8.1).

Protected areas

The renewed emphasis during the 1990s on protected areas grew out of the disillusionment of many actors with the formalized world of intergovernmental forest negotiations. By the mid 1990s, most NGOs considered that the forests

Box 8.1 Global Environment Facility

The GEF was created in 1991 and endorsed at the 1992 UNCED. Its implementing agencies are the World Bank, the United Nations Development Programme and the United Nations Environment Programme. The chair resides in the World Bank's Environment Department. The GEF is intended to mobilize new and additional financial resources to fund the incremental costs necessary to implement policies that yield global environmental benefits. These policies must be consistent with multilateral environmental agreements recognized by the GEF. Only states that have ratified such agreements are eligible for GEF funding. Until 2002, the GEF funded projects in four focal areas administered by the:

- Framework Convention on Climate Change, 1992;
- Convention on Biological Diversity, 1992;
- Regional and international waters agreements;
- Vienna Convention for the Protection of the Ozone Layer, 1985, and its Montreal Protocol.

During 2002, the GEF recognized two further conventions:

- Convention to Combat Desertification, 1994;
- Stockholm Convention on Persistent Organic Pollutants, 2001.

There is no GEF focal area on forests. However, forest projects that are consistent with the objectives of the CBD and the Convention to Combat Desertification may qualify for GEF funding (see Chapter 9). These projects are covered by two GEF operational programmes (OPs, not to be confused with World Bank operational policies):

- OP3 on Biodiversity;
- OP15 on Sustainable Land Management.

By 2004 the GEF had committed US$778 million for forest-related funding, which had leveraged an additional US$1995 million from other investors.[24] The GEF also has a Small Grants Programme that finances decentralized community-based projects, many of them on forests.

convention debate was deflecting attention away from practical conservation initiatives on the ground. They increasingly preferred to devote their scarce resources to working with partners within countries, rather than contesting policy at the intergovernmental level. During this time WWF was working with national and local partners towards the target of securing an ecologically representative network of protected areas covering at least 10 per cent of the world's forest cover. The World Bank's interest in protected areas made it a natural ally for the WWF.

Using protected areas to secure forest public goods clearly depends upon how effectively the areas are managed. Many protected areas, especially in remote forests, are not secured under effective management. They have been dubbed 'paper parks': while they exist on documents, such as maps and management plans, they suffer frequent incursions. The problem here is not so much local communities entering protected areas to reclaim traditional land-use rights. It is more actors from outside the forest engaging in illegal logging and converting forests to other land uses. So serious is this problem that the World Bank–WWF Alliance included the separate target of securing under effective management 50 million hectares of existing protected areas.

The alliance stimulated a range of partnership-based actions. In 1999, it promoted a summit of Congo Basin forest ministers in Yaoundé. This eventually contributed to the launch at the 2002 World Summit and Sustainable Development of the Congo Basin Forest Partnership (see Chapter 7). The Congo Basin Forest Partnership aims to promote 10 million hectares of effectively managed protected areas across six Congo Basin countries – Cameroon, Central African Republic, Republic of the Congo, Democratic Republic of the Congo, Equatorial Guinea and Gabon – including the Sangha Tri-National Conservation Area complex that straddles the first three of these countries.[26] Also launched at the World Summit on Sustainable Development was the Amazon Region Protected Areas initiative. The World Bank–WWF Alliance played a catalytic role in creating this initiative, which commits 50 million hectares (12 per cent of Brazilian Amazonia) to protected areas over a decade and will, if successful, triple the area of protected forests in Brazil.[27]

Forest certification

The WWF was one of the prime movers behind the creation of the FSC, and the World Bank's 1991 strategy expressed support for forest certification.[28] There was agreement between the WWF and the World Bank that protected areas can only be effective over the long term if they are surrounded by zones of sustainably managed forests. David Cassells, formerly of the World Bank's forest department, has illustrated this point with reference to Costa Rica. In many respects Costa Rica represents the best and the worst in protected forest area practice. Costa Rican forests are well secured and are managed effectively with local community involvement. But unless these areas are surrounded by sustainably managed forests, their long-term prognosis could be biotic impoverishment within a century.[29]

With the alliance advocating that protected areas be surrounded by sustainably managed forests, it was necessary for the World Bank and the WWF to agree an independent approach for assessing sustainable management. The solution chosen was forest certification. Given the closeness of the WWF to the FSC, the alliance's certification policy reflects FSC principles, although the alliance has not formally endorsed the FSC or any other certification scheme. The alliance has enabled previously unsustainably managed forests to meet independent certification standards through creating improved national-level policy environments. Examples include the alliance's assistance in preparing a new forest law in Peru[30] and in drafting protected areas legislation in Cambodia.[31]

The alliance achieved mixed results on its 2005 targets. The new protected areas target was exceeded; but the target of securing 50 million hectares of threatened protected areas under effective management was not met. The alliance fell well short of its target for certifying 200 million hectares of forests in World Bank-client countries (see Table 8.1).

The 2003 report of the alliance claimed that its 'ideals and principles on forest conservation and use are now internalized in the strategic thinking of both partner organizations.'[32] This thinking fed into the Bank's review of its 1991 forests strategy.

Table 8.1 *The World Bank–WWF Alliance for Forest Conservation and Sustainable Use: Targets, 1998–2005*

Targets announced in 1998	*Status as of 31 December 2005*
50 million hectares of new forests in protected areas	Target exceeded: 55 million hectares*
50 million hectares of existing, but highly threatened, protected areas to be secured under effective management	Target not met: 40 million hectares†
200 million hectares of production forests under independently certified sustainable forest management	Target not met: 25 million hectares‡

Notes:
* Many different actors are involved in the establishment of forest protected areas. While the World Bank–WWF Alliance supported the establishment of 55 million hectares, the total hectares of new forest protected areas created by all actors during this period was higher.
† In early 2006, the alliance was engaged in supporting improved management in 70 million hectares of forest protected areas. It was also working to define the criteria of 'effective management.'
‡ Not all forest certifications are due to the alliance. The view within the alliance is that it can legitimately claim involvement in bringing at least 25 million hectares under independent certification. (It may be noted that by 31 December 2005, the FSC had certified 68,125,087 hectares in World Bank-client countries.)

Source: Ken Creighton, WWF-US, email, 30 January 2006

The review of the 1991 forests strategy

In 1999, the World Bank initiated a review of its forests strategy. By this time the international forest policy dialogue had broadened to include all forest types, and the emphasis in the 1991 strategy on tropical forests was seen as discriminatory by developing countries and NGOs. Furthermore, the Bank's lending portfolio had changed. By the mid 1990s, half of the Bank's forest lending portfolio involved boreal and temperate forests, including to the former communist countries in Eastern Europe and the former Soviet Union. Forest loans to these countries were considered important since much of their logging equipment was outdated and environmentally destructive. Finally, certification and the criteria and indicator debates of the 1990s (see Chapter 6), along with the adoption of the ecosystem approach by the CBD (Chapter 9), had shifted the discourse on sustainable forest management and admitted the possibility that tropical forests could be conserved without a commercial logging ban.

As part of the review, the World Bank initiated a consultation process. A technical advisory group facilitated by the IUCN was formed comprising individuals from bilateral and multilateral lending agencies and environmental NGOs, including the World Resources Institute, the WWF and the Forest Peoples Programme. Early on in the consultation process, the Forest Peoples Programme, backed by some other NGOs, proposed that the World Bank's Operations Evaluation Department (OED) should carry out a review of the 1991 strategy.[33] The Bank agreed to this. The OED review, published in 2000, argued that the 1991 strategy had led to a more 'conservation-oriented' approach, but concluded that the focus on tropical moist forests ignored the importance of non-tropical biodiversity-rich forests.[34] The 1991 strategy had had a 'chilling effect' on World Bank forest investment.[35] Bank staff had become wary of possible reputational damage if they violated what came to be perceived as a general 'do no harm' rule.[36] The result was gradual disengagement from all forest investment as a psychological reaction, so that although the strategy only prohibited logging in primary tropical moist forests, the World Bank slowly withdrew from forestry investment in secondary tropical forests and non-tropical regions. The OED review concluded that the Bank should 'capitalize on its convening powers to facilitate partnerships that mobilize additional financial resources.'[37] In other words, the Bank should re-engage in forest investment.

It can be argued that this conclusion judges the World Bank on quantitative criteria, namely the money that is mobilized and invested, rather than on other, more qualitative, criteria, such as respect for cultural diversity and land rights.[38] It was strongly resisted by most NGOs. The main exception was the WWF. By entering into an alliance with the World Bank, the WWF now supported Bank investment in the forest sector in pursuit of its objectives on forest certification and the creation of ecologically representative forest protected areas. During the late 1990s, some other NGOs, including the World Rainforest Movement and Friends of the Earth in the UK, tried to persuade the WWF that while their objectives on certification and protected areas were important, supporting World Bank forest investments was not the best way of achieving these objectives. These efforts met with no success.[39] Some other NGOs supported Bank re-engagement in the forest sector, including Friends of the Earth-Brazil.[40]

An important factor in the 1991 decision to implement a logging ban was the concerted opposition of NGOs against Bank investment in tropical forests. A decade later a large majority of NGOs continued to hold to this position, arguing not only for the retention of the ban in the tropics, but its extension to old growth forests in the temperate and boreal zones. However, with a minority of NGOs opposing this position, World Bank forest staff were able to claim that NGOs were divided. The Bank argued that a total ban on commercial logging in primary moist forests could no longer be justified. Instead of protecting all primary tropical moist forests, protected areas should be chosen on merit. Admittedly, re-engagement carried with it social and environmental risks; but, World Bank staff argued, forest lending was so important that the Bank should re-engage despite the risks. The 1991 strategy had kept the Bank free of direct association with deforestation, but at the cost of neglecting its poverty alleviation mission. Africa was cited as one example. 'Forest lending has plunged in Africa, where the need for forest assistance is greatest and where the poor are overwhelmingly dependent on forest products and services.'[41] The general tenor of much of the OED review was that the Bank could act as a force for good by working with forest loggers. The choice was presented as a stark one between leaving loggers free to operate by their own standards, which are usually weak or non-existent, or working with commercial operators and pressuring them to adhere to the Bank's safeguards. While bans were justified in areas where severe logging had taken place or where the forest ecosystem was highly fragile, the notion that the World Bank should finance no logging in forest-rich areas created a barrier between the Bank and its client governments in the tropics.[42]

Another debate during this period centred on adjustment lending policy. Some NGOs argued that the macro-economic policies that the World Bank prescribed to borrowing countries through neoliberal lending conditionalities, such as cutting public expenditure and exporting natural resources to earn hard currency, had offset the benefits of the logging ban. The World Rainforest Movement argued that despite the 1991 strategy, the Bank 'continued being a major actor in forest destruction' as it imposed 'an economic model which resulted in even more serious social and environmental impacts.'[43] The NGOs in the technical advisory group argued that the revised forests strategy should take into account the forest-related effects of the World Bank's adjustment lending policy. Against this, the Bank insisted that the impacts on forests of adjustment lending required an amendment not to forest policy, but to adjustment lending policy. If, for example, adjustment lending policy was amended to take into account the effects on forests, then it would be inequitable not to do the same for other environments, such as coral reefs and wetlands. Hence, it was further argued, to avoid a piecemeal alteration to World Bank policy, a new adjustment lending policy should be prepared that took into account the environmental and social effects of lending on all sectors. This view prevailed, and the 2002 forests strategy makes no mention of the effects on forests of adjustment lending. IUCN and the NGOs in the technical advisory group then pressed the Bank to issue interim guidance on the forest-related effects of adjustment lending pending a full review of adjustment lending policy.[44] They were unsuccessful. However, as a result of these debates, the World Bank promised that a full review of adjustment lending policy would take place. We examine this later.

The 2002 forests strategy

In 2002, the executive board approved the 'revised' forests strategy (in effect, a completely new strategy). We first consider this strategy, and then the new safeguard policies that were prepared in parallel to it.

The strategy

The 2002 strategy is structured around three 'pillars of engagement,' each of which corresponds to one of the main objectives of the World Bank's 2001 Environment Strategy (see Table 8.2).

Whereas the key phrase of the 1991 strategy was that 'the Bank Group will not under any circumstances finance commercial logging in primary tropical moist forests,' the hallmark of the 2002 strategy is that 'Under no circumstances will the Bank Group invest in non-sustainable commercial logging or logging in environmentally or culturally critical forest areas.'[45] The World Bank emphasized that re-engagement did not mean the automatic approval of any commercial logging project. Instead, it signified the replacement of a blanket prohibition with 'targeted prohibitions on financing logging in critical forest areas in all forest types in all countries.'[46] By claiming that its new strategy

Table 8.2 *Main objectives of the World Bank's 2001 Environment Strategy and 2002 Forests Strategy*

World Bank Environment Strategy, 2001	***World Bank Forests Strategy, 2002***
Improving the quality of life	Harnessing the potential of forests to reduce poverty
Improving the quality of growth	Integrating forests within sustainable economic development
Protecting the quality of the regional and global commons	Promoting vital local and global environmental services and values

Source: World Bank website, www.lnweb18.worldbank.org/ESSD/envext.nsf/41ByDocName/Environment (accessed 13 May 2005); World Bank (2002) *A Revised Forest Strategy for the World Bank Group*, Washington DC: World Bank, pp.2–7

had the support of some NGOs, the World Bank was able to legitimize re-engagement, although the support from civil society for the 2002 strategy is minimal compared to a decade earlier when the logging ban enjoyed wholesale support from environmental and human rights NGOs.

The 2002 strategy noted the need to foster markets for ecological public goods. A recurring theme in the strategy is that many forest values fall outside the market and are thus not fully taken into account in policy. Whereas the 1991 strategy had merely noted the potential of new markets, the 2002 strategy pledged World Bank support at two levels: building and financing markets for international public goods, with particular reference to carbon and biological diversity; and assisting governments in establishing national markets for environmental goods and services.[47] The rationale for using market mechanisms to realize public goods was that countries would only be prepared to conserve forests if they were to receive revenue for doing so. However, the OED review had provided a caution; that support for 'developing carbon and other markets (certification, ecotourism, water) is not universal, and international willingness to pay for these services is questionable.'[48] The Bank also indicated an awareness of the risks of market mechanisms, particularly disruption of communal forest management systems, and increased competition between social groups 'resulting in restricted access by the poorest of the poor to essential forest products.'[49]

If the emphasis on new markets in the 2002 strategy is one area that has its intellectual origins in neoliberal ideology, another is the emphasis on increasing private sector investment flows.[50] The rapid expansion of Bank-led private sector lending was one of the features of the presidency of James Wolfensohn (1995–2005), with the Bank devoting an increasing share of its resources to guaranteeing private sector investment, rather than lending to governments. Environmental activist Bruce Rich has condemned this practice as 'little more than corporate welfare.'[51]

The policy of the World Bank–WWF alliance that sustainable forest management should be assessed through independent forest certification also appears in the 2002 strategy, as well as in the safeguard policies that accompany this strategy.

The World Bank's forest safeguard policies

World Bank strategies are operationalized with reference to the Bank's Operational Manual. Within this manual are operational policies (OPs), namely short statements on the Bank's specific obligations under a strategy, and Bank procedures (BPs), which guide Bank staff on the procedures for implementing an OP.[52] The relevant OPs for forests are known as OP 4.36, while the accompanying Bank procedures are BP 4.36. The World Bank's management has identified ten key areas within the Operational Manual that are 'critical to ensuring that potentially adverse environmental and social consequences are identified, minimized, and mitigated.'[53] These are known as the safeguard policies, to which particular attention is paid during project design and implementation. The safeguard policies cover ten areas, of which forests is one.[54] The forest safeguards apply only to project funding, and not to adjustment lending, which accounts for one-quarter of the Bank's lending portfolio.

The reversal of the commercial logging ban required new forest safeguard policies, and, in parallel to the preparation of the 2002 forests strategy, new drafts of OP 4.36 and BP 4.36 were prepared. The Bank consulted the technical advisory group when drawing up these safeguards. An early draft of BP 4.36 was circulated in 2002. The IUCN urged the World Bank to delete the word 'significant' from the passage 'Projects with the potential for significant conversion or degradation of natural forest or other natural habitats will receive a Category A rating.'[55] (Category A ratings are given to projects 'likely to have significant adverse environmental impacts that are sensitive, diverse, or unprecedented.'[56] Category A ratings are usually assigned to projects that could affect critical ecosystems or indigenous peoples, or which involve dam construction.) In the final draft of BP 4.36, the word 'significant' no longer appears before the word 'conversion,' as the IUCN had requested; but it was inserted later in the same sentence, which had been amended to read:

> *A project with the potential for conversion or degradation of natural forests or other natural habitats that is likely to have significant adverse environmental impacts that are sensitive, diverse, or unprecedented is classified as Category A.*[57]

Furthermore, the final draft of OP 4.36 stipulates: 'The Bank does not finance projects that, in its opinion, would involve *significant conversion or degradation* of critical forest areas or related critical habitats' [emphasis added].[58] Critical forest areas include protected areas, forests protected by traditional local communities, old growth forests and biodiversity rich forests.

The Bank had insisted on retaining the word 'significant' with respect to forest conversion and degradation in its forest safeguards. Why was this? Does this mean that conversion or degradation of critical forest areas that is less than 'significant' can be tolerated? The Bank wished to retain the word since to have deleted it would have ruled out investments that might lead to very small environmental impacts in forests. And it was from this absolutist position, to which it was bound by the 1991 strategy, that the Bank now wished to move. Without the word 'significant' the Bank would have been unable, for example, to finance construction of a schoolhouse if it required forest clearance.[59] The 'chilling effect' would have remained in place.

It should be noted that on plantations OP 4.36 reads: 'The Bank does not finance plantations that involve any conversion or degradation of critical natural habitats, including adjacent or downstream critical natural habitats.'[60] Here the word 'significant' does not define 'conversion or degradation,' as the Bank has committed itself not to finance plantation projects that can lead to biodiversity loss. It should also be noted that these semantic questions were not only discussed with external stakeholders. When the safeguard policies were considered by the executive board, there was considerable discussion on the word 'significant,' whether it should remain and, if so, how it should be applied.

The NGO members of the technical advisory group made several proposals on participation, biodiversity conservation, the rights of forest dwellers, indigenous peoples and implementation.[61] Most of these proposals were not adopted.[62] However, the NGOs did exert some influence. For example, the Forest Peoples Programme successfully argued that the World Bank should ensure that its strategy was consistent with international environmental law.[63] The Forests Peoples Programme also argued for the strategy to contain a commitment to respect international human rights law, although the Bank's lawyers refused to agree to this.[64] NGOs also lobbied for the safeguards to include strong provisions on indigenous peoples' rights. Bank staff responded that the safeguards could not refer extensively to indigenous peoples, as the Bank's internal policy-making systems are structured around separate, but consistent, safeguard policies, with separate Bank safeguards on indigenous peoples.[65] Despite this, BP 4.36 states that World Bank staff should improve 'the participation of indigenous peoples and poor people,' and take into account issues relevant to them in community-based forest management.[66]

Indigenous peoples are also mentioned with respect to certification policy. The Bank's commitment to forest certification demanded a clear position on certification standards. The part of OP 4.36 that addresses certification bears a distinct resemblance to the FSC principles for well-managed forests (see Table 8.3). Unlike the FSC's principles, the World Bank's certification policy does not mention plantations (although, as we have noted, the Bank does not finance plantations that would result in biodiversity loss). OP 4.36 stipulates that a forest certification system 'must require independent, third-party assessment of forest management performance.'[67] So striking was the similarity between the FSC principles and the Bank safeguards that at the executive board some delegates asked if the World Bank was binding itself too closely to the FSC. The Bank's reply was that it was committed to the standards contained in OP 4.36, and that it neither endorsed nor rejected any particular certification scheme.[68] The executive board also queried the policy of financing forest operations that had yet to be certified. The solution that the Bank eventually crafted was that in such circumstances industrial harvesting operations should agree and 'adhere to a time-bound action plan acceptable to the Bank for achieving certification.'[69]

The safeguards apply only to the International Bank for Reconstruction and Development and the International Development Agency, and not to other institutions within the World Bank Group. They were approved by the executive board in 2002. At this time, the Bank's management undertook to prepare a sourcebook that would clarify the meaning of terms such as 'significant impact' and 'critical forests.' However, by the end of 2005 the sourcebook had not been published, which, according to several NGOs, left the Bank free to make up the rules for projects that affect forests, the value of which is estimated at approximately US$3 billion.[70]

Table 8.3 *World Bank operational policies on forests and equivalent Forest Stewardship Council principles*

World Bank OP 4.36, paragraph 10	***Equivalent FSC principles***
To be acceptable to the Bank, a forest certification system must require:	
• compliance with relevant laws;	1
• recognition of and respect for any legally documented or customary land tenure and use rights, as well as the rights of indigenous peoples and workers;	2, 3
• measures to maintain or enhance sound and effective community relations;	4
• conservation of biological diversity and ecological functions;	6
• measures to maintain or enhance environmentally sound multiple benefits accruing from the forest;	5
• prevention or minimization of the adverse environmental impacts from forest use;	6
• effective forest management planning;	7
• active monitoring and assessment of relevant forest management areas; and	8
• the maintenance of critical forest areas and other critical natural habitats affected by the operation.	6, 9

Note: The ten FSC principles can be found in Box 6.2, Chapter 6, p.120, of this volume.

Source: World Bank (2002) *World Bank Operational Manual*, Section OP 4.36 – Forests (November 2002, revised August 2004), www.wbln0018.worldbank.org/Institutional/Manuals/OpManual.nsf/tocall/C972D5438F4D1FB78525672C007D077A?OpenDocument (accessed 1 February 2005)

The World Bank's adjustment lending policy

The World Bank's 2002 forests strategy and accompanying safeguards apply only to project lending and not to adjustment lending. Adjustment lending is pegged at 25 per cent of the Bank's lending portfolio, averaged over three years, with the remaining 75 per cent of Bank lending devoted to projects. Since the World Bank's forest safeguards contained no reference to adjustment lending policy, NGOs now insisted that the Bank honour its promise to adopt a new adjustment lending policy that took into account the effects on forests of adjustment lending.

Adjustment lending is a mechanism through which the World Bank leverages policy reforms in borrowing countries, and it is arguably the most controversial subject in the history of the World Bank. Along with its sister organization, the International Monetary Fund (IMF), the Bank has been involved in adjustment lending since the 1960s. Although the name has varied – austerity programmes, shock therapy, structural adjustment programmes (SAPs), adjustment lending policy and, more recently, development policy – the basics have remained essentially unchanged. ('Development policy' is now the preferred term in the World Bank; however, 'adjustment lending policy' is commonly used in the literature and is the term used here.) Adjustment lending takes place when the Bank agrees to make a loan to a country that is experiencing macro-economic difficulties – such as hyperinflation, a balance of payments deficit, a falling currency or a high level of external debt – providing that certain conditionalities are met that are intended to help the borrowing country achieve economic growth. Since the 1980s, Bank conditionalities have increasingly emphasized that market forces should be allowed to operate without government interventions in the borrowing country. The most common conditionalities, all of which have a clear neoliberal emphasis, are that the borrower needs to have implemented, or committed to implement, reforms such as privatizing state-owned economic sectors, cutting public expenditure, deregulation and abolishing subsidies that are considered trade distorting. Joseph Stiglitz, former chief economist of the World Bank, notes that during the 1980s a 'purge' took place within the Bank that helped to push the role of the market to the forefront of Bank policy. Government intervention came to be seen as something that prevented markets from functioning effectively. The purge – which could only have taken place with the support of the Bank's largest shareholder, the US government – helped to solidify the Washington consensus shared by the Bank, the IMF and the US Treasury Department and founded on what Stiglitz terms 'market fundamentalism.'[71]

The more dependent a country is on World Bank loans, the more vulnerable it is to neoliberal lending conditionalities. Cutting public welfare spending in tropical countries increases poverty, which can result in increased migration to forest areas as the rural landless poor seek land for subsistence agriculture. External debt repayment, which serves the interests of developed world governments and financial institutions, can fuel deforestation as trees are felled for export to earn hard currency through 'rip and ship' forest clearance. Not only are such policies bad for the forest, they have sometimes failed to achieve their stated objective of increasing hard currency earnings; increased timber exports increase the total world supply of timber, depressing prices and leading to reduced hard currency earnings per unit of export.

The World Bank has tried to take these criticisms into account. During the Wolfensohn presidency, the Bank sought to move away from adjustment policies that focused narrowly on debt repayment. There was considerable resentment within the Bank when the IMF called on Bank reserves to support massive adjustment lending deals with Indonesia, Thailand, South Korea and Russia in the wake of the financial crises of 1997 and 1998.[72] A consensus emerged that adjustment lending policies focusing solely on macro-economic factors were failing to address underlying issues such as corruption, and that a longer-term emphasis on development strategy was needed. This has been reflected in the rebranding of adjustment lending policy as 'development policy' as the Bank has tried to move towards conditionalities that promote broader benefits, such as poverty reduction, social stability and environmental conservation.[73]

But not all critics have been convinced by the World Bank's attempts to break from its previous narrow neoliberal approach to adjustment lending, and the review of what is now called 'development policy' was actively contested by environmental and human rights NGOs. The review of the relevant policies – OP 8.60 and BP 8.60 – was initiated in 2002.[74] The first draft of OP 8.60 issued in December 2003 did not even mention forests. This led to further NGO lobbying. The Forests Peoples Programme wrote to the World Bank arguing that the draft 'is in some measures weaker than the current policy... We urge that the draft policy be withdrawn and substantially revised.'[75] The policy that was finally adopted included the phrase: 'The Bank determines whether specific country policies supported by the operation are likely to cause significant effects on the country's environment, forests, and other natural resources.'[76] This weakly worded commitment leaves the Bank as the sole arbiter of what constitutes 'significant effects.'

The revised policy, which was approved by the executive board in 2004, also commits the World Bank to assessing the actions that borrowers will take to reduce any adverse environmental effects. If there are 'significant gaps in the analysis or shortcomings in the borrowers' systems,' the Bank will describe 'how such gaps or shortcomings would be addressed before or during programme implementation, as appropriate.'[77] This commitment uses the word 'would' rather than the stronger imperative 'should,' and is further softened by the caveat 'as appropriate.' Furthermore, there are no mentions of forests, environment or natural resources in BP 8.60.[78] Not surprisingly, the new policy has been interpreted by NGO activists as a breach of the promise made by the World Bank to integrate into its adjustment policy measures to redress the negative impacts on forests of adjustment lending.[79]

'Hard' and 'soft' conditionalities

The World Bank has used its agency to promote both neoliberal and conservation values. While there is a stronger conservation component in the 1991 forests strategy compared to that of 2002, the latter promotes stronger standards on forest certification, which, as a market-based tool, fits with the Bank's neoliberal bias. However, the Bank has not, and as pure neoliberalism would dictate, been content to let the market decide which forest certification scheme it wants. Many of the Bank's donor and client countries would have preferred safeguards that admitted all the major certification schemes; but in OP 4.36 the World Bank has effectively endorsed the most rigorous scheme currently available, the FSC. While the Bank's forest safeguards are intended as an internal guide for World Bank managers, they can, like Bank adjustment policy, have a broader influence by transmitting Bank policy preferences to other actors.

As with adjustment lending, the degree to which the World Bank can leverage forest policy reforms will vary according to the vulnerability of the borrowing country, in particular how heavily it is indebted to the Bank or dependent on it for future loans. The US and Canada do not, of course, depend on the Bank for finance. If the Bank did lend to these countries it would not fund the SFI and CSA forest certification schemes, which do not meet Bank safeguards, especially on indigenous peoples' rights. Similarly, the failure of most national PEFC schemes to include indigenous peoples' rights rules out Bank lending for such schemes.[80] Like the SFI and CSA, PEFC emerged from a region that is not dependent on the Bank for loans. The competitor schemes have tended to be weaker than the FSC, in part because they have emerged from economically strong countries that can ignore Bank certification policy.

However, countries in Eastern Europe that may depend on the Bank for loans can less easily afford to do this. For example, in Poland and Russia, to which the Bank lends, the FSC has so far prevailed over the PEFC. It remains to be seen whether PEFC can achieve the same dominance in Eastern Europe and the tropics as it has in Western Europe, or whether the Bank's role as a donor in these regions will provide agency for the FSC at the PEFC's expense. However, the Bank's promotion of standards that are FSC compliant does not always lead to the adoption of FSC certification. In tropical countries where the Bank is active as a lender, FSC certification has lagged well behind the developed countries, which have a more environmentally discriminating consumer base. But it is fair to conclude that the FSC would have had even less success in the tropics were it not for the Bank's safeguard policies on forests.

Those forest industries that choose FSC certification in countries where the Bank is active may do so voluntarily, but not necessarily because they want to. They may do so to avoid losing future World Bank investment in their country's forest sector. Even in countries where the forest industry is not dependent on Bank loans, other actors that are may pressure the forest sector to follow Bank certification policy. In this respect, certification in Bank client countries is not purely market driven. It is also driven by the pragmatic responses of some forest businesses to the structural economic power that the World Bank wields.

While the World Bank has used its agency to promote sustainable forest management, there is, as we have seen, a disconnect between the Bank's forests strategy and its adjustment lending policy. Which of these two is the strongest? Many of the criticisms of the Bank over the last 30 years are premised on the argument that the Bank's adjustment lending and macro-economic policies have more normative force than its environment policies. Many NGOs have questioned how hard the Bank pushes its forest standards on borrowing countries. The World Rainforest Movement (WRM) has distinguished between 'hard' and 'soft' conditionalities. Hard conditionalities include privatization and opening economies to foreign investment, while soft conditionalities include forest conservation and respect for indigenous peoples.[81] Hard conditionalities are neoliberal policy prescriptions that are central to adjustment lending and which the World Bank invariably insists on, whereas soft conditionalities are those on conservation and human rights, which, according to the WRM, the Bank is prepared to ignore if the borrowing country does not comply.[82]

The World Bank and private sector governance

Under the Wolfensohn presidency, the expansion of lending to the private sector was the fastest growing area of operational activity of the World Bank Group.[83] Growing engagement with business and the banking sector has led to the involvement of the World Bank in private sector governance. In 1998, the World Bank and the WBCSD brought together chief executive officers (CEOs) from some leading forestry businesses to create the CEOs Forum on Forests.[84] This was later expanded to include forest product retailers. It then stalled before being reactivated in 2001 as the Global Forest Industry CEO Forum. This body is now working with the WBCSD's Sustainable Forest Products Industry working group (formed in 1997) to promote 'sustainable solutions' for forest-based industries.[85] The CEO Forum and the Sustainable Forest Products Industry group are purely voluntary mechanisms to promote partnership formation within forest-based industries. Their actual contribution to global forest governance is difficult to gauge. Both groups aim for the continued growth of major forest industry corporations. Having launched the CEO process, the World Bank is no longer centrally involved with it, although it retains an active involvement with the WBCSD, including through The Forests Dialogue (see Box 6.5, Chapter 6).

The World Bank Group has also sought to promote voluntary private sector governance through the International Finance Corporation (IFC). The IFC aims to reduce poverty through private sector development. It has developed its own safeguard policies, which have fed into a global private sector governance initiative: the Equator Principles.

The forest safeguard policies of the International Finance Corporation

The IFC has an increasingly important role in World Bank Group lending in general, and forests in particular. During the1990s, the IFC adopted a set of OPs similar to those of the World Bank. Confusingly, the IFC's OPs on 'forestry,' adopted in 1998, bear the same reference number as those of the World Bank on 'forests,' namely OP 4.36. The content, however, is very different. The IFC safeguards state that the 'IFC does not finance commercial logging operations or the purchase of logging equipment for use in primary tropical moist forest.'[86] They pledge that the IFC 'involves the private sector and local people in forestry and conservation management' and that 'IFC's financing operations in the forest sector are conditional on the project sponsor's commitment to undertake sustainable management and conservation-oriented forestry.'[87] This includes undertaking 'social, economic, and environmental assessments of forests being considered for commercial use' and safeguarding 'the interests of forest dwellers, specifically their rights of access to and use of designated forest areas.'[88]

The current IFC forestry safeguards were designed to be consistent with the 1991 forests strategy and have not been revised since the adoption of the World Bank's 2002 strategy. The position of the IFC is that while it is legally bound by the 1998 safeguards, it will adhere to the spirit of the World Bank's 2002 strategy.[89] The next revision of the IFC safeguards will synchronize them with the 2002 strategy and will involve the replacement of the pledge not to invest in primary tropical moist forest with a commitment not to finance 'significant' damage in 'critical' forest areas.

The Equator Principles

In 2002, a series of discussions took place involving the IFC and some leading investment banks on the possibility of a set of guidelines that would enable investment banks to evaluate the environmental and social risks of development projects when considering funding. The motives of the banks involved varied, but included public pressure, reputational risk, shareholder expectations and client demands.[90] A consensus emerged that developing new guidelines and standards would be problematic and time consuming, as each individual bank would seek to negotiate standards that suited its own interests. Hence, a set of neutral 'off-the-shelf' principles was preferable. It was agreed that the IFC safeguard policies and other World Bank and IFC guidelines should be adopted. These were incorporated within what was eventually called the Equator Principles, so-called as they circle the globe and are intended to be common to both the northern and southern hemispheres.[91] The Equator Principles, which were adopted in Washington in 2003, apply to a range of industrial sectors, including forestry and wood industries (see Box 8.2).

Some investment banks that have not adopted the Equator Principles are finding that they have to follow them; for example, if they are involved in a consortium with another bank that has adopted them. The principles are thus acquiring the status of an industry-wide set of standards. However, the implementation and monitoring mechanisms of individual banks are far from transparent, as are the dispute settlement procedures. It is not clear what recourse individuals or groups have if they believe that the principles have been violated. The Equator Principles apply only to large projects with a total capital cost of more than US$50 million, so a large swathe of transnational investment flows are not covered. Despite the adoption by some major investment banks of the principles, their independent normative pull is unclear. The principles are very much a neoliberal product; they are purely voluntary, no other stakeholders were consulted, and there is no independent third-party monitoring. It is unlikely that the Equator Principles will lead to an effective system of private sector governance that rigorously upholds environmental and social standards.

Box 8.2 The Equator Principles

Investment banks that adopt the Equator Principles 'seek to ensure that the projects we finance are developed in a manner that is socially responsible and reflect sound environmentally management practices... We will not provide loans directly to projects where the borrower will not or is unable to comply with our environmental and social policies and processes.' The principles are linked to three exhibits:

1 *Exhibit I* is the environmental and social screening process of the World Bank Group, whereby proposed projects are categorized as A, B and C. A Category A proposal is 'likely to have significant adverse environmental impacts that are sensitive, diverse, or unprecedented.' A Category B proposal has potential adverse environmental impacts that 'are less adverse than those of Category A projects.' Category C proposals have 'minimal or no adverse environmental impacts.'
2 *Exhibit II* comprises the ten IFC safeguard policies, including OP 4.36 'Forestry.'[92]
3 *Exhibit III* consists of two World Bank Group guidelines. The first is the *World Bank Pollution, Prevention and Abatement Handbook*, which covers best practices for handling materials and chemicals for a broad range of industries, including agriculture, copper, pharmaceuticals, pulp, paper and wood preserving. The second is a set of IFC environmental, health and safety guidelines for industry, including forestry operations, wildland management and wood products.

Source: Equator Principles website, www.equator-principles.com/ (accessed 27 May 2004)

The renewal of the World Bank–WWF Alliance

The initial phase of the World Bank–WWF Alliance ended in December 2005. As we have seen, the alliance did not achieve all of its targets (see Table 8.1). Discussion began in 2004 on the renewal of the alliance. The possibility of admitting new partner organizations was considered, and different models for a new alliance were explored, including a core that involved the World Bank and the WWF, with other partners clustered around particular themes and issues.[93] Ben Gunneberg, the secretary-general of the PEFC, made it clear to World Bank forest staff that the PEFC would like to be involved in a new alliance.[94] PEFC membership could have had various effects, including a rapprochement between the FSC and the PEFC and a reappraisal of the World Bank's forest certification policy. Eventually it was agreed to renew the alliance with just the original two members, the WWF and the World Bank, an outcome that was welcomed by supporters of the FSC. A new goal was agreed; to reduce the rate of global deforestation by 10 per cent by 2010. Three new targets were also agreed:

1 25 million hectares of new forest protected areas;
2 75 million hectares of existing forest protected areas under improved management;
3 300 million hectares of forest outside strict protected areas under improved forest management, comprising three sub-targets:
   - 100 million hectares of forest independently certified consistent with the standards outlined in the World Bank's Operational Policy of Forests (OP 4.36);
   - 100 million hectares of forest progressing towards such independent certification;
   - 100 million hectares of forest land under community-based forest management agreements that improve local livelihoods.[95]

Concluding thoughts

Prior to 1978 the dominant view in the World Bank was that forests were an abundant but underutilized resource. Investment capital, not forests, was seen as the scarce resource; developing countries needed access to capital to develop their forests. On paper, the 1978 policy sought to balance commercial development, poverty alleviation and environmental objectives. However, the vast bulk of Bank lending between 1978 and 1990 went to commercial operations. This resulted in some disastrous Bank investments, which, along with growing international concern at deforestation, prompted a paradigm shift towards a preservationist stance. The 1991 forests strategy illustrates that the Bank has been prepared to swim against the prevailing ideological current when it believes this is necessary: a logging ban is a policy that has no basis at all in neoliberal thought. On the contrary, it can be seen as a massive market distortion. The ban suggested that tropical forests, not investment capital, should now be seen as the scarce resource. In effect, the 1991 strategy recognized the importance of forests to global public good provision. But the 'chilling effect' of the strategy led to the Bank tending to avoid all forest-related investments. The 2002 strategy thus reversed the preservationist emphasis of 1991 and sanctioned Bank re-engagement in primary forests. The lifting of the ban can be seen as a neoliberal reaction to 'excessive regulation' that interfered with free international trade and stifled investment opportunities for transnational capital.

The World Bank is sometimes seen as a blunt and uncritical neoliberal agent; but the evidence assembled in this chapter reveals that the Bank is also prepared to use its agency to promote environmental conservation. Certainly the World Bank has developed a strong neoliberal ethos that reflects the ideological orientation of its major shareholders, in particular the US and UK.

The neoliberal bias of the Bank is most clearly evident in its adjustment lending policy, although neoliberal values have helped shape Bank forest policy too. Hence the emphasis in the 2002 forests strategy that deforestation is due in part to the failure of markets to price environmental goods and services properly. It follows, therefore, that 'new markets' are needed for carbon, ecotourism and non-timber forest products. But the World Bank has not been uncritical of markets, and it recognizes their limitations.

The result has been a policy mix that is a hybrid of neoliberalism and conservationism. The World Bank actively promotes private sector investment in forests, new environmental markets and voluntary private sector regulation. But it has also supported interventions that interfere with markets, such as the logging ban, the creation of protected areas and strong certification standards. The Bank has promoted the creation of the CEOs Forum on Forests, but also entered into a strategic alliance with the WWF. It has agreed safeguards for forest project lending, but also adopted adjustment lending policies that drive deforestation. It is on this last point where the main problem lies. The World Bank has the most comprehensive and rigorous forest safeguard policies of any international financial institution; yet these safeguards do not take into account the macro-economic effects on forests of adjustment policy. The adjustment lending safeguards provide only one cursory mention of forests. A pressing task for the Bank must surely be to restructure its internal policy systems so that project lending safeguards are integrated within adjustment lending policy. Only then will environmental and human rights standards be an integral component of all areas of World Bank activity.

9

The International Forests Regime

This chapter develops the concept of the international forests regime. The most common definition of an international regime is a framework of 'norms, rules, principles and decision-making procedures around which actors' expectations converge in a given area of international relations.'[1] There are different ways of conceptualizing an international regime. It may be seen as a form of governance constructed around a single international legal convention and subsequent protocols. So, for example, the ozone regime is structured around the 1985 Vienna Convention, the Montreal Protocol and subsequent amendments. With no forests convention, there is, in this view, no international forests regime.[2] A second approach developed by European forest policy scholars and international lawyers is that an identifiable international forests regime has emerged since the mid 1990s.[3]

This chapter develops the second approach. It argues that a distinct forests regime has evolved that is gradually being expanded and strengthened as new areas of agreement emerge. It is founded on three sources. First, there are hard legal instruments. We include here multilateral environmental agreements with a forest-related mandate and human rights conventions that apply to forest-dwelling peoples, principally ILO Convention No 169 of 1989 (Convention Concerning Indigenous and Tribal Peoples in Independent Countries). Since the hard provisions of the regime are found under different legal covers, there is an overlay between the forests regime and other international regimes. Second, there is soft law on forests: the UNCED Forest Principles (1992); Chapter 11, 'Combating deforestation,' of Agenda 21 (1992); the IPF proposals for action (1997); the IFF proposals for action (2000); and the UNFF resolutions (since 2002). States appear to have opted deliberately for soft forest law, which contains political rather than legal commitments, in order to stress that although agreement on forests is incomplete, there should, nonetheless, be guiding principles.[4] Finally, the regime embraces private international law on forests For example, the legal chain of custody of the FSC and FSC forest management principles have status in private contract law.

Convention on Biological Diversity

Of the hard legal instruments that contribute to the regime, the CBD is the most important. So central is the CBD to forest use that some states have suggested that if there is to be an international legal forests instrument, it should be a protocol to the CBD rather than a separate convention. The three objectives of the CBD are 'the conservation of biological diversity, the sustainable use of its components and the fair and equitable sharing of the benefits arising out of the utilization of genetic resources' (Article 1). There has always been a delicate balance between these objectives, which embrace economic, social and environmental goals. The conference of the parties to the CBD meets approximately every two years, while the smaller Subsidiary Body on Scientific, Technical and Technological Advice (SBSTTA) meets between conferences. The main political dynamic in CBD negotiations is between developed countries and the biodiversity-rich countries, the so-called like minded megadiverse countries (LMMCs). There are 17 LMMCs; Bolivia, Brazil, China, Colombia, Costa Rica, Democratic Republic of the Congo, Ecuador, India, Indonesia, Kenya, Madagascar, Malaysia, Mexico, Peru, the

Table 9.1 *The world's 25 biodiversity hotspots by region*

North and Central America	***Africa***
Caribbean	Madagascar and Indian Ocean Islands
California Floristic Province	Eastern Arc Mountains and Coastal Forests
Mesoamerica	Guinean Forests of West Africa
	Cape Floristic Region
South America	Succulent Karoo
Tropical Andes	
Choco–Darien–Western Ecuador	***Asia–Pacific***
Atlantic Forest	Philippines
Brazilian Cerrado	Sundaland
Central Chile	Wallacea
	South-west Australia
Mainland Asia	New Zealand
Mountains of south-west China	New Caledonia
Indo–Burma	Polynesia and Micronesia
Western Ghats	
	Europe and Central Asia
	Caucasus
	Mediterranean Basin

Source: Conservation International website, Biodiversity Hotspots, www.biodiversityhotspots.org/xp/Hotspots (accessed 23 September 2004).

Philippines, South Africa and Venezuela. The LMMC group does not, however, include all the world's main biodiversity hotspots (see Table 9.1).

In 1998, the CBD adopted a programme of work on forest biodiversity to promote conservation, sustainable use, and access and benefit-sharing.[5] This programme was expanded in 2002. There are 12 goals to the expanded programme of work (see Box 9.1). There is considerable overlap between the IPF/IFF proposals for action and the CBD's Expanded Programme of Work on Forest Biodiversity. However, the proposals for action tend to straddle the international and national levels and are abstract in places, whereas the expanded programme of work bridges the national and sub-national levels, with more of an emphasis on practical action.

In 2002, the sixth conference of parties to the CBD adopted a strategic plan 'to achieve by 2010 a significant reduction of the current rate of biodiversity loss.'[6] What constitutes a 'significant reduction' is not defined; like the UNFF and the FLEG processes, the CBD has been unable to agree any quantified targets (see Chapters 5 and 7).

As a system of global governance, the forests regime – a mix of public and private, and hard and soft, provisions that provide an embryonic system of rights and obligations for states and other actors – is simultaneously coherent and fragmented. Coherence is achieved through a legal 'spillover effect.'[7] Any body of law, including soft law, can be precedent-setting, and principles adopted in one legal instrument may subsequently influence others. A principle may thus find expression in several legal codes. The fragmentation of the regime has two causes. First, the existence of several international institutions with forest-related mandates inevitably leads to inefficiencies, gaps and duplications. This fragmentation has, to some extent, been addressed by the Collaborative Partnership on Forests (CPF) (see Chapter 5). However, international organizations can only be coordinated by a mechanism such as the CPF when they share similar values, and then only to a limited degree. Without a shared value base, coordination is difficult, if not impossible. The CPF cannot eliminate all of the uncertainties and differences between its member organizations. This leads on to the second source of fragmentation, which has its origins in broader patterns of global governance: the regime seeks to secure both the long-term viability of the forest resource base and the continued exploitation of forests. Hence, while the regime promotes long-term forest public good enhancement, it also promotes continuing private good exploitation, including through new market mechanisms and new intellectual property rights that reflect neoliberal assumptions. A deep driver of deforestation is that international law promoting neoliberal values has been ascribed more coherence and greater normative force than international law promoting forest public goods (see Chapter 10).

In previous chapters we have seen that international forest institutions, despite their deficiencies and failures, have generated a measure of agreement on, for example, national forest programmes, criteria and indicators for sustainable forest management, forest certification and illegal logging. In this chapter we consider ten additional principles that guide actors' behaviour

Box 9.1 The 12 goals of the Convention on Biological Diversity's Expanded Programme of Work on Forest Biological Diversity

Programme element 1: Conservation, sustainable use and benefit-sharing

Goal 1: To apply the ecosystem approach to the management of all types of forests;
Goal 2: To reduce the threats and mitigate the impacts of threatening processes on forest biological diversity;
Goal 3: To protect, recover and restore forest biological diversity;
Goal 4: To promote the sustainable use of forest biological diversity;
Goal 5: Access and benefit-sharing of forest genetic resources.

Programme element 2: Institutional and socio-economic enabling environment

Goal 1: Enhance the institutional enabling environment;
Goal 2: Address socio-economic failures and distortions that lead to decisions that result in loss of forest biodiversity;
Goal 3: Increase public education, participation, and awareness.

Programme element 3: Knowledge, assessment and monitoring

Goal 1: To characterize and to analyse from forest ecosystem to global scale and develop general classification of forests on various scales in order to improve the assessment of status and trends of forest biological diversity;
Goal 2: Improve knowledge on and methods for the assessment of the status and trends of forest biological diversity, based on available information;
Goal 3: Improve understanding of the role of forest biodiversity and ecosystem functioning;
Goal 4: Improve the infrastructure for data and information management for accurate assessment and monitoring of global forest biological diversity.

Source: Secretariat of the CBD (2004) *Expanded Programme of Work on Forest Biological Diversity*, Montreal: Secretariat of the CBD.

in the forests regime. We examine the three guiding principles of the CBD: conservation; sustainable use; and fair and equitable benefit-sharing. We consider three principles that are central to indigenous peoples' rights: self-determination; free, prior and informed consent; and preservation of traditional knowledge. We also examine the principles of sustainable land management and the protection of endangered species. Two politically contentious principles are examined, namely the use of forests as carbon sinks, and trade liberalization of forest products.

The conservation of forest biodiversity

The CBD has established the principle of *in-situ* conservation: 'the protection of ecosystems, natural habitats and the maintenance of viable populations of species in natural surroundings' (Article 8(d)). The main mechanism for *in-situ* conservation is protected areas (Article 8(a)), a concept that is deeply embedded in the forests regime and other areas of global environmental governance. The concept of protected areas is embodied in the Convention on Nature Protection and Wild Life Preservation in the Western Hemisphere (1940), which emphasizes 'national parks,' 'national reserves' and 'strict wilderness reserves' (Article 1). Similarly, the emphasis in the Ramsar Convention on Wetlands of International Importance especially as Waterfowl Habitat (1971) on 'nature reserves' equates with protected areas. Globally, there are more than 100,000 protected areas of all types, including marine protected areas, covering approximately 10 per cent of the world.[8] The vast number of protected areas is necessary as incursions into nature have become so massive and routine.

No single international institution makes protected area policy. Protected areas formed the third theme of the G8 Action Programme on Forests and are central to the World Bank's forest policy (see Chapters 7 and 8). CBD protected area policy is being shaped continuously, much of it by institutions outside the CBD. A key actor in the preparation of the CBD's Programme of Work on Protected Areas was the IUCN. In 1994, the IUCN revised its categorization of protected areas, which had previously been conceived as state-controlled areas, to allow greater scope for alternative management and tenure regimes by local communities, indigenous peoples and the private sector (see Box 9.2). The IUCN World Commission on Protected Areas (WCPA) provides a policy advisory role to various actors, including the CBD. It comprises approximately 1000 leading specialists. Every decade since 1962, the WCPA has organized on behalf of the IUCN the World Parks Congress (formerly the World Conference on National Parks).

Local communities and protected areas

The exclusionary model of protected areas first developed during the 19th century in the US and subsequently adopted elsewhere has been criticized for denying the rights of indigenous peoples and local communities, and sometimes for requiring their forced relocation.[9] During the 1980s, Chico Mendes

Box 9.2 IUCN's Protected Areas Management Categories

Category Ia: Strict Nature Reserve: protected area managed mainly for science. Area of land and/or sea possessing some outstanding or representative ecosystems, geological or physiological features and/or species, available primarily for scientific research and/or environmental monitoring.

Category Ib: Wilderness Area: protected area managed mainly for wilderness protection. Large area of unmodified or slightly modified land, and/or sea, retaining its natural character and influence, without permanent or significant habitation, which is protected and managed so as to preserve its natural condition.

Category II: National Park: protected area managed mainly for ecosystem protection and recreation. Natural area of land and/or sea, designated to:

(a) protect the ecological integrity of one or more ecosystems for present and future generations;
(b) exclude exploitation or occupation inimical to the purposes of designation of the area; and
(c) provide a foundation for spiritual, scientific, educational, recreational and visitor opportunities, all of which must be environmentally and culturally compatible.

Category III: Natural Monument: protected area managed mainly for conservation of specific natural features. Area containing one, or more, specific natural or natural/cultural feature which is of outstanding or unique value because of its inherent rarity, representative or aesthetic qualities or cultural significance.

Category IV: Habitat/Species Management Area: protected area managed mainly for conservation through management intervention. Area of land and/or sea subject to active intervention for management purposes so as to ensure the maintenance of habitats and/or to meet the requirements of specific species.

Category V: Protected Landscape/Seascape: protected area managed mainly for landscape/seascape conservation and recreation. Area of land, with coast and sea as appropriate, where the interaction of people and nature over time has produced an area of distinct character with significant aesthetic, ecological and/or cultural value, and often with high biological diversity. Safeguarding the integrity of this traditional interaction is vital to the protection, maintenance and evolution of such an area.

Category VI: Managed Resource Protected Area: protected area managed mainly for the sustainable use of natural ecosystems. Area containing predominantly unmodified natural systems, managed to ensure long term protection and maintenance of biological diversity, while providing at the same time a sustainable flow of natural products and services to meet community needs.

Source: IUCN (1994) *Guidelines for Protected Areas Management Categories*, Cambridge, UK, and Gland, Switzerland: IUCN

and other activists developed the idea of extractive reserves, a concept that sought to fuse nature conservation with the needs of local peoples. Extractive reserves are intended to protect the Brazilian Amazon from deforestation while providing local communities with the means for livelihood. The concept was developed by rubber tappers, but has since been applied to those whose livelihoods depend on other non-timber forest products, such as Brazil nut gatherers. IUCN's 1994 revision of its protected area management categories subsequently introduced the livelihood needs of communities into mainstream international protected area policy.

But despite this, conflict between nature conservation and local livelihoods continues to dominate protected area management. An indigenous peoples' representative argued at the Vth World Parks Congress in 2003 that protected areas have:

> *... resulted in our dispossession and resettlement, the violation of our rights, the displacement of our peoples, the loss of our sacred sites and the slow but continuous loss of our cultures, as well as impoverishment... First, we were dispossessed in the name of kings and emperors, later in the name of State development and now in the name of conservation.*[10]

Running throughout this debate are two different notions of proprietorship: land ownership rights recognized by the state through legal titles and deeds, and collective notions of property upheld by local communities that may be legitimate as customary ownership forms but which do not always enjoy legal status. The concept of community conserved areas has been developed to overcome this dualism by making indigenous and local communities integral to protected areas policy.

Recognition of community conserved areas

The concept of community conserved areas originated from within the IUCN.[11] Central to the concept is the notion that communities relate culturally to ecosystems and species, that community management results in long-term conservation, and that primary decision-making should rest with communities.[12] We now track the process that led to the adoption of this concept by the World Parks Congress and the CBD.

In 2002, the CBD established a technical expert group on protected areas, and encouraged collaboration between this group and the World Parks Congress. The member of this group promoting the rights of indigenous peoples was Grazia Borrini-Feyerabend, who in her work with the IUCN has been central to the development and promotion of community conserved areas.[13] Her influence helps to explain why the 2003 report of the CBD technical expert group contains several mentions of community conserved areas.[14]

In September 2003, the World Parks Congress meeting in Durban received a draft report from the CBD technical expert group. The Durban Accord and Durban Action Plan subsequently issued by the congress urged new approaches on protected areas, including community conserved areas.[15] Two months later, the SBSTTA met to discuss its recommendations for the seventh conference of parties to the CBD. It considered the outputs of the technical expert group and the World Parks Congress and recommended that the CBD should 'Establish policies and institutional mechanisms to facilitate the legal recognition and effective management of indigenous protected areas and community conserved areas.'[16]

The CBD's seventh conference of the parties held in Kuala Lumpur in February 2004 formally endorsed the idea of community conserved areas, which features prominently in the CBD's Programme of Work on Protected Areas, also adopted in Kuala Lumpur.[17] The CBD also decided to appoint a working group on protected areas. To place this in context, it should be noted that before the seventh conference only three CBD working groups had been established: on access and benefit-sharing, Article 8(j) of the convention and, for parties to the Cartagena Protocol on Biosafety only, on liability and redress on biosafety issues.[18] The proposal to establish the protected areas working group was made by the EU. Canada countered by proposing a working group on implementation.[19] Eventually, both proposals were agreed.[20]

With community conserved areas now adopted by the World Parks Congress and the CBD, indigenous and local communities will feature more prominently in protected areas policy.[21] While the six IUCN categories have not changed in a formal sense, alongside them a widely accepted governance innovation centred on local communities is now developing that will significantly affect protected area management in the future.[22]

While the CBD is the main international legal instrument on protected area policy,[23] another that is growing in importance is the 1972 World Heritage Convention, administered by UNESCO.[24] This convention, with a mandate that includes the conservation both of nature and of cultural monuments, has been activated by the international forest policy community as a tropical forest conservation mechanism, with the Center for International Forestry Research (CIFOR) playing a prominent role. In 2000, the convention's World Heritage List included 33 tropical forest sites covering approximately 2.5 per cent of the world's closed tropical forests.[25]

Protected areas play an important role in public good provision. But, as Brian Child has argued with respect to African parks, 'we must question if parks are providing the correct public goods. In many cases parks are, as a consequence of tradition rather than principle, providing values suited to developed rather than developing countries.'[26] Child argues that the wilderness values of many African protected areas are written into management plans by consultants from outside Africa, part of a 'techno-bureaucratic elite' representing developed world interests rather than local rural communities. Protected areas that seek to exclude local communities are, rightly, criticized; a strict exclusionary approach to protected areas may safeguard some of the public goods of forests, such as biodiversity habitat, but will prevent local communities from enjoying local public goods, such as local ancestral sites, as well as private goods of a rival nature, such as fuelwood. The significance of the notion of community conserved areas is not solely that it moves towards recognizing the rights of local communities and their role in protected area management. It should also allow for a fairer distribution of forest public goods so that the distant users of protected areas do not benefit at the expense of the proximate users.

Protected areas remain politically contested spaces. While environmental and human rights values have influenced protected area policy, so too have neoliberal precepts. Many governments are selling off protected areas to private owners, thus relieving the state of an obligation while generating a source of income. Privatization is generally opposed by NGOs. A study by the Global Forest Coalition reports that many governments have handed over protected areas to the private sector on the basis that this will lead to more cost-effective management. However, private sector managers often exclude local communities and indigenous peoples, both from land and from policy.[27] Privatization also leads to the removal of national resource management as a public policy issue. A Friends of the Earth report argues that privatizing biodiversity management 'risks subjecting critical biodiversity resources and ecological functions to the vagaries of market pressures and corporate control,' whereas biodiversity and natural landscape should be seen as 'public assets that the state should hold in trust for its people.'[28]

The sustainable use of forest biodiversity

Sustainable use is defined in the CBD as the use of biodiversity 'in a way and at a rate that does not lead to the long-term decline of biological diversity, thereby maintaining its potential to meet the needs and aspirations of present and future generations.'[29] States agree to integrate conservation and sustainable use within national decision-making and to encourage cooperation between the private sector and government.

In 2000, the fifth conference of parties to the CBD adopted the ecosystem approach, which 'promotes conservation and sustainable use in an equitable way.'[30] The ecosystem comprises 12 principles (see Box 9.3). Whereas the concept of community conserved areas holds that protected areas cannot be managed in isolation from local communities, the ecosystem approach holds that protected areas cannot be managed in isolation from other landscapes and ecosystems. As Achim Steiner of the IUCN noted after the Vth World Parks Congress, there have never been as many protected areas as there are now, yet species continue to be lost. In the past, Steiner notes, protected areas were seen 'as islands of protection in an ocean of destruction. We need to learn to look on them as the building blocks of biodiversity in an ocean of sustainable human development.'[31] Steiner's emphasis is consistent with the CBD's ecosystem approach.

Discussion has taken place at the CBD and the UNFF on the compatibility of sustainable use (in the CBD and the ecosystem approach) and sustainable forest management (SFM) (see Chapter 6).[32] In 2004, CBD delegates agreed that the ecosystem approach and SFM are complementary and that 'SFM can be considered a means of applying the ecosystem approach in forests,'

Box 9.3 The 12 principles of the ecosystem approach of the Convention on Biological Diversity

1 The objectives of management of land, water and living resources are a matter of societal choice.
2 Management should be decentralized to the lowest appropriate level.
3 Ecosystem managers should consider the effects (actual or potential) of their activities on adjacent and other ecosystems.
4 Recognizing potential gains from management, there is usually a need to understand and manage the ecosystem in an economic context. Any such ecosystem management programme should:
   (a) Reduce those market distortions that adversely affect biological diversity
   (b) Align incentives to promote biodiversity conservation and sustainable use;
   (c) Internalize costs and benefits in the given ecosystem to the extent feasible.
5 Conservation of ecosystem structure and functioning, in order to maintain ecosystem services, should be a priority target of the ecosystem approach.
6 Ecosystems must be managed within the limits of their functioning.
7 The ecosystem approach should be undertaken at the appropriate spatial and temporal scales.
8 Recognizing the varying temporal scales and lag-effects that characterize ecosystem processes, objectives for ecosystem management should be set for the long term.
9 Management must recognize that change is inevitable.
10 The ecosystem approach should seek the appropriate balance between, and integration of, conservation and use of biological diversity.
11 The ecosystem approach should consider all forms of relevant information, including scientific and indigenous and local knowledge, innovations and practices.
12 The ecosystem approach should involve all relevant sectors of society and scientific disciplines.

Source: CBD Decision V/6, 'Ecosystem approach,' Annex, Fifth Conference of the Parties, Nairobi, May 2000

although the two concepts apply at different scales: 'the ecosystem approach can be applicable over large areas (landscape level), while SFM has historically emphasized forest management-unit levels of work at typically small spatial scales.'[33] However, the relationship between the two principles has proved more contentious at the UNFF. Dissimilarities were noted at the UNFF's fourth session. For example, a delegate from the Republic of Congo noted that while timber extraction is central to SFM, it was not clear whether it is consistent with the ecosystem approach.[34]

Under Article 10(c) of the CBD, states undertake to 'Protect and encourage customary use of biological resources in accordance with traditional practices that are compatible with conservation or sustainable use requirements.'[35] The secretariat of the CBD has concluded that 'customary use can only effectively occur within the framework of *in-situ* conservation'[36] and that 'it would seem appropriate to provide for customary uses of biological resources in accordance with traditional cultural practices within national laws.'[37] The concepts of sustainable use, customary use and *in-situ* conservation should thus be seen as inextricably interlinked. In order to comply with the provisions of the CBD and to promote the customary use of forest biodiversity, the Forest Peoples Programme has emphasized the importance of secure land rights. Indigenous communities should receive land title that is 'inalienable, unmortgageable and not subject to distraint' and the 'title so issued should be registered in the national land cadastre.'[38]

The principle of sustainable use illustrates the importance of indigenous peoples in forest policy. Such peoples include the Saami of Scandinavia, the Inuit of Canada and Alaska and numerous forest-dwelling peoples in South America, Africa and Asia. We now consider three rights salient to indigenous communities that should also be seen as principles of the international forests regime: self-determination; free prior informed consent; and preservation of traditional knowledge.

Self-determination

The right to self-determination, not to be confused with the right to secession or independence, applies under international law to all peoples, including indigenous peoples. The same formulation of the principle appears in the United Nations International Covenant on Economic, Social and Cultural Rights and the United Nations International Covenant on Civil and Political Rights, namely that by virtue of the right of self-determination, peoples 'freely determine their political status and freely pursue their economic, social and cultural development.'[39]

This formulation also appears in the Draft United Nations Declaration on the Rights of Indigenous Peoples.[40] The preparation of this declaration was initiated in 1993 by the UN Commission on Human Rights. It has yet to be concluded, which indicates that indigenous peoples' rights remain politically contentious between states (although the 2005 World Summit of heads and states and government agreed to finalize the draft declaration 'as soon as possible'[41]). Article 29 of the draft states that 'Indigenous peoples are entitled to the recognition of the full ownership, control and protection of their cultural and intellectual property.'[42] This implies the demarcation of indigenous territories, with secure land tenure rights.

ILO Convention No 169 does not mention self-determination, although the convention is consistent with the principle, with Article 7.1 stating that:

> *The peoples concerned shall have the right to decide their own priorities for the process of development as it affects their lives, beliefs, institutions and spiritual well-being and the lands they occupy or otherwise use, and to exercise control, to the extent possible, over their economic, social and cultural development.*[43]

Free, prior and informed consent

The principle of prior informed consent has different applications in international law. In the 1989 Basle Convention on the Transboundary Movements of Hazardous Wastes and Their Disposal, it denotes that countries importing wastes should be informed of the chemical composition in advance, and they should give their consent to accept such material. Prior informed consent is mentioned in the CBD, which states that 'Access to genetic resources shall be subject to prior informed consent of the contracting party providing such resources' (Article 15.5). The emphasis in both these instruments is on consent between states.

A different version of the principle – free, prior and informed consent – is crystallizing in international human rights law. Free, prior and informed consent is the right of indigenous peoples to participate in decision-making on development activities that affect their lands, territories, resources and rights. Consent should be *free* (that is, freely given or withheld), *prior* (that is, before implementation) and *informed* (that is, based on a full understanding of how any development activities will affect traditional lands).[44] Among the bodies that have issued declarations consistent with this principle are the Inter-American Commission on Human Rights, the United Nations Sub-Commission on the Promotion and Protection of Human Rights and the now defunct United Nations Centre for Transnational Corporations.[45] One of the criteria for principle 3 of the FSC is consistent with free, prior and informed consent: 'Indigenous peoples shall control forest management on their lands and territories unless they delegate control with free and informed consent to other agencies.'[46] The World Commission on Dams explicitly endorsed free, prior and informed consent,[47] although the World Commission on Forests and Sustainable Development did not.

The World Bank has resisted entreaties from civil society groups to include the principle in the World Bank's operational policies on indigenous peoples.[48] An early version of the Draft Declaration on the Rights of Indigenous Peoples stated that indigenous peoples 'have the right to require that States obtain their free and informed consent prior to the approval of any project affecting their lands, territories and other resources.'[49] If accepted, this formulation could have subordinated government policy to indigenous peoples. It proved controversial to some states, and in negotiations held in 2003 the word 'obtain' was placed in square brackets, indicating that some states disputed it, and was replaced with 'seek,' a softer formulation that would require only that states try to obtain approval from indigenous peoples, but not necessarily to secure it.[50]

Preservation of traditional knowledge

This principle that the traditional knowledge of indigenous peoples should be respected and preserved was recognized in many of the UNCED outputs.[51] The strongest formulation of the principle appears in Article 8(j) of the CBD, where states agree to 'respect, preserve and maintain knowledge, innovations and practices of indigenous and local communities embodying traditional lifestyles.'[52] The principle has been reiterated by other international institutions, including the IPF, IFF and UNFF, although agreement on its application has not always proved possible (see Chapters 2, 4 and 5). The Convention to Combat Desertification of 1994 requires states to 'protect, promote and use in particular relevant traditional and local technology, knowledge, know-how and practices' (Article 18.2). While the Ramsar Convention on Wetlands of International Importance does not mention traditional knowledge, parties to the convention have agreed that wetlands management should 'incorporate the traditional knowledge and management practices of indigenous peoples and local communities.'[53]

While indigenous peoples' rights are endorsed in international law, they have often been elaborated in instruments that have yet to attract support from large numbers of states. For example, while ILO Convention 169 is frequently cited as an authoritative legal source by indigenous peoples, it has received just 14 ratifications.[54] The strongest standards are contained in the Draft Declaration on the Rights of Indigenous Peoples, although the normative force of this draft is limited given that it has not been finalized; furthermore, even if, as intended, it is adopted by the UN General Assembly, it will be a non-binding instrument.[55] The CBD, which illustrates that environmental law and human rights law are overlapping jurisprudences, is the principal international environmental instrument upholding indigenous peoples' rights, although its emphasis is on encouraging states rather than obliging them.

The role of indigenous peoples in the UN system was enhanced in 2002 with the creation of the Permanent Forum on Indigenous Issues (PFII), a subsidiary body of the ECOSOC. It comprises 16 people, eight elected by ECOSOC government delegates and eight nominated by indigenous peoples' groups. All members serve as experts, and not as representatives of governments or indigenous peoples. The forum's mandate is to make recommendations to the ECOSOC and to promote the integration of indigenous issues within the UN system.

Equitable sharing of the benefits from the use of forest biological resources

While the CBD provides for the fair and equitable sharing of the benefits from genetic resource use, it does not indicate a formula by which the benefits should be shared. A complication is that the history of this subject has seen different groups of states use different international institutions to promote their interests. In addition to an environmental institution (the CBD), there is also a trade institution (the WTO, which has responsibility for the Agreement on Trade-Related Aspects of Intellectual Property Rights, or TRIPS) and an intellectual property institution, namely the World Intellectual Property Organization (WIPO). WIPO is currently working towards a single internationally harmonized system of international law on patent rights. This could have ramifications for biological resource patents, and the harmonized system is likely to incline more towards the TRIPS interpretation of intellectual property than that of the CBD, which recognizes traditional knowledge (see Chapter 4).[56] WIPO has lost its former role as the leading international intellectual property rights forum and now plays a subservient role to TRIPS, such as assisting developing countries to become TRIPS compliant.[57] The IPF, IFF and UNFF have also taken an interest in this area. However, at the UNFF's fourth session, negotiations for a resolution on traditional forest-related knowledge collapsed since the G77 did not wish to agree a UNFF resolution that could later be invoked against developing country interests in access and benefit-sharing at the CBD (see Chapter 5). The UNFF is unlikely to play a future role in this issue area.

The international politics of benefit-sharing is thus played out between the CBD, WIPO and the WTO. The tripartite dynamics between these institutions reflect the different claims on who legitimately holds, and should benefit from, knowledge on the properties of biological resources. Under TRIPS, the benefits from biodiversity use accrue to patent holders. For a patent to be granted, an applicant must demonstrate that something new has been created and that no other party has developed the knowledge used in the creation. A patent will be denied if evidence exists in written form that what is claimed as original was previously known. In distinction, traditional knowledge is often passed verbally from generation to generation, hence no codified evidence may exist.[58] The TRIPS agreement, indigenous peoples' groups argue, favours corporations that patent traditional knowledge at the expense of the original knowledge holders.

The inclusion of equitable benefit-sharing in the CBD has brought intellectual property rights to the heart of international environmental policy, and elaborating the details of benefit-sharing, which cannot be considered in isolation from access rights to biodiversity, has occupied considerable time at CBD negotiations. Benefit-sharing is supported by two main sets of actors. The governments of biodiversity-rich countries wish the benefits from biodiversity use to flow to the national level of the country of origin, while indigenous peoples' groups argue that the benefits should flow to the local level. There is thus some measure of agreement on the principle, but not on its practical application.

The CBD has created two open-ended working groups relevant to this area, both of which allow NGO participation. In 1998, the working group on the implementation of Article 8(j), which is tasked with developing protection for the knowledge of indigenous and local communities, was created.[59] Two years later the working group on access and benefit-sharing (ABS) was created.[60] The second group was initially tasked with preparing a set of voluntary guidelines, which were adopted in 2002 as the Bonn Guidelines on Access to Genetic Resources and Fair and Equitable Sharing of the Benefits Arising out of their Utilization. The guidelines advise governments on the development of national access and benefit-sharing regimes to protect traditional knowledge and propose a prior informed consent system for access to biological resources, including 'an internationally recognized certificate of origin system as evidence of prior informed consent.'[61]

The Bonn Guidelines meet, in part, the demands of developing countries and indigenous peoples' groups. However, they are non-legally binding, do not present a serious challenge to the TRIPS agreement and provide no guidance on how the benefits from biodiversity use should be shared among claimant groups. The LMMC group, viewing the guidelines as the first stage of an evolutionary process, pressed for stronger provisions. Just six months after the adoption of the Bonn Guidelines, the biodiversity-rich states successfully negotiated into the plan of implementation of the 2002 World Summit on Sustainable Development a call for states to:

> *Negotiate within the framework of the Convention on Biological Diversity, bearing in mind the Bonn Guidelines, an international regime to promote and safeguard the fair and equitable sharing of benefits arising out of the utilization of genetic resources.*[62]

The seventh conference of the parties to the CBD agreed in 2004 that the negotiation of this regime will take place within the ABS working group, with input from the working group on Article 8(j). It is not clear what form the regime will take. The 'provider' countries, represented principally by the LMMC group, support a legally binding regime, possibly a protocol. The 'user' countries, such as the EU, Canada, Japan and Australia, have argued that a regime could comprise several different instruments operating at different levels.[63] Meanwhile, indigenous peoples groups, represented by the International Indigenous Forum on Biodiversity (IIFB), have emphasized that equitable benefit-sharing does not mean simply a slice of the profits from pharmaceutical and biotechnology corporations. Protecting traditional knowledge under existing arrangements such as TRIPS is seen by some NGOs as unethical since it would hasten the enclosure and commodification of such knowledge.[64] The IIFB has stated that the ABS regime could violate indigenous peoples' rights and that 'a precondition of the regime must be consistency with international human rights law.'[65] Many indigenous peoples' groups want the regime to recognize the rights of indigenous and local communities to decide who has access to their resources, who does not and what form any benefits should take.

Sustainable land management

The main international instrument for sustainable land management is the Convention to Combat Desertification (CCD), the objective of which is 'to combat desertification and mitigate the effects of drought in countries experiencing serious drought and/or desertification, particularly in Africa.'[66] Three of the convention's four regional annexes mention forests, including that for Africa, which notes the need to ensure 'integrated and sustainable management of natural resources, including ... forests ... and biological diversity.'[67] All CCD projects are in dry arid zones that are poor economically, and all are expected to address poverty alleviation. The sixth conference of parties to the CCD, held in Havana in 2003, agreed that the CCD secretariat would promote joint activities with the Tehran Process on Low Forest Cover Countries (see Chapter 4).[68]

We have argued that coherence in the forests regime is achieved by establishing conceptual connections across institutions. The following account on the status of forest and land issues within the Global Environment Facility (GEF) illustrates how language agreed in one forum can be invoked in others.[69] The October 2002 meetings in Beijing of the GEF Council and the GEF Assembly agreed to add land degradation as a focal area eligible for GEF funding (Box 8.1, Chapter 8).[70] The following year a draft of GEF Operational Programme 15 (OP15) on Sustainable Land Management was produced, but it made no mention of forests. During informal discussions on the margins of the May 2003 meeting of the GEF Council, US negotiator Jan McAlpine suggested to other delegates that, in line with the GEF decisions of the previous year, OP15 should include language on forests. Cameroon, Iran, Nigeria and the Philippines supported this position and the GEF Council subsequently agreed to amend OP15. The revised list of activities eligible for GEF funding includes:

> *improvement of forest health, controlling damaging invasive species, strengthening forest inventory, monitoring assessment and sustainable harvesting practices, [and] establishment of community woodlots to provide fuel wood as an alternative source to natural forests and woodland.*[71]

Six months later at the fourth session of the UNFF, McAlpine reported on behalf of the US delegation that the GEF now included a list of forest activities eligible for funding under OP15. However, the progress report[72] prepared for the GEF Council later that month (May 2004) made no mention of these activities, many of which appear in the UNFF's plan of action. McAlpine suggested that this was a failure of UNFF member states, which needed to consider how to integrate the UNFF plan of action into GEF activities.[73] The GEF observer at the UNFF, Kanta Kumari, responded that the US's comments would be relayed to the GEF Council.[74]

This sequence of events again illustrates that the decisions of one institution can spill into others. But spillover does not happen 'automatically.' It requires the exercise of human agency. In this case, the connections were drawn by an experienced negotiator with membership on US delegations to the GEF and UNFF. Without such vigilance, there would probably today be a lower profile for forest activities under the GEF's operational programme on sustainable land management.

The use of forests as carbon sinks

Forests play important roles in climatic regulation as carbon sinks and carbon sources. A forest acts as a sink when it takes up carbon dioxide from the atmosphere. When the carbon stored is released, through, for example, dieback or fire, a forest becomes a carbon source. The principle of enhancing the sink function of forests to offset a country's greenhouse gas emissions has legal status under the 1997 Kyoto Protocol of the Framework Convention on Climate Change (FCCC) that entered into legal effect in February 2005 following ratification by Russia, and under which states agree to the 'protection and enhancement of sinks and reservoirs of greenhouse gases not controlled by the Montreal Protocol ... promotion of sustainable forest management practices, afforestation and reforestation' (Article 2). The protocol allows countries to take into account 'removals by sinks' when calculating their net carbon dioxide equivalent emissions (Article 3.7). It establishes the Clean Development Mechanism (CDM) under which the Annex I countries (that is, developed countries and countries with economies in transition, most of which have agreed to reduce their greenhouse gas emissions) can offset against their emissions mitigation activities in developing countries. So, for example, if an Annex I country establishes a forest in another country, the carbon taken up by that forest can be credited against the emissions of the Annex I country.

Since the Kyoto Protocol was negotiated, the politics of negotiating the fine details of carbon sequestration forestry have proven complex and divisive. At the sixth conference in The Hague in 2000, the US, supported by Australia, Canada and Japan, argued that countries should be allowed to offset against their greenhouse gas emissions the sink functions of their existing domestic forests. The US demanded a 300 million tonne credit against its emissions. The UK attempted to act as a mediator between the US and the EU, and by the end of the negotiations had negotiated the US down to 75 million tonnes. However, the EU then refused to accept even the reduced US figure, and the negotiations collapsed without agreement.[75]

The strongest argument against using forests for carbon sequestration centres on the question of permanence. The notion of carbon sequestration holds that carbon that was previously stored underground and which has been extracted and burned as fossil fuels can be re-fixed in trees. But carbon that was stored under the Earth's surface existed in a stable state; it did not form part of the biosphere; in other words, those parts of the Earth's oceans, surface and atmosphere where life exists. Once this carbon has been extracted and burned as fuel, it enters the biosphere as carbon dioxide. It cannot be permanently fixed in trees. A carbon forest can, like any other forest, come under conversion pressures from the rural landless poor, from illegal loggers and so on. Even well-managed and well-protected carbon forests can be subject to natural pressures, such as fire, pest attack and storm damage. The viability of carbon forests as a long-term fix to greenhouse gas emissions is thus highly questionable. There will always be a substantial risk that carbon sequestered in trees will re-enter the atmosphere through social or natural causes. Carbon sequestration is thus a non-permanent solution that, it can be argued, violates the principle of intergenerational equity; future generations will either have to devote time and resources to re-fixing sequestered carbon that has been re-emitted, or suffer the climatic and environmental consequences.[76]

One issue that has arisen in international climate negotiations is additionality. When a country calculates its net carbon dioxide emissions, it can only count carbon stored in a new carbon forest or plantation when this is additional to what would otherwise have been stored. This leads onto the question of historical responsibility. For example, should a country that has deforested in the past be allowed to establish carbon plantations on former forestland and offset the sink value of the plantations against its greenhouse gas emissions? According to the World Rainforest Movement, most plantations have replaced natural forests, which are more effective as carbon sinks than plantations. Hence, plantations should be considered a net source of carbon rather than a sink.[77]

Additionality was debated at the ninth conference of parties to the FCCC in Milan in December 2003, dubbed the 'forests conference' since it addressed many forest-related items. The politics of the negotiations involved the buyers of carbon sequestration credits, namely the OECD countries, and the sellers, principally the G77. The OECD countries can be sub-divided into two groups. Australia, Canada, Japan and New Zealand strongly advocated forest sink projects while tending to ignore their socioeconomic and environmental ramifications. The EU, Norway and the so-called Environmental Integrity Group of Mexico, South Korea and Switzerland pushed for an agreement that would address climate change without leading to negative socioeconomic and environmental consequences.[78] Although the US is somewhat sidelined in the climate negotiations by its refusal to ratify the Kyoto Protocol, it continues to be a significant actor and is most closely associated with the first group.

The deal struck allows forest and plantation owners to sell carbon credits to polluting countries and businesses. Given that most forests in the developing world are publicly owned (see Chapter 1), carbon credits provide a potentially lucrative source of revenue for forest-rich governments, many of which, including Brazil, support the idea. Other supporters are Northern businesses, including the UK Confederation of British Industry, which sees carbon trading as preferable to an energy tax, and the Japanese Federation of Economic Organizations.[79] Northern governments can use carbon trading to gain a measure of control over the policies and forests of developing countries. In what can be seen as one of the world's first 'debt-for-carbon' swaps, the Canadian International Development Agency has agreed to write off US$680,000 of the external debt of Honduras in exchange for commitments to promote tree plantations in the country.[80] FERN has argued that carbon sequestration amounts to a new form of colonialism – 'CO_2lonialism' – which will tie up large areas of forest lands in the developing world so that developed countries can continue polluting.[81]

The Milan conference agreed that any emission reductions claimed under the CDM should be independently audited and certified. A distinction was drawn between temporary certified emission reductions (*t*CERs) and long-term certified emission reductions (*l*CERs).[82] However, this distinction overlooks the fundamentally non-permanent nature of carbon sequestration forestry. Sink accounting under the CDM remains controversial. There is scientific uncertainty as to how much carbon is sequestered by trees of different species, ages, sizes and so on. While young trees fix carbon, older trees may leach it. Global warming may lead to increased leaching of carbon from trees as temperatures rise. There is also the question of determining when carbon sequestration is due to human intervention in forests and when it is due to natural factors. Carbon accounting thus allows ample scope for future political disagreement.

The Milan conference failed to establish a direct link between the work of the CDM and the CBD. This was despite a decision by the fifth conference of parties to the CBD urging parties to the FCCC and Kyoto Protocol to ensure that carbon sequestration is 'consistent with and supportive of the conservation and sustainable use of biological diversity.'[83] The forests work of the CDM is focused on greenhouse gases to the almost total neglect of other forest values. Greenpeace has criticized the CDM for failing to prohibit GM trees, for allowing large-scale monoculture plantations such as eucalyptus and pine, for prioritizing a global forest function at the expense of local commons, and for failing to include provisions on the rights of indigenous peoples.[84] These concerns have been reiterated by the scientific community, with CIFOR urging that CDM projects should include mandatory social assessments and provide economic benefits to local communities, particularly the poor.[85]

The US now acts as a free rider in the global climate regime. Buying emission rights and funding carbon forests in other countries would prove expensive for the high polluting US. This, coupled with a disinclination towards any kind of multilateral commitment, led the US to repudiate the Kyoto Protocol shortly after Bush junior took office, with one member of the administration asking: 'Why should we pay a billion dollars to the Russian mafia to keep a car plant open in Chicago?' [86]

Protection of endangered species

The principle that endangered forest species should be protected through trade restrictions is provided for in the Convention on International Trade in Endangered Species of Wild Fauna and Flora (CITES). The principle has been tempered by sovereignty; CITES is authorized to restrict and monitor the international trade of species that face extinction, but has no mandate on trade within countries. There are three CITES annexes. Species listed on Appendix

I are banned from international trade. Appendix II species can be traded internationally with a CITES export permit. Such species are monitored to ensure that international trade 'will not be detrimental to the survival of that species.'[87] Appendix III species are monitored, although listing takes place through unilateral action by a range state, and not through a conference decision, as for Appendices I and II. The interest of CITES in forest species includes, in addition to tree species, plants and fauna.

CITES listing of tree species

CITES has a mixed record on listing tree species. The first Appendix I tree species was Parlatore's podocarp (*Podocarpus parlatorei*), listed in 1974. The following year the first Appendix II listings were made, namely Aji (*Caryocar costaricense*), Holywood lignum vitae (*Guaiacum sanctum*), Caribbean walnut (*Oreomunnea pterocarpa*) and Mexican mahogany (*Swietenia humilis*).[88] There were no new listings in the 1980s (although some species moved between appendices). In 1992, CITES was reactivated as a tree conservation mechanism with the Appendix I listing of Brazilian rosewood (*Dalbergia nigra*). That year, three listings on Appendix II were made; Commoner lignum vitae (*Guiacum officinale*), Afromosia (*Pericopsis elata*) and American mahogany (*Swietenia mahagoni*).[89] In 1994, CITES created a Timber Working Group, which includes representation from the International Tropical Timber Organization (ITTO), to provide recommendations on listings.

Throughout the 1990s, concern grew about the long-term sustainability of populations of Bigleaf mahogany (*Swietenia macrophylla*). The high quality, durability and beauty of the species make it highly sought after. The range in South America has become increasingly fragmented in Brazil, Bolivia and Peru leading to growing CITES interest. The Costa Rican population of Bigleaf mahogany was listed on Appendix III in 1995.[90] After a proposal from Brazil to list the species on Appendix II was rejected in 1997,[91] the Brazilian and Bolivian governments added their populations to Appendix III in 1998. In 2000, the 11th conference of parties to CITES accepted a Brazilian proposal to create a Mahogany Working Group, which, like the CITES Timber Working Group, includes ITTO representation. The group noted discrepancies in the trade data of Bigleaf mahogany between importing and exporting countries, which suggested a significant illegal international trade.[92] CITES finally agreed to list Bigleaf mahogany on Appendix II in 2003. This was the first time a high-volume, high-value tree species was listed on Appendix II. Previous listings were for rarer species traded in lower volumes. In 2004, another high-volume, high-value species – Ramin (*Gonystylus* spp), a species targeted by illegal loggers in Indonesia – was listed on Appendix II.[93]

CITES listing of forest fauna

The forest-related work of CITES encompasses species other than trees, including forest-dwelling fauna. All surviving species of tiger and rhinoceros were placed on Appendix I at the first CITES conference. Of the great apes, the gorilla, chimpanzee and bonobo (all Africa) and the orang-utan (Southeast Asia) are also listed on Appendix I. Many of the threats to the great apes are the same as those to forests, such as commercial and illegal logging, road building, increased migration and land clearance, with the net result being severe fragmentation of the range.[94] In Africa, the great apes face an additional threat: the demand for bushmeat, which is an important protein source in the Congo Basin. Bushmeat is not only consumed by rural populations. People living in the urbanized areas of Central and Western Africa often prefer bushmeat to agricultural meat, as it is reminiscent of traditional village life.[95] Logging operations attract labourers to the forests, many of whom prefer bushmeat as food. The role that CITES can play in addressing the bushmeat crisis is limited since its mandate is restricted to international trade, and most bushmeat is consumed within range countries.

The CBD has established a liaison group on non-timber forest products, including bushmeat. This group includes CITES representation.[96] CITES–CBD cooperation has increased since the signing of a memorandum of cooperation between the secretariats of the two conventions in 1996. This commits the secretariats to encouraging 'effective conservation and promoting the sustainability of any use of wildlife as part of the biological diversity of our planet.'[97] By adopting the language of sustainability and biological diversity, CITES has moved conceptually closer to the CBD. Although the two organizations have different mandates – CITES focuses on individual species, whereas the CBD has an ecosystem approach – their roles are complementary. The signature of the memorandum has resulted in the involvement of the CITES Plants Committee in the CBD's Global Strategy for Plant Conservation.[98] But despite overlapping membership – most states that have ratified one convention have ratified the other – and despite their mutually supportive mandates, CITES-CBD collaboration has resulted only in the modest linking of work programmes. Cooperation takes place principally at the secretariat level rather than on project implementation.

Trade liberalization of forest products

International trade has always been a controversial subject in environmental politics. While trade can provide an important means of livelihood for forest communities, unregulated access to forests by domestic businesses and transnational corporations has undermined and degraded local common regimes, impoverished communities and resulted in the erosion and destruction of forest public goods.[99]

CITES is one of the few multilateral environmental agreements to allow restrictions to international trade on environmental grounds. Another is the Montreal Protocol on ozone depletion, which aims to phase out the consumption and use of chlorofluorocarbons (CFCs) and other ozone-depleting substances. However, the compatibility of the trade restriction measures of CITES and the Montreal Protocol with the WTO agreements, which promote international trade liberalization through the removal of tariff and non-tariff barriers, is unclear. Neither the WTO nor, before the WTO's creation in 1995, GATT has challenged the trade restrictions of a multilateral environmental agreement. Were a challenge to be made, trade restrictions on environmental grounds could be ruled in breach of international trade law.

The CBD contains no trade restriction measures, although some issues have a trade-related aspect, such as access and benefit-sharing.[100] The CBD stipulates that the convention 'shall not affect the rights and obligations of any contracting party deriving from any existing international agreement, except where the exercise of those rights and obligations would cause a serious damage or threat to biological diversity' (Article 22.1). This implies that the CBD could implement a trade restriction measure if not to do so would seriously damage or threaten biodiversity. But such a measure could be ruled WTO illegal. Where there is an inconsistency between two pieces of international law, the most recent instrument, which has post-dated the older instrument, usually prevails. The WTO agreements post-date most multilateral environmental agreements, including the CBD.[101]

At the end of the 1990s, work was initiated on a WTO forest products agreement, which would have promoted trade liberalization through tariff reduction on forest products. The American Forest and Paper Association has long advocated the elimination of tariffs on forest products by US trading partners in order to improve the market opportunities for the US forest industry in these countries,[102] and the US government was a mover behind the proposed agreement.[103] Throughout 1999, NGOs lobbied against the proposed agreement, arguing that tariff reduction would lead to increased logging and deforestation. To date, no WTO forest products agreement has been concluded. The lack of transparency with which the WTO operates makes it difficult to assess why this is so, although the negotiations for the agreement appear to have been abandoned, in part, due to the Seattle street demonstrations at the WTO 1999 ministerial conference, since when the WTO agenda has been dominated by other issues, particularly agricultural subsidies.[104] But tariff elimination on forest products has not been defeated as an idea. The US is pursuing tariff liberalization on forest products and other natural resources under the Free Trade Area of the Americas (FTAA) agreement.[105]

Expanding the international timber trade and conserving the forest resource base can be seen as mutually exclusive, although both objectives have appeared in the International Tropical Timber Agreements negotiated in 1983, 1994 and 2006. Of these two objectives, trade expansion has proved the strongest at the ITTO. In 1990, the WWF urged the ITTO to seek a waiver from any GATT clauses incompatible with forest conservation. However, the ITTO producer countries have opposed such a measure. International trade law prohibits discrimination between 'like products' on the basis of the process and production methods used in their manufacture, which means that states cannot discriminate between timber imports from sustainably managed forests and those from unsustainable sources, such as clear-felled forests. Under international trade law the principle of equal market access for 'like products' takes precedence over environmental degradation.[106] This is consistent with the neoliberal logic of removing barriers to international trade, but acts against sustainable forest management.

Amending international trade law to allow discrimination against timber from unsustainable sources would face both technical challenges, such as agreeing a clear definition of sustainable forest management, and political challenges, particularly from states with an important forest products sector. A further complication is that a trade restriction provision designed to protect the environment could be used in a protectionist way; a state could ban timber imports from a country, claiming that the timber was harvested from 'unsustainable sources' when its real intention is to exclude timber that would compete with domestic industries. Any system allowing discrimination between like products would thus need clear criteria for distinguishing between environmental protection and environmental protectionism.[107]

Rather than addressing such issues, the WTO has avoided them, thus sidelining environmental considerations in international trade law, which has a stronger normative force than international environmental law in two respects. First, most states attach a higher priority to trade than to the environment. International law on the former is invoked more frequently than that on the latter, and international trade institutions receive better resourcing from states than international environmental institutions. International law is thus *de facto* stronger than international environmental law. Second, states have provided the WTO with the authority to require changes to national law consistent with WTO rules on pain of sanctions. No environmental treaty has such powers.[108] Trade law is thus *de jure* stronger than international environmental law. The strength of the WTO in relation to international environmental law forms an integral part of what Stephen Gill has termed disciplinary neoliberalism, namely the ascendancy of the rights of businesses and investors over other rights, such as human rights or the right to a clean environment.[109] Robyn Eckersley develops the concept of disciplinary neoliberalism when arguing that conferences of parties to environmental agreements have become increasingly self-censorious, avoiding measures that may not survive a WTO challenge and opting only for 'cool' interpretations of any trade restriction measures on environmental grounds.[110] States favouring minimal trade restrictions on environmental grounds, principally the US, Australia and many developing countries, have prevailed over those states that would grant more flexibility, such as the EU, Norway and Switzerland.[111]

Concluding thoughts

The international forests regime overlaps with other international environmental regimes, particularly the CBD. However, in the absence of a forests convention the consensus on forest-related issues is fragmentary and incomplete. Several international institutions deal with the different dimensions of forest conservation and use, and obvious connections between these institutions are often not made. International forest policy-making remains scattered among an array of institutions, and the parties to one legal instrument are not bound in any formal or legal sense by the decisions of any other. Hence, the Framework Convention on Climate Change, which is the most 'isolationist' of the multilateral environmental agreements, has paid no heed in its work to the CBD, despite entreaties from the CBD's secretariat. Numerous calls from government delegates to the UNFF for the various forest-related institutions to 'exploit areas of synergy' have gone largely unheeded. The result has been a creeping *ad hoc* incrementalism. The international forests regime is disconnected and multicentric; it has developed at different speeds and in different directions, rather than strategically and holistically along a common front.

To develop this point we may recall the very different principles of the forests regime. Examples of environmental principles include the conservation of biological diversity and the protection of endangered species. Human rights principles include self-determination and free, prior and informed consent. Some regime principles are grounded in both environmental and human rights norms, such as sustainable use and the preservation of traditional knowledge. All of these principles have a strong public goods element. Others promote the private goods that forests provide. The trade liberalization of forest products and the contentious principle of access and benefit-sharing each aim explicitly at the continued exploitation of forests for private gain. These principles are not constrained by the conservationist and human rights principles of the regime (as in, say, access to biological resources may be granted only when effective biodiversity conservation measures are in place and when the right to free, prior and informed consent is respected). The result is that those regime principles promoting private goods compete directly with, and are unchecked by, those that promote public good provision. This reflects a broader and more fundamental tension in global governance, with different international institutions operating largely in isolation from each other while promoting different, and sometimes conflicting, values and objectives. Examples include the conflict between the WTO and multilateral environmental agreements, and between those international institutions that promote different intellectual property rights. We have argued that where such conflicts occur, those international institutions promoting neoliberal objectives will prevail. In the final chapter we develop this argument in greater depth.

10

The Crisis of Global Governance

The preceding eight chapters have documented the failures, and few successes, of global forest politics. There have been some notable developments, including the activation of CITES to restrict the international trade in endangered tree species, the Low Forest Cover Countries initiative and the Forest Stewardship Council. But many of the negotiated outputs on forests, such as the IPF and IFF proposals for action and the UNFF resolutions, are weak and ambiguous. And there have been significant failures, such as the World Commission on Forests and Sustainable Development and the inability of states to agree a convention to promote sustainable forest management. This does not augur well for the future of the world's forests and suggests a more fundamental political problem. But to condemn states and intergovernmental organizations would be to miss the heart of the problem. The problem – of deforestation and of democratic governance – is the penetration of publicly accountable organizations, and of nature and common resources worldwide, by business corporations. This has been facilitated by a neoliberal economic order dedicated to promoting the expansion of capital into new spaces. The bulk of this chapter develops this argument before advocating an alternative model of global governance. A fundamentally different type of politics is proposed, one that is founded on a vibrant public domain and the democratization of politics at all levels of global governance.

A crisis of public accountability

Much of the problem lies with decision-making procedures at the UN. After the 1992 UNCED, it took three years to build up sufficient confidence for regular intergovernmental forest meetings to be held. No delegation has been prepared to risk this fragile consensus. The unwritten rules of diplomacy mean that the environmentally destructive policies of governments and their corporate allies are rarely criticized, except in veiled form. There has been no sustained dialogue on the deep political, social and economic drivers of forest degradation. To compound the problem, policy responses have been framed within, and thus delimited by, the core assumptions of neoliberalism.

Few delegates are genuinely satisfied with UN forest-related institutions, although few express their frustrations in formal government interventions.

Many delegates are, as individuals, committed to forest conservation; but all, in one way or another, must work to governmental instructions. In informal discussions many express personal support for positions that they cannot endorse as delegates.[1] The unwillingness of delegates to take important issues to a vote has empowered intransigent states. Decisions have been made by consensus; while it needs all to say yes, it takes only one to say no. The result has been a diluted politics of the lowest common denominator, resulting in suboptimal outcomes.

From this it might be concluded that politically accountable bodies – such as intergovernmental organizations and the states that comprise them – can no longer govern in the common interest. Such a view might be articulated like this. Due to economic globalization, intergovernmental organizations and states can no longer control transnational economic processes, with the result that their capacity to govern effectively is weakened. New governance mechanisms in which non-state actors are central should be created to fill this vacuum. Indeed, it can be argued, a restructuring of global governance is already under way. John Ruggie argues that a reconstituted global public domain is emerging, 'an increasingly institutionalized transnational arena of discourse, contestation and action concerning the production of global public goods, involving private as well as public actors.'[2] What Paul Wapner has termed a global civic politics is emerging in which civil society players act in the public interest.[3] In this new public domain, partnerships between a diversity of actors drive global public standard-setting.

Such dynamics are apparent in global forest governance. While the regional FLEG processes have initially been driven by states, a central role is envisioned for business and NGOs in promoting governance reforms to tackle illegal logging. Business and NGOs have been the driving forces behind forest certification. The FSC has garnered several advantages by prohibiting governments from membership. By excluding governments, the FSC forestalls allegations that it is a tool of particular governmental interests. Since it cannot be construed as an intergovernmental organization, it has so far avoided charges that it is illegal under international trade law. Finally, the absence of governments prevents the FSC from becoming another forum dragged down to lowest common denominator politics.[4] Significantly, some governments have successfully submitted public forests for FSC certification (see Chapter 6). There is a paradox here: NGOs and business are producing new rules that are then adopted by public authorities as public standards. In some respects, the state is now a taker, rather than the maker, of standards.

But a question that needs to be asked is whether a politics in which the state and intergovernmental organizations lack the willingness, the capacity or both to govern in the public interest, leaving a void that other actors must fill, is desirable. This chapter will argue that it is not, and that despite the fundamental weaknesses of international forest policy which this book has documented, governmental and intergovernmental politics, far from being abandoned, should be reclaimed, revitalized and democratized.

Neoliberalism and deforestation

Neoliberalism, which is dedicated to the promotion and expansion of global capital, has proved itself an adaptable and resilient ideology. Neoliberalism has not rejected environmental ideas *per se*, although it has rejected those that challenge its core assumptions, such as privatization, reduced state regulation, voluntary governance and market solutions. According to Steven Bernstein's theory of liberal environmentalism, ideas are selected according to their fitness with liberal norms. Ideas with a poor fit are rejected, as with the radical challenge of deep green ecologism, or reshaped to render them acceptable. Liberal environmentalism is thus the result of environmental ideas interacting with the norms of a liberal world economic order, what I refer to as neoliberalism.[5] To Bernstein, states have failed to agree on the need for a forests convention since the regulatory nature of such an instrument would conflict with key norms of liberal environmentalism.[6]

Neoliberal policies have failed to halt deforestation as they have both failed to address its root causes and, by supporting the expansion of global capital, have promoted further deforestation. Bernstein's theory helps to explain the penetration of the forests regime by neoliberal principles, such as promoting international trade in forest products and enhancing private sector forest investment. These principles feature prominently in the IPF and IFF proposals.[7] They have legitimized increased corporate access to forests. Intergovernmental organizations which, it will be recalled, are public institutions, now play an increasing role in promoting private sector investment. In 2003, the World Bank invited 150 senior executives to a forest investment forum to explore opportunities for forestry investments.[8] In 2006, the International Tropical Timber Organization (ITTO) co-hosted an international tropical forest investment forum with private sector companies prominently represented.[9]

Most global environmental problems, such as acid rain, global warming and deforestation, are caused by what Julian Saurin terms the 'normal and mundane practices' of modern capitalism, such as industrial production and natural resource consumption.[10] Environmental problems are not, therefore, exceptional or accidental; they are the cumulative result of routine social actions. Deforestation is generated by the massive daily consumption of pulp, paper, timber and agricultural produce, the production of which places severe pressure on forest spaces. Deforestation is thus a crisis that is driven by global capitalism.

Deforestation is also exacerbated by crises *of* global capitalism, when the viability and legitimacy of the capitalist system itself is questioned. Such crises occur during periods of dynamic economic disequilibrium leading to falls in financial and currency markets that can be controlled only with huge interventions from international financial institutions, if they can be controlled at all.[11] The immediate crisis is then followed by a period of economic depression and unemployment. Most national economies are affected in varying degrees. Since World War II, core countries have been able to insulate themselves from the worst effects of capitalist crises by using IMF and World Bank structural adjustment policies that emphasize debt repayment and ensure that the costs of economic adjustment fall principally on peripheral countries.[12] Peripheral countries cannot easily ignore the IMF, which sets the international credit ratings that determine whether a country can raise loans from the international financial sector.[13] The most recent crisis, caused largely by speculation, was the capital flight from Indonesia, Malaysia, Russia and South Korea in 1997 and 1998, leading to sharp falls in currencies followed by economic depression.

One result of the 1997 Indonesia crisis was an increase in deforestation. The IMF recommended that Indonesia earn hard currency to repay its debt by restructuring its forest sector. This led to the harvesting of younger trees and of more tree species.[14] The IMF letter of intent to the Indonesian government also recommended removing 'all formal and informal barriers to investment in palm oil plantations.'[15] This led to increased forest fire burning in Kalimantan and Sumatra to clear forestland for palm oil plantations.[16] Friends of the Earth argued that in order for Indonesia to repay its debts to international bankers, the costs of economic adjustment fell on nature and the poor.[17] Victor Menotti subsequently suggested a causal relationship between falling currencies and falling trees.[18]

The Suharto government, which was heavily involved with unscrupulous timber barons, fell during the crisis, and the subsequent democratization of Indonesia has weakened the patron–client relations between forest corporations and politicians, although it has not eliminated them completely. Indeed, close corporate–political relationships characterize most countries. In the developed world, G8 politicians have pressed for the opening of state-administered tropical forests to private sector investment (see Chapter 1). One way in which they have done so is by promoting public–private partnerships between tropical governments and forest corporations. There are several mentions in the IFF proposals of 'public–private partnerships,' a concept that has assumed a prominent place in neoliberal discourse and which captures the idea that private money should be mobilized to replace declining public expenditure.[19] However, the concept is problematic with respect to democracy. It assumes shared interests and equality between the public and private sectors, whereas in a democracy public authorities should regulate and monitor business. This requires a hierarchical relationship in which the public sector oversees the private sector.[20]

The fact that corporations have attained a position of equality with states, or at least that the idea is intelligible and widely accepted, indicates the enormous economic power that is concentrated in corporations. The per capita consumption of forest products in developed countries far exceeds that in the developing world. To Kees van der Pijl, deforestation is explicable by the structural dependence of developed countries on transnational timber and paper corporations.[21] Neoliberal economic policy has enabled the growth of a forest industry oligopoly based in the US, where forest industry corporations have grown enormously following successive mergers and acquisitions. The largest forest sector corporation in terms both of annual revenue and annual wood consumption, International Paper, has expanded through acquiring Hammermill Paper in 1986 and merging with the Federal Paper Board in 1996.[22] In 1995, Scott Paper merged with Kimberley-Clark, the world's largest tissue paper manufacturer, which owns the Kleenex brand. In 1998, Jefferson-Smurfit and Stone Container merged to yield Smurfit-Stone. In 1999, Weyerhaeuser acquired Macmillan Bloedel. All of these forest industry corporations, which are among the world's largest, are members of the AFPA, an influential trade association that enjoys close contacts with the US State Department and a seat on US delegations to international forest negotiations. The increased influence of corporations in international negotiations has led to what may be termed the 'privatization' of the United Nations, in which international agreements reflect the preferences of the business sector.[23] The International Tropical Timber Agreement of 2006 can be interpreted in this light, with most delegations to the ITTO including timber trade advisers.

If forest corporations are growing in size and political influence, then how can local communities regain democratic control over them? Before we can approach this question we first need to examine the foundations of corporate power.

The legal basis of corporate power

The modern business corporation is a fairly recent type of organization that was conceived to serve public needs. The legal right of a corporation to exist and raise money through issuing shares is provided by public authorities in a charter. In principle, corporations that fail to meet the public needs for which they were created can be dissolved. In 1886, a US court case, *Santa Clara County v. Southern Pacific Railroad*, ruled that a corporation should be considered a legal person with the same constitutional rights as a US citizen.[24] Since this ruling a jurisprudence has developed, first in the US and then globally, that defines and upholds corporate rights. The sole responsibility of the corporation – its fiduciary duty – is to act in the interests of its shareholders. This is usually interpreted as the maximization of shareholder value. In principle, the shareholders own the corporation and have the authority to set policy. In practice, there is no genuine owner authority of the corporation. Corporate charters are granted easily on payment of a fee and are rarely revoked, even for corporations that break the law.

Neoliberalism has enabled corporations to gain ownership of previously public forests for profit, while deregulation has freed corporations from public oversight. The combination of these two processes has fuelled deforestation, which is a symptom of a broader pattern of commons enclosure, both of land and, through patents, of biological resources. David Bollier documents in *Silent Theft: The Private Plunder of our Common Wealth* how corporations are penetrating a range of formerly public spaces, including common water resources in developing countries, publicly funded academic research, mineral resources on public lands and public sector broadcasting.[25] As David Harvey has argued, the neoliberal agenda has led, in effect, to commons enclosure becoming an objective of the state.[26] In the neoliberal era the aims of the corporation and the state have become fused.

The case of the US energy corporation Enron illustrates this. In 1995, Enron lobbied the US Congress for further privatization and the curtailing of state regulation.[27] Enron was a major donor to Bush junior's election campaigns for the Texas governorship. Bush subsequently endorsed a state energy deregulation bill for which Enron had lobbied. Enron then donated to Bush's first presidential campaign. After his election, Enron pressed Bush to appoint a new head of the Federal Energy Regulation Commission after the previous head refused Enron's entreaties for further deregulation. Policies sought by Enron were included in the recommendations of an energy task force convened by Vice President Cheney.[28] Enron had a poor environmental record. It promoted deforestation in the Amazon by co-funding a pipeline and service roads through Bolivian and Brazilian forests. Enron opposed the rerouting of the pipeline around ecologically sensitive areas on cost grounds.[29] The corporation collapsed bankrupt in 2001 following systematic corruption and embezzlement. While Enron's spectacular collapse was exceptional, in other respects the corporation was typical. The lack of attention to Enron's environmentally destructive activities from its shareholders is emblematic of the low priority that financial investors attribute to environmental degradation. The lack of public oversight of the corporation is again typical. Indeed, public accountability of Enron only became an issue after its collapse, and then only because investors lost money due to depressed US stock prices.[30]

The lobbying of federal politicians for corporate-friendly policies is a daily occurrence in Washington, DC, and the Bush–Enron relationship is typical of the mutual support that US politicians and corporations give each other. The Bush administration's domestic forest policy, the Healthy Forests Initiative, can also be explained in terms of corporate influence. Its stated aim is to reduce the risk of fire on public lands by selective logging.[31] The Sierra Club, a leading American NGO, charges that the initiative will promote private logging of wilderness forests that are miles from communities at risk from fire, and that the fire issue is being used 'to cut the public out of the public lands management decision-making process.'[32] Private logging in US public forests dates back 200 years. Richard Behan argues that the *Santa Clara County v. Southern Pacific Railroad* case facilitated increased exploitation of public natural resources by US timber and mining corporations.[33] Many private logging operations have been directly subsidized from public funds, encouraging logging that would otherwise be unprofitable and, in effect, promoting the conversion of national forests into corporate wealth. A hidden form of subsidies is forest road building to enable better access for logging companies. In a five-year period during the 1990s, over US$387 million of US public funds was spent building forest roads.[34]

The globalization of corporate power

Private investment in forests is driven, in part, by increased demand for forest products and, in part, by the need of corporations to secure new investment opportunities. The instability in global financial markets can be explained by an increasingly desperate search on the part of organized capital for new areas of investment to soak up surplus investible funds in order to maintain the value of stock capital.[35] The neoliberal state has aided this process by opening new spaces for private investment through privatization. To further facilitate investment, core countries used the OECD during the late 1990s as the forum for negotiating a multilateral agreement on investment (MAI). The first draft of the MAI was prepared by the US Council for International Business, founded in 1945 to represent US corporations within the UN.[36] The intent behind the MAI was to internationalize and strengthen the rights that corporations enjoy in developed countries, especially in the US, and to empower corporations to sue governments if these rights were breached. The rights proposed in the draft MAI were so blatant that they alienated normally pro-business politicians, and the MAI foundered after some US politicians opposed it asking: 'Why would the US willingly cede sovereign immunity and expose itself to liability for damages?'[37]

The defeat of the MAI does not, however, mean the end of corporate demands for international investors' rights. Some OECD governments are looking to the WTO to promote such rights.[38] Even before the MAI was proposed, corporations had gained the right to sue governments under the North American Free Trade Agreement (NAFTA).[39] In one case, the US waste disposal corporation Metalclad had opened a hazardous waste facility in Mexico. After a geological survey showed that the facility would contaminate groundwater, the local state governor declared the area an ecological zone. Metalclad sued the Mexican government, arguing that its investment had been appropriated. In 2002, Metalclad was awarded US$16,685,000 by a NAFTA tribunal.[40] NAFTA, which imposes no legal responsibility on investors to avoid environmental damage, indicates the sort of rights for which corporations are pressing. At the time of writing, the latest drafts of the Central American Free Trade Agreement and the Free Trade Area of the Americas agreement allow investors to sue governments. Meanwhile, the US government is promoting investors' rights in bilateral investment treaties. A former US trade representative uses the term 'competitive liberalization' to denote how countries compete to offer the best bilateral terms to US business in exchange for trade and investment deals.[41]

Here we may recall arguments developed earlier in this book. We argued that international trade and investment law has a stronger normative force than international law on the environment and human rights (see Chapter 9). We also contended that World Bank policy conditionalities on macroeconomic adjustment are more important to Bank shareholders and managers than conditionalities on the environment and human rights (see Chapter 8). To develop this point we may introduce a framework proposed by the philosopher John McMurtry. McMurtry distinguishes between two global value sets: a dominant life-blind value set, in which everything, including people and nature, is subordinated to capitalist accumulation; and a life-ground ethic, in which life values governs the economy so that nature is conserved and the social, cultural and spiritual needs of humans are satisfied. McMurtry stresses that for life values to be met, there needs to be 'an international economy *within a constitutionally governed, democratically accountable framework*' [emphasis in original].[42] This framework is consistent with Charles Derber's argument that we are living through a 'constitutional moment,' an ideological battle between two constitutional traditions. The first is a business-based constitutionalism expressed through the IMF, World Bank and WTO in which corporate rights and capitalist expansion are central, what Stephen Gill has termed a 'new constitutionalism' based on disciplinary neoliberalism.[43] In Derber's second tradition human rights are central. This tradition is based on the principles of the US Bill of Rights, the 1948 Universal Declaration of Human Rights and European social democracy.[44] Since the 1970s, the struggle between these two constitutional visions has become increasingly polarized.[45]

The proponents of a business-based constitutionalism – developed world governments, investment banks and transnational corporations – have ensured that all significant international legal instruments promoting neoliberal principles have been consolidated under the auspices of one organization: the WTO. In distinction to the forests regime, which is scattered among several institutions, the WTO agreements are administered by a single organization with strong enforcement and compliance mechanisms. For example, all WTO member states are required to accept the Agreement on Trade-Related Aspects of Intellectual Property Rights (TRIPS). The WTO agreements have collectively evolved as a coherent body of law that enables the access of investors to the economies and resources of developing countries. Any inconsistencies between individual WTO agreements can be resolved decisively through the WTO's dispute resolution mechanisms. In distinction, no single institution has the authority to resolve conflicts and inconsistencies between different bodies of international law on forests or, for that matter, on any other environmental or social issue.

The WTO agreements have evolved separately from international environmental and human rights law by design, and not by accident. There is no reason why instruments on, say, forest conservation and indigenous peoples' rights cannot come under the auspices of the WTO in the same way as other 'trade-related' agreements. The reasons this has not happened are political rather than legal. The corporate and political interests that promote trade and investment wish any international instruments on the environment and human rights to be kept soft and outside the purview of the WTO. These instruments, which lack enforcement powers comparable to those of the WTO, are kept entirely separate from international trade law, thus ensuring that they neither compromise neoliberal objectives, nor are subject to WTO compliance mechanisms. The case of forests is illustrative. In 1999, the US, supported by

Brazil, backed a WTO forests products agreement to promote the international trade in forest products through tariff elimination (see Chapter 9).[46] This agreement would have been limited solely to trade-related issues. However, both the US and Brazil have long opposed a global forests convention in which trade liberalization would, inevitably, become entangled with conservationist and human rights issues.[47]

This illustrates an increasingly bifurcated strategy of the US government towards international law. Trade and investment treaties are promptly ratified, and the US government uses a variety of incentives to induce other states to become parties to such treaties. Violating or laggard states can expect to incur costs, such as lost trade deals or a cooling of diplomatic relations with the US. However, on occasion, the US itself will derogate from such treaties when this suits US economic interests. One such instance was the temporary imposition of steel tariffs, in contravention of WTO rules, by the US government in 2002. This reveals an occasional tension between the two main roles that the US government plays in the global economy. As the key representative of capitalism, the US government uses its economic power to leverage open new spaces for investment. But as the representative of the US economy, it uses its power to maintain, or improve on, US relative advantages in international trade and finance.

This tension has existed since the US attained the position of global hegemon at the end of World War II. What is now different is the increasing ease with which the US abandons multilateralism. This leads onto the second US approach to international law, which has been used with some environmental and human rights treaties. The US will partake in the negotiations and exert its negotiating power to bargain for provisions that are as closely aligned with US interests as possible. However, when the final text is considered unacceptable by US political elites, the US will not become a party to the treaty. In some cases, treaties are not signed. An example here is the 1998 Rome Statute for the International Criminal Court.[48] Alternatively, the US will become a signatory, but the treaty will subsequently fail to be ratified by Congress. Examples include the CBD and the Kyoto Protocol.[49] Either way, the result is a treaty that has been weakened to accommodate the US, but which the US does not adopt.[50] These examples are part of a more disturbing turn towards unilateralism from the Bush administration, of which the illegal invasion of Iraq is a particularly disconcerting example.[51] US support for multilateralism is now increasingly conditional. This raises a more fundamental question: how effective can international cooperation on forests – or any other issue – be when the multilateral system itself is increasingly under threat from the world hegemon?[52]

Neoliberalism is a rationalizing ideology for US political and economic elites, who will abandon it when it no longer promotes their interests. But for now neoliberalism is a powerful ideological framework that plays an essential role in framing international environmental policy. We now consider how neoliberalism has helped to shape policy in two areas considered earlier in this book: certification and corporate social responsibility (CSR).

Forest certification

Forest certification schemes closely follow neoliberal premises. They are market based, aim at greater efficiency in resource use and are voluntary alternatives to state regulation and intergovernmental regimes. By targeting the supply chain of a private good, namely timber, certification schemes have contributed to the provision of forest public goods. They represent a clear demonstration that market mechanisms can promote sustainable forest management. However, certification has been effective on only a limited scale, proving more popular in developed countries than the tropics. As with CSR, some businesses have reservations about certification, which can still be seen as a regulatory burden, albeit one that comes in a new, voluntaristic guise. Opponents claim that the role of external stakeholders in certification schemes interferes with the rights of landowners to manage their resources as they wish.

To develop this point a digression is necessary. Will Hutton differentiates between American and European property rights traditions.[53] In the US, property rights tend to be seen as absolute. Landowners are essentially free to use their land as they want. This tradition is most strongly expressed through the anti-regulatory Wise Use movement, which is closely associated with US forest industries.[54] The movement cites the fifth amendment of the US constitution, which includes the phrase 'nor shall private property be taken for public use, without just compensation.'[55] This amendment has enabled landowners to claim compensation for restrictions to their property. When public authorities have passed legislation to conserve vulnerable species or ecosystems, some forest owners have claimed that the restrictions are so severe as to constitute a 'taking' of property from the owner, as the forest is less profitable than previously.[56] The agenda of the Wise Use movement is to increase the costs to public authorities of regulation, thus effectively promoting deregulation, and to ensure that when regulation does restrict forest use, public authorities pay compensation.[57]

To Hutton, the European tradition is a more collectivist one; owners have obligations to the public, and property should be managed consistent with the common good. In Europe, owning property is considered a privilege that carries with it reciprocal obligations.[58] This is expressed in the German constitution, which states: 'Property imposes duties. Its use should also serve the public weal.'[59] Forest certification schemes are more consistent with the European tradition; while forest owners have the right to profit from their forests, other citizens have a stake in how forests are managed. Forest certification schemes rest less easily within the American property rights tradition, although they have taken root in the US. The voluntary nature of the schemes means that they evade Wise Use charges that they are imposed regulatory forms that restrict the rights and profitability of forest owners.

The distinction between different property rights traditions in Europe and the US may help to explain the different levels of acceptance on the two continents of CSR, to which we now turn.

Corporate social responsibility

Corporations now routinely claim that there is no need for public regulation, as businesses are socially responsible actors. This strategy has led to CSR, the adoption by a corporation, either singly or with others, of voluntary principles and standards. The voluntarism of CSR is consistent with neoliberalism. Following the failure of the UN to agree a code of conduct for transnational corporations (see Chapter 2), some businesses presented to the UNCED the Business Charter for Sustainable Development, a voluntary declaration of principles.[60] In 2000, UN secretary-general Kofi Annan launched the United Nations Global Compact of ten voluntary principles to which endorsing corporations pledge commitment (see Box 10.1).[61] Today, CSR is the dominant discourse in business (self-)regulation. While it originated from business, CSR has been accepted by many developed governments and by the EU.[62]

Opponents claim that CSR is a contradiction in terms. Under the law, the corporation has no social responsibility, only the fiduciary duty to maximize its shareholders' interests. Corporations adopt CSR schemes only when it is in their interests; for example, to avoid harder forms of regulation or to provide predictability in an uncertain policy environment, as with the Equator Principles (see Chapter 8). Corporations, it is charged, often evade the voluntary commitments that they make.[63] Proponents respond that CSR helps to fill a global regulatory vacuum by exercising an independent normative pull that raises standards. Corporations that take a lead in CSR can set new standards that laggards will adopt to avoid reputational damage.[64] To John Ruggie, schemes such as the Global Compact signify 'the emergence of a new advocate for a more effective global public sector: business itself.'[65] The Global Compact can be seen as a compromise between, on the one hand, those who advocate international regulation of business and, on the ohter, transnational corporations seeking to avoid the extra costs that this would impose.

How has the Global Compact been received by the forest and paper businesses? In terms of annual revenue and/or wood consumption 14 forest corporations qualify as one of the top ten corporations in the world (see Table 1.3 in Chapter 1). Of these, seven are North American corporations, of which five are based in the US and two in Canada; none of these corporations has signed the Global Compact although, significantly, most have joined the AFPA, which promotes forest business interests. Only six of the major forest corporations shown in Tabe 1.3 have endorsed the Global Compact, all of them from Europe and Japan. They are Metsälitto (Finland), Nippon Paper (Japan), Norske Skogindustrier (Norway), Oji Paper (Japan), Stora-Enso (Finland) and UPM-Kymmene (Finland).[66]

Neoliberal policies in context

Voluntary CSR schemes and voluntary certification schemes can contribute to standard-raising, although such mechanisms alone cannot ensure the long-

Box 10.1 The ten principles of the United Nations Global Compact

Human Rights

Principle 1 Businesses should support and respect the protection of internationally proclaimed human rights; and

Principle 2 make sure that they are not complicit in human rights abuses.

Labour Standards

Principle 3 Businesses should uphold freedom of association and the effective recognition of the right to collective bargaining;

Principle 4 the elimination of all forms of forced and compulsory labour;

Principle 5 the effective abolition of child labour; and

Principle 6 the elimination of discrimination in respect of employment and occupation.

Environment

Principle 7 Businesses should support a precautionary approach to environmental challenges;

Principle 8 undertake initiatives to promote greater environmental responsibility; and

Principle 9 encourage the development and diffusion of environmentally friendly technologies.

Anti-corruption

Principle 10 Businesses should work against all forms of corruption, including extortion and bribery.

These principles are derived from:

- Universal Declaration of Human Rights, 1948
- Rio Declaration on Environment and Development, 1992
- International Labour Organization's Declaration on Fundamental Principles and Rights at Work, 1998
- United Nations Convention Against Corruption, 2003.

Source: UN Global Compact website, 'The Ten Principles,' www.unglobalcompact.org/AboutTheGC/TheTenPrinciples/index.html (accessed 8 June 2006)

term future of forests. In a neoliberal policy environment, it makes perfect sense to degrade forests for private gain. While this can be criticized on moral grounds – costs and risks are shifted onto future generations while the benefits are internalized by an elite of the present generation – it is rational for utility-maximizing corporations. Admittedly, not all forest corporations engage in forest degradation; but in the absence of regulation to the contrary, market players have every incentive to subordinate environmental sustainability to profit.

The neoliberal argument that privatization and voluntary regulation leads to more efficient resource use and promotes resource conservation can thus be rejected. This view is derived from neoclassical economics, which provides much of the theoretical and intellectual foundation of neoliberalism. But as Robert Nadeau points out, the argument confuses part–whole relationships. Neoclassical economics focuses on the relationships between economic actors (parts) and economic systems (the whole), and holds that over the long term markets tend towards equilibrium. But for the global environment, the relevant relationships are entirely different, involving organisms (parts) and ecosystems (the whole).[67] Markets seek out the most profitable outcomes, rather than the most environmentally sustainable.

In neoclassical economic theory a tree does not have economic value until it is felled and becomes a commodity. Despite the huge losses of forest cover over recent decades timber prices have remained relatively stable. This reveals a foundational problem: neoclassical economics does not factor in the value of natural capital as it does for economic capital. If the stock market value of a corporation falls, buyers and sellers on international stock exchanges will engage in a flurry of activity as they seek to protect their positions. Policy within the corporation will be revised as directors and managers seek to restore shareholder value. Similarly, an announcement that a country's GDP has fallen will influence the policies of a range of private and public actors, leading, say, to interest rate changes and spending reviews. But news that a large area of rainforest has been burned or a certain species of mammal has become extinct will have no such effect. There is simply no feedback loop in neoclassical economics that factors natural resource depletion into economic decision-making.[68] The World Commission on Forests and Sustainable Development drew attention to this problem by proposing a forest capital index, which, if adopted, would have tracked changes in the natural capital value of forests over time. Even if this proposal had been adopted, it would have made little difference in the absence of a feedback mechanism so that the index registered on decision-makers (see Chapter 3).

The benefits of those forest policies that are consistent with neoliberalism, such as CSR and certification, are more than negated by the environmentally damaging practices that neoliberalism promotes through a combination of deregulation and increased private sector investment in forests. If neoliberalism is part of the problem, what should the response be? In the next section we argue that for deforestation to be arrested, publicly accountable bodies need to regain the authority and power to regulate forest use for the common good.

The reassertion of publicly accountable politics

We may distinguish between two approaches for addressing the social and environmental problems caused by corporate globalization.[69] The first accepts globalization but seeks to democratize it, building new democratic global institutions that hold corporations accountable. The second approach may be termed localization: according to this view, globalization is the cause of most environmental problems; hence the solution is to reverse it by re-grounding politics at the local level. Only by doing this, it is claimed, can social, cultural and biological diversity be maintained.[70]

These two approaches should not be seen as mutually exclusive. Even the best local responses will ultimately fail if they take place within a neoliberal economic system. To be effective, local responses need to be embedded within an international framework dedicated to the enhancement of what McMurtry calls life values. The democratization of globalization and the democratization of local spaces should thus be seen as mutually reinforcing processes. Such an endeavour requires a fundamental restructuring of the international regulatory environment. This section proposes a model for achieving this. Central to the model is the democratic regulation of the corporation. The social and environmental consequences of business practices should no longer be subordinated to the quest for profit. The fiduciary duty of corporations to maximize shareholders' interests – a duty that from a legal standpoint currently trumps all moral and ethical considerations – should be replaced by an explicit legal responsibility for the corporation to act *pro bono publico* (for the public good). To infuse accountability into corporate practices, the notion should be revived that the right of the corporation to operate is conditional on its satisfying public needs as stipulated in public charters. However, there are some important differences between the charter system that currently exists and that proposed here. In particular, the final arbiters on what constitutes the public interest and the standards that corporations should adhere to should be local public bodies.

Under the system of international law that has emerged over the last four centuries, states both make international law and are subject to it. However, the state is no longer the main international actor. To ensure minimum global standards a new body of international law is called for that regulates corporations rather than states. States, as the legitimate representatives of national publics in international politics, will negotiate this corpus of law, which will stipulate the standards to which corporations should adhere in order to operate and trade internationally. This new body of law should be agreed free of interference from corporations, which, as regulatees rather than regulators, would be denied access to the legislative process.

The first step would be the negotiation of a Convention on Transnational Corporations outlining the responsibilities and duties of corporations.[71] This would set a raised plateau of tough obligations, standards and conditions that would apply to all corporations in all countries. Whereas an individual state adopts a treaty by signing and ratifying it through the domestic legislature, an individual corporation would adopt the Convention on Transnational Corporations by endorsing it through its board of directors and shareholders' meetings. No corporation could trade or invest internationally without first adopting the convention.[72] By doing so corporations would recognize that their right to engage in transnational economic activity is conditional on the observance of obligations to the global public. These obligations would, at a minimum, include commitments to uphold environmental quality, respect the precautionary principle, undertake environmental and social impact assessments, allow independent financial and environmental auditing, observe a duty of care towards the public, and respect community rights and traditions. Corporations endorsing the convention would receive an international charter permitting them to trade internationally. In terms both of intent and content, the convention would be the opposite of the aborted multilateral agreement on investment, which outlined corporate rights rather than duties.

To invest in a country, a corporation would then need to obtain a country-level charter from a public authority in the host country. This would stipulate the terms and standards that the corporation should observe. The granting or withholding of country-level charters would be a matter of public debate, with the final decision being made by the national government. The conditions stipulated in a country charter could be stronger, but not weaker, than those outlined in the convention. A corporation would need a separate charter for every country in which it operated. It would be unlikely that the charters issued to different corporations would be the same. Since different corporations would be admitted to different countries to meet different public needs, then inevitably the contents of charters would vary.

A corporation with a country-level charter would then be free to find a locality in which to invest. Publicly accountable groups at the sub-state level would have the right to determine which corporations invested in their space. The conditions would be stipulated in a local charter. The provisions contained in a local charter could strengthen, but not weaken, those laid out in the country-level charter. Actors at the sub-state level charged with upholding the public interest will come in many guises, reflecting the rich diversity of local cultures, economies and commons regimes that are to be found in most countries. They will include, for example, democratically elected local councils and more traditional local community governance structures. The size of the territory for which sub-state authorities are responsible would vary from case to case, depending on population size, population density, the local economy, culture, social conditions, and so on. Irrespective of the form of governance at the local level, a local group or authority would be under no obligation to issue a charter if it did not consider this to be in the interests of its public. Under the system of nested and differentiated governance proposed here, countries and communities would not adopt identical social and environmental rules, although they would adopt minimum rules. The system would promote universal values and standards while respecting the diversity and right to self-determination of countries and communities.[73]

We have proposed three levels at which charters are issued, but there could be more. Regional bodies such as the EU may wish to issue charters. Federal states usually have more layers of government than unitary states. The precise model of publicly accountable governance would thus vary according to region and country, in line with local needs and conditions. But the local level cannot entirely dominate. Global targets should be set on public good provision. Targets would be necessary on, for example, forest cover and greenhouse gas emissions. Multilevel coordination between the different layers of governance would then be necessary to ensure that these targets were met. Agreement would need to be reached at the intergovernmental level on the responsibilities of individual countries with respect to global targets. Similarly, within countries, coordination will be necessary between the state and sub-state authorities on the responsibilities that local spaces should make to global public good provision.

A Convention on Transnational Corporations is necessary so that countries in need of investment can resist pressure from unscrupulous businesses seeking to erode environmental and social standards through bargaining with the host government. Businesses adopting the convention would be obliged to ensure its implementation by all subsidiary companies. They would also undertake to trade only with businesses that respect and implement the convention. This would reinforce standards among market players. As forest certification has demonstrated, standards adopted by one player can be passed along the supply chain. Supply chains can thus act as transmission belts along which rules are passed and enforced. Insisting that a corporation know its supply chains is not unreasonable, but it would call for mandatory international product certification schemes. The experiences of the FSC and other ISEAL organizations, which aim to ensure that production processes take place within an ethical framework that respects environmental and social principles, would be relevant here (see Chapter 6). If businesses are required to know the practices of their suppliers, then the defence, currently a common one in commerce, that a corporation simply did not know that a product was manufactured, for example, using environmentally unsustainable practices or child labour, will be rendered void. Obliging corporations to be accountable for the practices of their subsidiaries, contractors and trading partners will generate mutually reinforcing monitoring networks. All businesses will be mindful of what their associates are doing lest they themselves risk opprobrium or financial censure.

A corporation would be called on to answer for its actions before an International Environmental Court if evidence were to emerge that it had violated the Convention on Transnational Corporations.[74] Actions in this court could be brought by various plaintiffs: international organizations, a host state, local authorities or civil society groups. If the court were to find that the convention had been violated then the corporation would be liable to penalties, such as fines, suspension of the international charter or, in severe cases, the dissolution of the corporation and the seizure of its assets. Similarly, if a corporation breaks a national or local charter it would answer to national and local authorities, and may suffer financial penalties or a loss of charters.

When applying for any new charters anywhere in the world, a corporation and its directors would be obliged to disclose all previous malfeasance that resulted in fines or charter revocation.

How might this system of governance operate in the forest sector? At the local level a public authority may include in a charter issued to a logging corporation the commitment to maintain the natural capital stock value of the forest concession. So if the corporation were granted, say, a 20-year concession, at the end of this period it would be obliged to ensure, as far as possible, that the forest was in the same natural condition as at the start of the concession. (There would, of course, be technical difficulties in measuring changes to the capital stock value over time.[75] Some causes of forest decline could be due to factors outside the concession, such as acid rain or global warming. But these points do not negate the general principle.) Where the natural capital stock value has declined due to unsustainable, unscrupulous or negligent practices, then the corporation would be required to make good on the damage caused and to pay a financial penalty. Should the corporation have insufficient funds to pay these costs, the courts should have the power to dissolve the corporation and seize its assets.

The courts should also be able to seize the private assets of the company's directors. Some will argue that this is draconian. In response, three points should be made. First, in many countries the law can be used to pursue assets where individuals and businesses have engaged in illegal and immoral activities, such as drug smuggling and prostitution. There is no reason why those who enrich themselves from the unsustainable exploitation of natural assets should be treated differently. Second, the risk of financial penalties accruing to corporations and directors would act as a strong incentive for long-term environmental sustainability.[76] Third, financial penalties would co-opt the world's financial markets into promoting environmental public goods. Financial investors want profitable businesses. If knowledge emerged that a forest corporation had engaged in unsustainable practices that violated a charter, the result would be a fall in the value of the corporation's stock as the markets reacted in anticipation of heavy fines or the dissolution of the corporation. Financial investors would thus have an incentive to monitor corporate environmental practices. This would help to repair the problem of the missing feedback loop whereby the depletion of natural capital simply does not register with corporations and international financial markets.

Towards a democratic and publicly accountable global governance

One can expect various protests to the governance system proposed here. It may be claimed that international rules outlining the obligations of corporations are impractical. Even if the model proposed here is not feasible at present, it has the clear advantage of signposting the governance reforms that are necessary if communities are to regain control over nature. In any case, business leaders and their political allies have consistently pressed for international rules on trade, investment and intellectual property rights. On what basis can it be claimed that international rules for environmental conservation and human rights are, in some way, impractical?

It may be argued that people in local communities would not be motivated to engage in debates on corporate charters. Certainly, there is a high degree of public apathy within contemporary representative democracies; many people believe that they will make no difference through becoming politically involved. However, the revitalization of the charter as an active instrument for accountability will stimulate citizen engagement and participatory democracy. It may help to overcome one of the main barriers to political involvement, what David Held has termed *nautonomy*, namely 'the asymmetrical production and distribution of life chances which limit and erode the possibilities of political participation.'[77] Citizens' views would now count. Corporations would have to adjust to the needs of local communities rather than, as now, communities being forced to adjust to intrusions from big business playing by international trade rules. In a post-neoliberal world, a new mode of doing business would emerge based on the needs of the local economy. This will benefit not only those communities who wish to retain local forests, but those who wish to conserve local fishing and water resources and those who do not want large superstores eroding the distinctiveness and diversity of the local economy. Furthermore, if local communities exercise greater control over their economies then a greater share of the income from resource use will flow to the local level where it can be used to fight poverty.

It may be claimed that a multilevel system of charters would be 'bureaucratic' and 'protectionist,' both of which are pejorative terms for neoliberals. But if bureaucratization is the price of accountability, then let these charges be made, as the social and environmental costs of allowing corporations to penetrate new spaces without effective public oversight have already proved too large. And it is hardly protectionist to point out that local and, sometimes, national businesses are at a significant disadvantage when competing with transnational corporations. A system of public charters would help to equalize the huge power asymmetries between local communities and big corporations. Indeed, on what basis should anyone seek to deny public authorities the right to decide which corporations can have access to their economies and which should not?

It may be argued that the public interest, which lies at the heart of the proposal made here, is an essentially contested concept. Indeed it is. The problem, however, is that at present contestation usually takes places away from the public gaze, with corporate interest groups lobbying politicians behind closed doors. Democratic governance would see debate on the public interest and the economy taking place openly, transparently and inclusively. It may also be claimed that if environmental and social standards are set 'too high,' then corporations will stop investing in local communities. But given that a key driving force of environmental degradation is investment that is not constrained by strong safeguards, this argument should be dismissed. Irrespective of the regulatory environment, corporations will continue to search for investment opportunities. They will soon learn to adjust to the needs of local communities if the alternative is not to invest at all. With effective public safeguards in place, only environmentally and socially irresponsible investors need be concerned.

Clearly, much depends on the role of the state and its capacity to regulate the corporation (as an agent) and capital accumulation (as a process). As Richard Falk has noted, under neoliberalism the state has become an instrument of business, leading to 'a loss of capacity and will to promote the public good, in general, and its environmental aspects, in particular.'[78] It may be argued that it is utopian, even naive, to argue that corporate-driven environmental degradation can be reversed using the same institutions – the state and intergovernmental organizations – that so far have failed. In response it should be asked that if publicly accountable bodies are not going to represent the interests of citizens and promote public goods, then who, precisely, will do this? The problem is not so much with publicly accountable bodies, but with their penetration by corporate interests.[79] The solution is to reform and strengthen public bodies rather than to accede to the neoliberal view which, premised on the notion that a weakened state and ineffective intergovernmental regimes are 'inevitable' under globalization, searches for new forms of governance that hand more power to business, such as CSR and public–private partnerships.

From where might a democratic post-neoliberal politics emerge? Civil society organizations, such as indigenous peoples' groups and NGOs, will continue to play a vital role in highlighting environmental degradation and human rights abuses. Indeed, such groups have become increasingly important precisely because many public bodies at the international, national and local levels are failing to uphold the public interest with which they are entrusted, often entering into close relationships with environmentally destructive businesses. As Geoffrey Underhill argues, an effective public domain requires an authority that will not win or lose from market conditions and transactions.[80] Charles Derber makes a similar point; robust democracy requires a 'firewall' between government and business.[81] Rather than promoting business opportunities, a post-neoliberal state would be charged solely with governing in the common good. Corporations and other interest groups would be excluded from the law-making process and from donating to political parties, which should be funded from the public purse.

There are signs that some state institutions are starting to act against corporations on behalf of their citizens. In 2006, a Nigerian court ordered Shell to pay US$1.5 billion as damages for polluting the Niger delta, including compensation to the Ijaw people and other local communities whose crops and fishing grounds have been degraded.[82] The same year the Bolivian president, Evo Morales, a vocal critic of the neoliberal model, announced that the country's corporate owned gas and oil supplies would be brought under public ownership (although the exact form that public ownership would take was not clear as this book went to press).[83] Prior to his election Morales was one of the leaders of the widespread protests in Bolivia against water privatization that had resulted in transnational corporations imposing price increases that took water supplies out of the reach of the poor. Examples of governments standing up to corporations are currently few in number; most politicians in developing countries recognize that challenging corporations may deter private investment and risk a chilling of diplomatic relations with the powerful developed states that protect corporate interests. However, under the governance system proposed here a power shift would take place from the private sphere to the public. All corporations would have to play by the same international rules, and such concerns would no longer apply.

Will change emerge from the world's hegemonic power: the United States? The current Republican strain of American politics has a profound aversion to corporate regulation and of government spending on social needs. Under Bush junior spending on environmental protection has fallen and the main environmental regulatory body, the Environmental Protection Agency, is involved in fewer prosecutions. Given the US turn to unilateralism and the unwillingness of the American political establishment to participate in strong international regimes on the environment and human rights, a fundamental shift in federal politics is necessary if the US is to become a force for an environmentally sustainable global governance. The activist Susan George views the Bush government as an 'adversary' that is opposed to international treaties, ecological sustainability and social responsibility.[84] However, environmentally responsible policies by individual states in the US may create political space for change at the federal level. In the absence of action from Washington, some individual states are assuming leadership on environmental issues. For example, in 2005 nine states in the north-eastern US agreed to reduce carbon dioxide emissions from power plants by 10 per cent by 2020.[85]

Among developed world governments, the seeds of a democratically accountable and environmentally sustainable global governance are more likely to come from Europe than the US. The introduction by the EU of demand-side policies on illegal logging, which the US has rejected, is indicative of a broader trend (see Chapter 7). On a range of issues – such as climate change, chemical weapons, GM food and landmines – the EU is now the international leader.[86] Over the last 20 years, the EU has progressively raised standards by tightening environmental policy legislation. Whereas US environmental policy has a marked neoliberal hue, relying principally on market-based incentives, the European tradition of governance lays more emphasis on shared rules and accountability.[87] But if Europe is to present a coherent alternative to neoliberalism, then it will need support from other countries. The world's largest democracy, India, is one possible ally.[88] In fact, it may not be too fanciful to suggest that the developing countries, which bear most of the social costs of environmental degradation, could generate an alternative to neoliberalism without the EU, especially if they can recapture the solidarity that characterized their attempts to form a New International Economic Order (NIEO) in the 1970s.[89] Certainly developing countries are growing increasingly intolerant of the use of the WTO to promote rules that favour developed countries and their corporate allies; hence the difficulties that the WTO Doha Round negotiations have experienced since 2003. Even if the developing countries do not act in unison, smaller coalitions could provide a regional locus that challenges neoliberalism.

A revitalized and democratized UN is essential. Multi-stakeholder dialogue has failed to promote genuine accountability, and the Global Compact is indicative of a serious malaise at the heart of the UN system. While the Global Compact enhances the status of the UN secretary-general, it completely bypasses, and thereby diminishes, intergovernmental bodies within the UN system. It presumes that the era of public regulation is drawing to a close, supplanted by a new era of global privatized governance. As David Coleman has argued, by appealing directly to corporations to adopt global standards, and, in so doing, inferring that states and international organizations are the laggards at standard-setting, the Global Compact has subverted public decision-making and passed global regulation to the private sector.[90] The model of corporate accountability proposed here is the antithesis of the Global Compact. It would render obsolete voluntary CSR and private sector regulation and restore full political authority to a dynamic, inclusive and democratically accountable global public domain.

Concluding thoughts

This book has examined global forest policy from mid 1995 to early 2006. Throughout this period there was unprecedented activity on forests in international organizations; yet massive deforestation continued, especially in the tropics. It is not simply a case of policy development being too slow to match the pace and scale of forest degradation. Often the wrong policies have been pursued within broader political and economic structures that routinely generate deforestation. It is now clear that piecemeal and incremental policy shifts from forest-related international organizations will continue to fail as they do not address the deep causes of deforestation, principally predatory and unregulated corporations that profit from the degradation and destruction of forest public goods. The politicians of the neoliberal state have aided and abetted this process in various ways, including through political alliances, often corrupt, with forest industries, and by promoting international trade, investment and intellectual property rights law that has a stronger normative force than international law on the environment and human rights. In this respect, deforestation is symptomatic of a broader structural crisis of global governance.

Arguments that the regulatory capacity of the state has been fatally weakened are premature. In many countries the state apparatus has been harnessed to considerable effect in creating corporate investment opportunities and in promoting national and international law that favours business. In one respect, this is good news; the state remains influential. If rendered accountable and working in concert with civil society organizations and publicly accountable bodies at the supra-state and sub-state levels, the state can play an essential role in public good enhancement through ensuring that corporations act in the common public interest, harnessing their economic power for the benefit of all citizens. There will, admittedly, be huge costs to corporations if they are to adopt more environmentally sustainable and socially responsible practices. But the environmental and social costs that have arisen from predatory and exploitative corporate practices – including massive displacement of communities, species loss and climate change – have already been vast. Simply put, nature and communities are currently paying the costs of corporate enrichment. In any case, the governance reforms advocated here will be minimal and small scale compared with the massive interventions and curbs on corporate activity that will be necessary in, say, another half century if global environmental degradation continues unchecked.

To accept the values and agenda of a neoliberal capitalist order, knowing that this system has driven deforestation and other environmental and social problems, yet to continue to insist that the solutions can be found within the premises of this system, would be both an abrogation of responsibility and a failure of imagination. There is nothing inevitable or permanent about the current neoliberal world. There are alternatives. The alternative advocated here is a democratic global politics that restores the primacy of publicly accountable authorities over business and which insists that economies are managed for the benefit of their communities. A truly democratic politics will value all the goods and services that forests provide: public as well as private. Forests would be valued for the timber, nuts, fruits, berries and other products that they can provide in perpetuity, if managed sustainably. They would be appreciated for the social and cultural services they provide to people, including recreation and spiritual fulfilment. They would be conserved for their life-supporting services, such as biodiversity habitat, climatic regulation and water provision. And they would be properly valued for their role in soil conservation and flood prevention, which, returning to where we started, would help to prevent further human tragedies such as the catastrophic mudslides that, in December 1999, shattered the lives of thousands of people living on the mountains and in the valleys of northern Venezuela.

Notes

Preface

1 Humphreys, David (1996) *Forest Politics: The Evolution of International Cooperation*, London: Earthscan.
2 Intergovernmental Panel on Climate Change (2001) 'Climate Change 2001: Synthesis Report – Summary for Policymakers,' IPCC Plenary XVII, London, 24–29 September 2001.
3 Thomas, Chris D. et al (2004) 'Extinction risk from climate change,' *Nature*, Vol.427, 8 January, pp.145–148.
4 United Nations University news release (2004) 'Two Billion People Vulnerable to Floods by 2050...,' 13 June.
5 Soares-Filho, Britaldo Silveira et al (2006) 'Modelling conservation in the Amazon basin,' *Nature*, Vol. 440, 23 March, pp.520–523.
6 Cox, Robert W. (1981) 'Social forces, states and world orders: Beyond International Relations Theory,' *Millennium: Journal of International Studies*, Vol. 10, No. 2, pp.126–155. For an application of Cox's distinction between problem-solving approaches and critical theory to global environmental problems see Elliott, Lorraine (1998) *The Global Politics of the Environment*, Basingstoke: Macmillan, pp.242–248.

1 Forests as Public Goods

1 I have recalled from memory this scene from visits made to La Guaira and Caracas between 1979 and 1982 as a merchant seaman on the cruise ship *Queen Elizabeth 2*.
2 On the Venezuela mudslides of December 1999 see: www.disasterrelief.org/Disasters/000322Venezuela9/ and www.disasterrelief.org/Disasters/001121venezuela/ (both accessed 9 March 2004). On the relationship between deforestation and flooding see Food and Agriculture Organization (2005) *Forests and Floods: Drowning in Fiction or Thriving on Facts? – RAP Publication 2005/03*, Rome: Food and Agriculture Organization.
3 On global public goods see Kaul, Inge, Grunberg, Isabelle and Stern, Marc A. (eds) (1999) *Global Public Goods: International Cooperation in the 21st Century*, Oxford: Oxford University Press/United Nations Development Programme; Kaul, Inge, Conceição, Pedro, Le Goulven, Katell and Mendoza, Ronald U. (eds) (2003) *Providing Global Public Goods: Managing Globalization*, Oxford: Oxford University Press/United Nations Development Programme.

4 On the categorization of environmental goods according to rivalness (a similar concept to rivalry) and excludability see, for example: Barkin, J. Samuel and Shambaugh, George E. (1996) 'Common pool resources and international politics,' *Environmental Politics*, Vol. 5, No. 4, pp.429–447; Barkin, J. Samuel and Shambaugh, George E. (eds) (1999) *Anarchy and the Environment: The International Relations of Common Pool Resources*, Albany: State University of New York Press; Vogler, John (2000) *The Global Commons: Environmental and Technological Governance* (second edition), Chichester: John Wiley and Sons. Buck uses a similar typology, categorizing goods according to rivalness and subtractability: Buck, Susan J. (1998) *The Global Commons: An Introduction*, London: Earthscan.
5 Hence such goods are sometimes called toll goods.
6 See, for example, Bowers, John (1997) *Sustainability and Environmental Economics: An Alternative Text*, Harlow: Prentice Hall, pp.33–40.
7 See, for example, Kaul, Inge and Mendoza, Ronald U. (2003) 'Advancing the concept of public goods,' in Kaul, Conceição, Le Goulven and Mendoza, n.3 above, pp.78–111; Musgrave, Richard A. (1959) *The Theory of Public Finance: A Study in Public Economy*, New York: McGraw-Hill.
8 Perrings, Charles and Gadgil, Madhav (2003) 'Conserving biodiversity: reconciling local and global benefits,' in Kaul, Conceição, Le Goulven and Mendoza, n.3 above, p.532.
9 Stockholm Declaration on the Human Environment, 1972, principle 21; Rio Declaration on Environment and Development, 1992, principle 2; Non-legally binding authoritative statement of principles for a global consensus on the management, conservation and sustainable development of all types of forests, 1992, para.29(a).
10 Forest Peoples Programme and FERN (2004) 'Indigenous peoples' rights, state sovereignty and the Convention on Biological Diversity.' This paper was circulated at the seventh conference of parties to the Convention on Biological Diversity in February 2004.
11 McKean, Margaret A. (2000) 'Common property: What is it, what is it good for, and what makes it work?,' in Gibson, Clark C., McKean, Margaret A. and Ostrom, Elinor (eds) *People and Forests: Communities, Institutions, and Governance*, Cambridge MA: MIT Press, p.27.
12 Hardin, Garrett (1968) 'The tragedy of the commons,' *Science*, No. 162, pp.1243–48.
13 Ostrom, Elinor (1990) *Governing the Commons: The Evolution of Institutions for Collective Action*, Cambridge: Cambridge University Press; The Ecologist (1993) *Whose Common Future?: Reclaiming the Commons*, London: Earthscan. Hardin recognized the validity of many of the criticisms made in response to his original essay, which he later said should have been called 'The tragedy of the unmanaged commons.'
14 See, for example, The Ecologist, n.13 above; Bollier, David (2003) *Silent Theft: The Private Plunder of Our Common Weal*, New York: Routledge; Prashad, Vijay (2002) *Fat Cats and Running Dogs: The Enron Stage of Capitalism*, London: Zed Books.
15 Geores, Martha E. (2003) 'The relationship between resource definition and scale: considering the forest,' in Dolšak, Nives and Ostrom, Elinor (eds) *The Commons in the New Millennium: Challenges and Adaptations*, Cambridge MA: MIT Press, p.84.
16 Geores, n.15 above.

17 Westoby, Jack (1989) *Introduction to World Forestry: People and their Trees*, Oxford: Basil Blackwell, p.57.
18 Scott, James C. (1998) *Seeing Like a State: How Certain Schemes to Improve the Human Condition Have Failed*, New Haven: Yale University Press, p.2.
19 Scott, n.18 above, pp.2–3.
20 Scott, n.18 above, pp.11–13.
21 Scott, n.18 above, pp.13–15.
22 Scott, n.18 above, Chapter 1, especially pp.18–22. See also Lipschutz, Ronnie D. (2004) *Global Environmental Politics: Power, Perspectives, and Practice*, Washington DC: CQ Press, pp.195–197.
23 Keal, Paul (2003) *European Conquest and the Rights of Indigenous Peoples: The Moral Backwardness of International Society*, Cambridge: Cambridge University Press; Banuri, Tariq and Marglin, Frédérique Apffel (1993) *Who Will Save the Forests?: Knowledge, Power and Environmental Destruction*, London: Zed Books.
24 See, for example, Sabine, George H. and Thorson, Thomas L. (1973) *A History of Political Theory* (Fourth Edition), Hinsdale Il: Dryden Press, pp.485–487.
25 Jaramillo, Carlos Felipe and Kelly, Thomas (1997) 'Deforestation and property rights in Latin America,' p.7. Available online at: www.iadb.org/sds/doc/1411eng.pdf (accessed 14 June 2005).
26 The Ecologist, n.13 above, p.15.
27 UN document E/CN.17/IFF/1998/8, 'Programme element II.d(ii), Valuation of forest goods and services; economic instruments, tax policies and land tenure; future supply of and demand for wood products and non-wood products; and rehabilitation of forest cover,' 19 June 1998, para. 17, p.6.
28 Anand, P. B. (2002) 'Financing the provision of global public goods,' *United Nations University/World Institute for Development Economic Research (UNU/WIDER) Discussion Paper* No. 2002/110; Kaul, Inge and Le Goulven, Katell (2003) 'Financing global public goods: A new frontier of public finance,' in Kaul, Conceição, Le Goulven and Mendoza, n.3 above, p.336.
29 Sato, Jim (2000) 'People in between: Conversion and conservation of forest lands in Thailand,' in Doornbos, Martin, Saith, Ashwani and White, Ben (eds) *Forests: Nature, People, Power*, Oxford: Blackwell, pp.153–154.
30 The six countries not included, for which there was insufficient or unreliable data, were Angola, Congo, Mozambique, Paraguay, Venezuela and Zambia.
31 White, Andy and Martin, Alejandro (2002) *Who Owns the World's Forests: Forest Tenure and Public Forests in Transition*, Washington DC: Forest Trends and Center for International Environmental Law.
32 White and Martin, n.31 above, p.2.
33 Saad-Filho, Alfredo and Johnston, Deborah (2005) 'Introduction,' in Saad-Filho, Alfredo and Johnston, Deborah (eds) *Neoliberalism: A Critical Reader*, London: Pluto, p.3.
34 As Bakan explains, the relevant US case law is *Dodge v. Ford*, where the court found against the Ford Motor Company after it cancelled a share dividend and passed the money onto customers in the form of lower pricing. The judge ruled that 'a business corporation is organized and carried on primarily for the profit of the stockholders' and not 'for the merely incidental benefit of shareholders and for the primary purpose of benefiting others.' See Bakan, Joel (2004) *The Corporation: The Pathological Pursuit of Profit and Power*, London: Constable, pp.34–36.
35 Bakan, n.34 above; Korten, David (1995) *When Corporations Rule the World*, London: Earthscan.

36 'G8 action programme on forests.' Available online at: http://birmingham.g8summit.gov.uk/forfin/forests.shtml (accessed 19 November 2002).
37 White and Martin, n.31 above.
38 Dauvergne, Peter (1997) *Shadows in the Forest: Japan and the Politics of Timber in Southeast Asia*, Cambridge MA: MIT Press; Dauvergne, Peter (2001) *Loggers and Degradation in the Asia Pacific: Corporations and Environmental Management*, Cambridge: Cambridge University Press.
39 World Rainforest Movement and Forests Monitor (1998) *High Stakes: The Need to Control Transnational Logging Companies – A Malaysian Case Study*, Montevideo and Ely: WRM/Forests Monitor.
40 Greenpeace (1999) *Facing Destruction: A Greenpeace Briefing on the Timber Industry in the Brazilian Amazon*, Amsterdam: Greenpeace International; Environmental Investigation Agency and Telapak (2005) *The Last Frontier: Illegal Logging in Papua and China's Massive Timber Theft*, London: EIA.
41 SKEPHI (1996) *Asian Forestry Incursions: Indonesian logging in Suriname, Report on NV MUSA Indo-Surinam*, Amsterdam: SKEPHI; Colchester, Marcus (1995) 'Asia logs Suriname,' *Multinational Monitor*, Vol. 16, No. 11. Available online at: http://multinationalmonitor.org/hyper/mm1195.05.html (accessed 25 September 2002).
42 Global Witness (1999) 'The untouchables: Forest crimes and the concessionaires – can Cambodia afford to keep them?,' a briefing document by Global Witness.
43 Jamie Aviles, cited in Ross, John (1998) 'Big pulp vs. Zapatistas,' *Multinational Monitor*, Vol. 19, No. 4, pp.9–12. Available online at: http://arts-sciences.cua.edu/pol/faculty/foley/Ross.htm (accessed 14 June 2005).
44 See, for example, Goldman, Michael (ed) (1998) *Privatizing Nature: Political Struggles for the Global Commons*, London: Pluto.
45 Hirst, Rob (2003) *Willie's Bar and Grill: A Rock 'n' Roll Tour of North America in the Age of Terror*, Sydney: Picador Pan Macmillan, p.49. See also Dodshon, Mark (2004) *Beds are Burning: Midnight Oil – The Journey*, Melbourne: Viking Penguin, pp.49–50.
46 Friends of Clayoquot Sound, Spring 2003 newsletter. Available online at: www.focs.ca/1newsroom/sprnl20031.html (accessed 6 October 2004); Common Ground (2003) 'Clayoquot was defining protest of our time.' Available online at: www.commonground.ca/iss/0308145/clayoquot.shtml (accessed 6 October 2004).
47 Friends of the Earth (2005) *Friends of the Earth International*, Issue 107, January, special issue on 'Privatization: nature for sale – the impacts of privatizing water and biodiversity.'
48 Lovera, Miguel; Avendaño, Tatiana Rosa and Torres, Irene Vélez (eds) (2005) *The New Merchants: Life as Commodity*, Bogotá: Global Forest Coalition/Censat Agua Viva, p.84.
49 Lovera, Avendaño and Torres, n.48 above.
50 Toke, Dave (2000) *Green Politics and Neo-liberalism*, Basingstoke: Macmillan.
51 UN document E/CN.17/IFF/1998/4, 'Transfer of environmentally sound technologies to support sustainable forest management,' 19 June 1998, para.29.
52 Shiva, Vandana (2001) *Protect or Plunder: Understanding Intellectual Property Rights*, London: Zed, pp.57–58.
53 Perry, David (undated) 'The global distribution of AIDS pharmaceuticals.' Available online at: www.scu.edu/ethics/publications/submitted/Perry/aids.html (accessed 23 June 2005); Quaker United Nations Office (undated leaflet), 'How

strong patent protection affects access to medicines – Patents, trade and health,' New York.

54 Bollier, n.14, pp.77–78.

55 The patenting of genetic sequences extends to the human genome, with nearly a fifth of all human genes now the subject of a patent, the majority owned by biotechnology corporations: Jensen, Kyle and Murray, Fiona (2005) 'Intellectual property landscape of the human genome,' *Science*, No. 310, 14 October, pp.239–240.

56 Prashad, n.14 above, pp.151–188. Bollier argues that the use of intellectual property rights in agriculture amounts to the corporate enclosure of the agricultural commons (n.14 above, p.77).

57 Shiva, n.52 above, p.49. (Shiva used the term *terra nullius* rather than *res nullius*.)

58 Convention on Biological Diversity 1992, Article 8(j).

59 The figure of 25 per cent of anthropogenic emissions of carbon dioxide is provided in: Food and Agriculture Organization (2005) *Global Forestry Resources Assessment*, Rome: FAO. Between 10 per cent and 25 per cent of global human-induced greenhouse gas emissions are attributable to tropical deforestation: Santilli, Márcio; Moutinho, Paulo; Schwartzman, Stephan; Nepstad, Daniel; Curran, Lisa; and Nobre, Carlos (2005) 'Tropical deforestation and the Kyoto Protocol,' *Climatic Change*, No. 71, pp.267–276. According to the Intergovernmental Panel on Climate Change (IPCC) information on emissions from forest fire burning is poor. See IPCC (2000) *IPCC Special Report on Emissions Scenarios*, section 3.5.2, 'Carbon dioxide emissions from anthropogenic land-use change.' Available online at: www.grida.no/climate/ipcc/emission/077.htm (accessed 15 September 2005).

60 Yamin, Farhana (ed) (2005) *Climate Change and Carbon Markets: A Handbook of Emissions Reductions Mechanisms*, London: Earthscan; Michaelowa, Axel and Butzengeiger, Sonja (eds) (2005) *EU Emissions Trading Scheme: A Special Issue of the Climate Policy Journal*, London: Earthscan.

61 World Rainforest Movement (2001) 'Convention on Climate Change: Privatising the atmosphere?' Available online at: www.wrm.org.uy/publications/briefings/CCC.html (accessed 19 February 2004).

62 *Guardian*, 28 June 2003, p.15; *Guardian*, 4 September 2004, p.17.

2 Intergovernmental Panel on Forests

1 '1892–1992, 100 years of IUFRO'(undated) International Union of Forest Research Organizations, Vienna, brochure.

2 Humphreys, David (1996) *Forest Politics: The Evolution of International Cooperation*, London: Earthscan, Chapters 2 and 3. The Tropical Forestry Action Plan was renamed the Tropical Forests Action Programme in 1990.

3 Kufuor, Edward, quoted by Chatterjee, Pratap (1991) 'G77 rejects "stewardship",' *Crosscurrents*, UNCED PrepCom 3, No. 8, 28–29 August, p.4.

4 In full, 'Non-legally binding authoritative statement of principles for the management, conservation and sustainable development of all types of forests.'

5 On the UNCED forest negotiations see Humphreys, n.2 above, Chapter 4; and Davenport, Deborah S. (2005) 'An alternative explanation for the failure of the UNCED forest negotiations,' *Global Environmental Politics*, Vol. 5, No. 1, pp.105–130.

6 On the negotiation of the International Tropical Timber Agreement of 1994 see Humphreys, n.2 above, Chapter 5; and Poore, Duncan (2003) *Changing Landscapes: The Development of the International Tropical Timber Organization and its Influence on Tropical Forest Management*, London: Earthscan, Chapter 9.
7 Jag Maini, interview, fifth session of the United Nations Forum on Forests, New York, 24 May 2005.
8 'UK Proposal, January 1995, proposals for EU approach towards forests issues for CSD 1995,' para. 13.
9 The other five issues were integrated management of land resources, combating desertification, sustainable mountain development, sustainable agriculture and rural development, and biodiversity.
10 *Earth Negotiations Bulletin*, Vol. 5, No. 27, pp.4–5.
11 *Earth Negotiations Bulletin*, Vol. 5, No. 27, p.5.
12 'Statement by the Swedish Ministry of Industry and Commerce, Dr Sten Heckscher,' pp.3–4.
13 FAO document COFO-95/REP, 'Report of the twelfth session of the Committee on Forestry, Rome, Italy, 13–16 March 1995,' para.30, p.4.
14 'Ministerial session, draft declaration of the European Union, 15.3.95,' p.2.
15 FAO (1995) 'Ministerial meeting on forestry, Rome, 16–17 March 1995, Rome Statement on Forestry,' para.4. The meeting was attended by 123 countries, 53 of which had ministerial level representation.
16 As reported by *Earth Negotiations Bulletin*, Vol. 13, No. 1, p.4.
17 Forest Principles, para. 12(d); Chapter 11 of Agenda 21, para. 11.14(d); Convention on Biological Diversity, Articles 8(j) and 17.2.
18 UN document E/CN.17/IPF/1995/2, 'Commission on Sustainable Development, Programme of work of the Intergovernmental Panel on Forests,' 16 August 1995, p.3.
19 Rosendal, G. Kirstin (2001) 'Overlapping international regimes: The case of the Intergovernmental Forum on Forests (IFF) between climate change and biodiversity,' *International Environmental Agreements: Politics, Law and Economics*, Vol. 1, No. 4, pp.452, 460.
20 Colchester, Marcus and Lohmann, Larry (1990) *The Tropical Forestry Action Plan: What Progress?*, Penang/Sturminster Newton, Dorest: World Rainforest Movement/The Ecologist; Winterbottom, Robert (1990) *Taking Stock: The Tropical Forestry Action Plan After Five Years*, Washington DC: World Resources Institute.
21 'Urgent calls for a full review of past failures of forest management in the Asia Pacific Region,' mimeo, March 1995.
22 Holdgate's status during this period is an interesting one. He was retired both from the British civil service and from the position of IUCN director-general. In order to serve as chair of the CSD Intersessional Working Group on Sectoral Issues of 1995 and of the IPF from 1995 to 1997 Holdgate was reemployed by the UK government on a per diem basis. Sir Martin Holdgate, interview, Kirkby Stephen, Cumbria, 17 August 2004.
23 Rodríguez was nominated for the post of 'southern' co-chair by Brazil: Manuel Rodríguez, interview, New York, 10 September 2004.
24 Jag Maini, interview, fifth session of the United Nations Forum on Forests, New York, 24 May 2005.
25 Initiatives sponsored by governments have been known variously as government-led initiatives, national initiatives and country-led initiatives. Initiatives sponsored

by an organization or organizations not attached to a particular government are sometimes called organization-led initiatives. The term intersessional initiatives is favoured throughout this book.

26 Sir Martin Holdgate, briefing paper to the UK government, 'Intergovernmental Panel on Forests, first meeting, New York, 11–15 September 1995,' paras. 5, 18, 20.

27 UN press release ENV/DEV/413, 'Global forest convention could ensure comprehensive, sustainable management, Commission on Sustainable Development told,' 10 April 1997. Available online at: www.un.org/News/Press/docs.1997/19970410.endev413.html (accessed 20 August 2004).

28 *Earth Negotiations Bulletin*, Vol. 5, No. 27, p.4.

29 ITFF information leaflet, 'Interagency Task Force on Forests,' undated (circa 1999).

30 The functions of these task forces vary from case to case. Some are established to promote coordinated strategies and policies, such as the UN Inter-Agency Task Force on Disaster Reduction, which was established by the General Assembly in 2000. Some are established for a fixed period to carry out research and to report to the UN secretary-general, such as the Inter-Agency Task Force on Sport for Development and Peace which was formed in 2002 and reported in 2003. The Inter-Agency Task Force on Gender and Trade was established in 2003: chaired by the UNCTAD it has nine member organizations, including the WTO, and aims to mainstream gender issues into international trade.

31 'German proposals for the organisation of the work of the Intergovernmental Panel on Forests, Ref.: Draft proposed programme of work of the CSD secretariat, 10 August 1995.'

32 This was agreed by the CSD when establishing the IPF. See UN document E/CN.17/IPF/1995/2, n.18 above, para. 19, p.19.

33 Credit belongs to the FAO, which responded to the criticisms of the TFAP and was largely responsible for the new conceptualization of national forest programmes (NFPs). See UN document E/CN.17/IPF/1996/14, 'Programme element I.1: Progress in national forest and land-use plans,' 16 August 1996.

34 The NFP concept has become particularly well established in Europe. COST Action E19, a major pan-European research programme carried out between 1999 and 2003, took the IPF proposals for action as its point of departure. The results of COST Action E19 can be found in: a special issue of *Forest Policy Economics*, Vol. 4, No. 4; Glück, Peter, Mendes, Américo M.S. Carvalho and Neven, Ine (eds) (2003) *Making NFPs Work: Supporting and Procedural Aspects. Report on COST Action 'National Forest Programmes in a European Context,' Publication Series of the Institute of Forest Sector Policy and Economics* 48, Vienna: Institute of Forest Sector Policy and Economics; Humphreys, David (ed) (2004) *Forests for the Future: National Forest Programmes in Europe, Country and Regional Reports from COST Action E19*, Luxembourg: European Communities.

35 On the controversy that surrounded this principle during the UNCED negotiations see Humphreys, n.2 above, pp.96–8. The principle appears in the final report of the IPF under the conclusions for programme area V: UN document E/CN.17/1997/12, 'Report of the ad hoc Intergovernmental Panel on Forests on its fourth session (New York, 11–21 February 1997),' 20 March 1997, para. 143.

36 Personal notes taken by co-chair Sir Martin Holdgate during the second session of the IPF. I am indebted to Sir Martin, who generously handed over to me his personal notes from the IPF negotiations, as well as other personal papers that have been invaluable in preparing this chapter.

37 Rio Declaration on Environment and Development, 14 June 1992, principle 22.
38 UN document E/CN.17/IPF/1997/6, 'Results of the international meeting of indigenous and other forest-dependent peoples on the management, conservation and sustainable development of all types of forests, Leticia, Colombia, 9–13 December 1996,' 17 January 1997, p.13.
39 UN document E/CN.17/1997/12, n.35 above, para. 36.
40 UN document E/CN.17/1997/12, n.35 above, para. 32.
41 I first made this argument in Humphreys, David (2004) 'Redefining the issues: NGO influence on international forest negotiations,' *Global Environmental Politics*, Vol. 4, No. 2, pp.51–74.
42 UN document E/CN.17/1997/12, n.35 above, para. 35; Convention on Biological Diversity, Article 8(j).
43 UN document E/CN.17/1997/12, n.35 above, paras. 47–50.
44 *Earth Negotiations Bulletin*, Vol. 13, No. 13, p.1.
45 *Earth Negotiations Bulletin*, Vol. 13, No. 3, p.2.
46 Advance unedited copy, UN document E/CN.17/1997/.../Add.1, 'Report of the High-Level Advisory Board on Sustainable Development for the 1997 review of the Rio commitments,' 18 February 1997, para. 15, p.6.
47 *Earth Negotiations Bulletin*, Vol. 13, No. 14, p.7.
48 *Earth Negotiations Bulletin*, Vol. 13, No. 7, p.2.
49 An important forum at this time for these issues was the CSD Intersessional Ad Hoc Working Group of Experts on Finance and Technology Transfer.
50 'Statement by Ambassador Daudi N. Mwakawago, Permanent Representative of the United Republic of Tanzania, Chairman of the Group of 77 and China, at the opening session of the ad hoc Intergovernmental Panel on Forests – fourth session, New York, 11 February 1997,' p.3.
51 On the unsuccessful efforts to negotiate a code of conduct see the account from Sidney Dell, the former executive director of the United Nations Centre on Transnational Corporations: Dell, Sidney (1989) 'The United Nations code of conduct on transnational corporations,' in Kaufmann, Johan (ed) *Effective Negotiation: Case Studies in Conference Diplomacy*, Dordrecht: Martinus Nijhoff/ Kluwer Academic, pp.53–74.
52 Bruno, K. (1992) 'The corporate capture of the Earth Summit,' *Multinational Monitor,* July/August. Available online at: www.essential.org/monitor.hyper/ mm0792.html (accessed 15 March 1997).
53 Humphreys, David (2001) 'Environmental accountability and transnational corporations,' in Gleeson, Brendan and Lowe, Nicholas (eds) *Governing for the Environment: Global Problems, Ethics and Democracy,* London: Palgrave, pp.88–101.
54 World Wide Fund for Nature, 'First meeting of the Intergovernmental Panel on Forests of the Commission on Sustainable Development, September 11–15 1995, New York – Forests for Life, Priorities for the IPF,' pp.9–10.
55 In January 1995 the Business Council for Sustainable Development merged with the World Industry Council for the Environment to create the World Business Council for Sustainable Development.
56 IUCN, 'Forum summary paper – *Not* seeing the forest for the trees: Focusing on realistic outputs for the IPF, 13 September, 1995,' p.3.
57 UN document E/CN.17/IPF/1997/2, 'Report of the ad hoc Intergovernmental Panel on Forests on its third session (Geneva, 9–20 September 1996),' 11 December 1996.

58 For the heavily bracketed text issued at the end of the Panel's third session see UN document E/CN.17/IPF/1997/2, n.57 above. For the co-chair's draft tabled at the Panel's fourth session see UN document E/CN.17/IPF/1997/3, 'Elements of a draft report, Note by the co-chairmen of the ad hoc Intergovernmental Panel on Forests,' 3 January 1997. For the Panel's final report see UN document E/CN.17/1997/12, n.35 above.

59 Sir Martin Holdgate, interview, Kirkby, Stephen, Cumbria, 17 August 2004.

60 UN document E/CN.17/IPF/1997/3, n.58 above, para. 60, p.24. This document had just six pairs of square brackets remaining on five issues: TFRK, voluntary codes of conduct, global CRI, trade sanctions and sustainable forest management, and certification.

61 UN document E/CN.17/1997/12, n.35 above, para. 69(a), p.26.

62 UN documents graduate proposals and recommendations according to different priorities. The verbs 'requested,' 'stressed,' 'urged' and 'called for' indicate a higher priority than 'recommended' and 'encouraged,' which in turn represent a higher priority than 'invited' and 'suggested.' See Pülzl, Helga and Rametsteiner, Ewald (2002) 'Evaluating the implementation of the IPF and IFF proposals for action & how to approach it (sic) In the experience of Austria and the European Community,' Institut für Sozioökonomik der Forst- und Holzwirtschaft,' Nr: P/2002-2, Vienna, p.8.

63 *Earth Negotiations Bulletin*, Vol. 13, No. 14, p.7.

64 Letter from Joshua Bishop of IIED to the Environmental Policy Department of the Overseas Development Administration, 25 February 1996. The IIED report was entitled *Economic Evaluation of Tropical Forest Land Use Options: A Review of Methodology and Applications*, London: IIED, December 1994. For the IPF paper that drew heavily from this publication see UN document E/CN.17/IPF/1996/7, 'Programme element III.1(b): Methodologies for proper valuation of the multiple benefits of forests,' 14 February 1996.

65 Haas, Peter M. (1990) 'Obtaining international environmental protection through epistemic consensus,' *Millennium: Journal of International Studies*, Vol. 19, No. 3, pp.347–363.

66 Smouts, Marie-Claude (2003) *Tropical Forests, International Jungle: The Underside of Global Ecopolitics*, London: Palgrave/Macmillan, pp.67–71. See also Angelsen, Arild and Kaimowitz, David (1999) 'Rethinking the causes of deforestation: Lessons from economic models,' *The World Bank Research Observer*, Vol. 14, No. 1, pp.73–98.

67 UN document E/CN.17/IPF/1996/7, n.64 above, p.29.

68 *Earth Negotiations Bulletin*, Vol. 13, No. 14, p.8.

69 *Earth Negotiations Bulletin*, Vol. 13, No. 9, p.1.

70 UN document E/CN.17/IPF/1997/2, n.57 above, para. 153(b), p.63.

71 This question dominated deliberations under IPF programme area V on international organizations and multilateral institutions and instruments. The Panel was aided in its deliberations on this subject by an intersessional initiative co-sponsored by the governments of Switzerland and Peru on international organizations and multilateral institutions. See UN document E/CN.17/IPF/1996/26, 'Swiss Peruvian Initiative on Forests, Report of the Independent Expert Group,' 21 August 1996.

72 I first outlined these points in Humphreys, David (2005) 'The elusive quest for a global forests convention,' *Review of European Community and International Environmental Law (RECIEL)*, Vol. 14, No. 1, pp.1–10.

73 EU document COM(96) 569 final, 'Communication from the Commission, A Common Platform: Guidelines for European Union preparation for the United Nations General Assembly Special Session to be held in New York in June 1997, to review Agenda 21 and related outcomes of the United Nations Conference on Environment and Development held in Rio de Janeiro in June 1992,' Brussels 12 November 1996, p.6; EU document 12713/96, ENV 407, DEVGEN 119, 'Council Conclusions, guidelines for European Union preparation for the United Nations General Assembly Special Session to be held in New York in June 1997,' Brussels, 10 December 1996, para. 9.

74 Manuel Rodríguez, interview, New York, 10 September 2004.

75 Which helps explain why the US Congress has yet to ratify the Convention on Biological Diversity.

76 *Earth Negotiations Bulletin*, Vol. 13, No. 20, p.2. See also 'Canadian Pulp and Paper Association calls for a legally-binding international forest agreement,' New York, April 13, 1995. Available online at: www.ee/lists/infoterra/1995/04/0039.html (accessed 29 July 2003).

77 Colchester, Marcus et al (2006) *Justice in the Forest: Rural Livelihoods and Forest Law Enforcement – Forest Perspectives 3*, Bogor: CIFOR, pp.45–46; Society of American Foresters (2003) 'International trade in forest products, a position statement of the Society of American Foresters.' Available online at: www.safnet.org/policyandpress/psst/international.cfm (accessed 24 March 2006).

78 Juliette Williams, Environmental Investigation Agency, interview, London, 24 July 1997.

79 Kolk, Ans (1996) *Forests in International Environmental Politics: International Organizations, NGOs and the Brazilian Amazon*, Utrecht: International Books, p.162.

80 Bill Mankin, interview, fourth session of the United Nations Forum on Forests, Geneva, 11 May 2004. The Global Forest Policy Project was an advocacy project of the National Wildlife Federation, Sierra Club and Friends of the Earth-US.

81 'International citizen (sic) declaration against a global forest convention,' 10 February 1997, New York. This was endorsed by 81 NGOs, namely 3 from Africa, 17 from Asia and Russia, 14 from Europe, 14 from Latin America and Caribbean, 28 from North America, 2 from Australia and 3 international NGOs.

82 'NGO declaration for IPF IV,' 21 January 1997.

83 Environmental Investigation Agency (1996) *Corporate Power, Corruption and the Destruction of the World's Forests: The Case for a New Global Forest Agreement*, London: Environmental Investigation Agency.

84 The FAO's Committee on Forestry is dominated by foresters and discussion on forestry. It could not have fulfilled the political role that the IPF assumed.

85 Manuel Rodríguez, interview, New York, 10 September 2004.

3 World Commission on Forests and Sustainable Development

1 United Nations General Assembly Resolution 38/161 of 1983 proposed the terms of reference for the World Commission on Environment and Development. The UN secretary-general subsequently invited Gro Harem Brundtland of Norway to chair the commission.

2 Independent Commission on International Humanitarian Issues (1986) *The Vanishing Forest: The Human Consequences of Deforestation*, London and New Jersey: Zed Books.
3 Independent Commission on International Humanitarian Issues (1987) *Indigenous Peoples: A Global Quest for Justice*, London and New Jersey: Zed Books.
4 World Commission for Peace and Human Right Council website: http://greenfield.fortunecity.com/swallowtail/408/ (accessed 19 November 2004).
5 World Commission on Global Consciousness and Spirituality website: http://globalspirit.org/ (accessed 19 November 2004).
6 Commission on International Development (1969) *Partners in Development*, New York: Praeger.
7 Independent Commission on International Development Issues (1980) *North–South: A Programme for Survival*, London: Pan Books.
8 Brandt Commission (1983) *Common, Crisis, North–South Co-operation for World Recovery*, London: Pan Books.
9 Independent Commission for Disarmament and Security Issues (1982) *Common Security: A Blueprint for Survival*, New York: Simon and Schuster.
10 Independent Commission on International Humanitarian Issues, n.2 above.
11 Independent Commission on International Humanitarian Issues, n.3 above.
12 World Commission on Environment and Development (1987) *Our Common Future*, Oxford: Oxford University Press.
13 World Commission on Culture and Development (1992) *Our Creative Diversity*, Oxford & IBH Publishing Company.
14 Commission on Global Governance (1995) *Our Global Neighbourhood*, Oxford: Oxford University Press.
15 Independent World Commission on the Oceans (1998) *The Ocean: Our Future*, Cambridge: Cambridge University Press.
16 World Commission on Forests and Sustainable Development (1999) *Our Forests, Our Future*, Cambridge: Cambridge University Press.
17 World Commission on Dams (2000) *Dams and Development: A New Framework for Decision-Making*, London: Earthscan.
18 World Commission on the Social Dimensions of Globalization (2004) *A Fair Globalization: Creating Opportunities for All*, Geneva: International Labour Organization.
19 Cited in a briefing from the Global Forests Policy Project to the Carter Center, 1993, undated.
20 Kilaparti Ramakrishna, Woods Hole Research Center, email 1 September 1995. (Ramakrishna coordinated the WCFSD organizing committee.)
21 Ullsten, Ola, Nor, Salleh Mohammed and Yudelman, Montague (1990) *Tropical Forestry Action Plan: Report of the Independent Review*, Kuala Lumpur: FAO, p.48.
22 Fax from George M. Woodwell, Woods Hole Research Center, Massachusetts to Jimmy Carter, The Carter Center, Atlanta, Georgia, 27 March 1992. This quote is reproduced in italics on pages 53–54 of this volume. The emphasis appears as plain text.
23 The organizing committee met on two subsequent occasions (November 1992, Ottawa and April 1993, New Delhi).
24 'Summary of GFPP activity re WCFSD as of December 8, 1993.'
25 Mankin, William E. (1998) 'Entering the fray – International forest policy processes: An NGO perspective on their effectiveness,' *Policy That Works for Forests and People: Discussion Paper* 9, London: International Institute for Environment and Development, p.36.

26 'Alert, To: All forest activists and NGOs, From: Global Forest Policy Project (of the National Wildlife Federation Sierra Club and Friends of the Earth-US, Re: Urgent input needed on forest user-group leaders, April 10, 1993.'
27 Letter from Bekki J. Johnson, Assistant Director of Operations for Programs, Carter Center, Atlanta, Georgia to George M. Woodwell, Woods Hole Research Center, Massachusetts, 22 April 1993.
28 Letter from Bekki J. Johnson, n.27 above.
29 Fax from George M. Woodwell, Woods Hole Research Center to Bekki J. Johnson, Carter Center, 5 May 1993.
30 World Commission on Forests and Sustainable Development (1993) 'Possible mandate, key issues, strategy and work plan: Issued under the responsibility of the organizing committee for the establishment of an independent World Commission on Forests and Sustainable Development, June 1993,' p.5; Fax from George M. Woodwell, Woods Hole Research Center, Massachusetts to Jimmy Carter, The Carter Center, Atlanta, Georgia, 27 March 1992.
31 I am grateful to this campaigner, who wishes to remain anonymous, for providing a verbal briefing and for allowing me access to several important documents.
32 Letter to Boutros Boutros-Ghali, secretary-general of the United Nations, from Environmental Defense Fund, National Audubon Society, Sierra Club, Friends of the Earth-US and National Wildlife Federation, dated 7 June,1993.
33 Letter to Boutros Boutros-Ghali, n.32 above.
34 Letter from President Jimmy Carter to Boutros Boutros-Ghali, 15 June, 1993.
35 Letter from Boutros Boutros-Ghali to Jimmy Carter, 26 July 1993.
36 In November 1992 in New York following the second meeting of the organizing committee: World Commission on Forests and Sustainable Development, n.30 above, p.3.
37 Manuel Rodríguez, WCFSD commissioner, interview, New York, 10 September 2004.
38 Funding was raised from a number of government agencies, international organizations and private foundations.
39 Copies of the letter sent to Ullsten and Salim can be found in 'World Commission on Forests and Sustainable Development: Proposed work programme, August 1995,' pp.27–33.
40 The letterhead under which the invitations to Ullsten and Salim were sent named Ullsten as a member of the InterAction Council. Other members at that time included Raul Alfonsin (Argentina), James Callaghan (UK), Malcom Fraser (Australia), Mikhail Gorbachev (Russia), Tadeusz Mazowiecki (Poland) and José Sarney (Brazil).
41 World Commission on Forests and Sustainable Development, n.16 above, p.xv.
42 World Commission on Forests and Sustainable Development, 'Potential areas of collaboration with the UNCSD Intergovernmental Panel on Forests,' 11 September 1995.
43 Manuel Rodríguez, WCFSD commissioner, interview, New York, 10 September 2004.
44 UN document E/CN.17/IFF/2000/8, 'Summary report of the World Commission on Forests and Sustainable Development,' 24 January 2000.
45 Manuel Rodríguez, WCFSD commissioner, interview, New York, 10 September 2004. The Yaoundé meeting was sponsored by the European Commission and the African Development Bank.
46 David Pearce, WCFSD commissioner, email, 29 September 2004.

47 Marcus Colchester, Forest Peoples Programme, email, 11 March 2006.
48 World Commission on Forests and Sustainable Development, n.16 above, pp.128–130.
49 Marcus Colchester, Forest Peoples Programme, email, 11 March 2006.
50 World Commission on Forests and Sustainable Development, n.16 above, pp.127–128, 141.
51 World Commission on Forests and Sustainable Development, n.16 above, p.33.
52 World Commission on Forests and Sustainable Development, n.16 above, pp.31–44.
53 World Commission on Forests and Sustainable Development, n.16 above, p.71.
54 World Commission on Forests and Sustainable Development, n.16 above, p.157.
55 WCFSD Secretariat (1997) 'Our Forests: Our Future, internal discussion draft,' Geneva, 30 June 1997, para. 355, p.64.
56 WCFSD Secretariat, n.55 above, para. 489, p.100.
57 World Commission on Forests and Sustainable Development, n.16 above, p.66.
58 World Commission on Forests and Sustainable Development, n.16 above, p.66.
59 *Official Journal of the European Communities*, C295, Vol. 33, 26 November 1990, 'Resolution on measures to protect the ecology of tropical forests,' 25 October 1990, para. 8, p.196.
60 For the full draft of the GLOBE draft convention see Humphreys, David (1996) *Forest Politics: The Evolution of International Cooperation*, London: Earthscan, Annex D, pp.246–275.
61 European Parliament document A3-0024/92, 'Report of the Committee on the Environment, Public Health and Consumer Protection on the need for a convention on the protection of forests, Rapporteur: Mr Hemmo Muntingh,' 24 January 1992. The document acknowledges that 'the rapporteur has borrowed from the GLOBE texts,' p.8.
62 David Pearce, WCFSD commissioner, email, 13 May 2005.
63 Angela Cropper is a former executive secretary of the Convention on Biological Diversity. She undertook most of the writing of the Commission's final report.
64 Manuel Rodríguez, interview, New York, 10 September 2004; David Pearce, email, 29 September 2004. I am grateful to Professors Rodríguez and Pearce for kindly taking the time to answer my questions on the Commission. Two other commissioners and a senior member of the WCFSD secretariat did not reply.
65 Bass, Stephen and Thomson, Koy (1997) 'Forest security: Challenges to be met by a global forest convention,' *Forest and Land Use Series No. 10*, London: International Institute for Environment and Development, p.iii.
66 World Commission on Forests and Sustainable Development, n.16 above, pp.142–143.
67 World Commission on Forests and Sustainable Development, n.16 above, p.76.
68 Global Forest Watch (2003) 'What is Global Forest Watch?' Available online at: www.globalforestwatch.org.english/about/index.htm (accessed 28 July 2003).
69 World Commission on Forests and Sustainable Development, n.16 above, p.146.
70 World Commission on Forests and Sustainable Development, n.16 above, p.146.
71 At a meeting at the Commonwealth offices in London on 23 March 2000 I asked a panel of five commissioners which countries had expressed interest in the proposed forest security council. Ola Ullsten replied that the government of Canada had expressed support. Ullsten noted that the Commission was also approaching other (unnamed) governments to canvass support for the proposal.

72 World Commission on Environment and Development, n.12 above, pp.49–52, 82–3.
73 Bernstein, Steven (2000), 'Ideas, social structure and the compromise of liberal environmentalism,' *European Journal of International Relations*, Vol. 6, No. 4, pp.464–512; Bernstein, Steven (2002) 'Liberal environmentalism and global environmental governance,' *Global Environmental Politics*, Vol. 2, No. 3, pp.1–16.

4 Intergovernmental Forum on Forests

1 Food and Agriculture Organization and United Nations Development Programme (1999) *Practitioner's Guide to the Implementation of the IPF Proposals for Action*, Eschborn: Gesellschaft für Technische Zusammenarbeit (GTZ).
2 FERN (2000) 'Implementation of the proposals for action of the Intergovernmental Panel on Forests: European report,' April. The ten countries surveyed were Austria, Belgium, Denmark, Finland, France, Germany, Ireland, Netherlands, Sweden and the UK. Two countries, Portugal and Greece, did not respond.
3 Verolme, Hans J., Mankin, William E., Ozinga, Saskia and Ryder, Sophia et al (2000) *Keeping the Promise? A Review by NGOs and IPOs of the Implementation of the UN Intergovernmental Panel on Forests 'Proposals for Action' in Selected Countries*, Washington DC: Biodiversity Action Network and Global Forest Policy Project, pp.1–2.
4 UN document E/CN.18/2000/14, 'Report of the Intergovernmental Forum on Forests on its fourth session,' 20 March 2000, para. 42.
5 UN document E/CN.18/2000/14, n.4 above, para. 41(a).
6 All quotations or reference to statements made at the fourth session of the IFF that are not individually referenced were witnessed by the author while observing the negotiations and committed to note form at the time. I am grateful to the United Nations Committee on Environment and Development for the UK (UNED-UK), now the Stakeholder Forum, for kindly accrediting me to visit the IFF.
7 Verolme, Hans J. and Moussa, Juliette (1999), *Addressing the Underlying Causes of Deforestation and Forest Degradation: Case Studies, Analysis and Policy Recommendations*, Washington DC: Biodiversity Action Network, p.i.
8 Verolme and Moussa, n.7 above, p.3.
9 UN document E/CN.18/2000/14, n.4 above, para. 58.
10 Verolme and Mousa, n.7 above, p.3.
11 Verolme and Mousa, n.7 above, p.4.
12 UN document E/CN.18/2000/14, n.4 above, para. 28.
13 UN document E/CN.17/IFF/1999/4, 'Programme element II.a, matters left pending and other issues arising from the programme elements of the Intergovernmental Panel on Forests process, the need for financial resources,' 18 February 1999, para. 54, p.12. The authorship of this report is not clear, although presumably it was prepared by one of the ITFF member organizations. The lead ITFF agency for financial resources was the United Nations Development Programme.
14 UN document E/CN.18/2000/14, n.4 above, para.30(c). For other references to private sector investment in this document see paras. 21, 22, 28, 30(c), 50, 56(b) and 111.
15 'Statement by Mrs Nandhini I. Krishna, Counsellor, Permanent Mission of India, at the fourth session of the Intergovernmental Forum on Forests under Category III issues,' 1 February 2000.

16 UN document E/CN.18/2000/14, n.4 above, para. 29.
17 UN document E/CN.18/2000/14, n.4 above, para. 47.
18 See, for example, the section of the IFF's final report on the transfer of environmentally sound technologies: UN document E/CN.18/2000/14, n.4 above, especially paras. 56(j) and (k).
19 Berkes, Fikret (1999) *Sacred Ecology: Traditional Ecological Knowledge and Resource Management*, Philadelphia: Taylor and Francis, p.4.
20 Berkes, n.4 above, p.5.
21 The term local knowledge is preferred in a paper prepared for the Intergovernmental Panel on Forests. See Antweiler, Christoph and Mersmann, Christian (1996) 'Local knowledge and cultural skills as resources for sustainable forest development, IPF programme element 1.3, traditional forest-related knowledge,' Eschborn, August.
22 Banuri, Tariq and Marglin, Frédérique Apffel (1993) 'A systems-of-knowledge analysis of deforestation, participation and management,' in Banuri, Tariq and Marglin, Frédérique Apffel (eds) *Who will Save the Forests: Knowledge, Power and Environmental Destruction*, London: Zed Books; Berkes, n.19 above, p.176. Berkes cites Capra, Fritjof (1996) *The Web of Life*, New York: Anchor/Doubleday.
23 See, for example Banuri and Marglin, n.22 above; and Shiva, Vandana (1993) *Monocultures of the Mind: Perspectives on Biodiversity and Biotechnology*, London: Zed Books, p.21.
24 Goldman, Michael (1998) 'The political resurgence of the commons,' in Goldman, Michael (ed) *Privatizing Nature: Political Struggles for the Global Commons*, London: Pluto, p.2.
25 ILO Convention 169, 1989, Article 7.
26 The UPOV acronym is derived from the French name for the convention, the Union internationale pour la Protection des Obtentions Végétales. The convention has been updated three times since 1961; in 1972, 1978 and 1991.
27 International Convention for the Protection of New Varieties of Plants (UPOV convention), 1961, revised 1972, 1978 and 1991, Article 5. Available online at: www.upov.int/en/publications/conventions/1991/act1991.htm (accessed 19 July 2004).
28 UPOV convention, n.27 above, Article 14.
29 Flitner, Michael (1998) 'Biodiversity: Of local commons and global commodities,' in Goldman (ed), n.24 above, p.152.
30 International Undertaking on Plant Genetic Resources, 1983, Annex II.
31 International Undertaking on Plant Genetic Resources, 1983, Annex I and Annex II.
32 Kloppenburg, Jack R. and Kleinman, Daniel Lee (1988) 'Seeds of controversy: national property versus common heritage,' in Kloppenburg, Jack R. (ed) *Seeds and Sovereignty: The Use and Control of Plant Genetic Resources*, Durham NC: Duke University Press, cited in Flitner, n.29 above, p.153.
33 media-g-eng-l@mailser.fao.org, 29 June 2004, 'International plant genetic treaty becomes law.'
34 Barrero, Joaquín Molano (2004) 'Colombia: The wealth of the jungle and the logic of expropriation,' in Lovera, Miguel, Avendaño, Tatiana Rosa and Torres, Irene Vélez (eds) *The New Merchants: Life as Commodity*, Bogota: Global Forest Coalition and Censet Agua Viva, p.17.
35 'Frequently asked questions about TRIPS in the WTO.' Available online at: www.wto.org/english/tratop_e/trips_etripfq_e.htm (accessed 26 October 2004).

36 Purdue, Derrick (1995) 'Hegemonic Trips: World trade, intellectual property and biodiversity,' *Environmental Politics*, Vol. 4, No. 1, pp.88–107.
37 'Agreement on Trade-Related Aspects of Intellectual Property Rights,' Article 27.3 (b).
38 As noted on the WTO website, 'TRIPS: Reviews, Article 27.3(b) and related issues': www.wto.org/english/tratop_e/trips_e/art27_3b_background_e.htm (accessed 26 October 2004).
39 WTO document WT/MIN901)/DEC/1, 'Ministerial Declaration: Adopted on 14 November 2001,' 20 November 2001, para. 19.
40 'The Thammasat Resolution.' Available online at: www.greens.org/s-r/16/16-13.html (accessed 3 November 2004). In the Thai language the word 'thammasat' has two meanings: 'knowledge of nature,' and 'justice.'
41 For an overview of TRIPS, the CBD and the differences between them, see Duttfield, Graham (2000) *Intellectual Property Rights, Trade and Biodiversity*, London: Earthscan, pp.75–90.
42 UN document E/CN.17/2000/14, n.4 above, paras. 68–75.
43 UN document E/CN.17/2000/14, n.4 above, para. 56(k).
44 Bob Watson, head of the World Bank's Environment Department, interview, Geneva, 25 February 1999. (Dr Watson was being interviewed for an Open University TV programme on climate change for the course 'Environmental Policy in an International Context.')
45 UN document E/CN.17/IFF/1998/8, 'Programme element II.d(ii), valuation of forest goods and services; economic instruments, tax policies and land tenure; future supply of and demand for wood products and non-wood products; and rehabilitation of forest cover,' 19 June 1998, pp.4–6.
46 UN document E/CN.17/IFF/1998/8, n.45 above, para. 15, p.6; UN document E/CN.17/IFF/1999/13, 'Issues that need further clarification: Economic instruments, tax policies and land tenure,' 18 February 1999, para. 40(b).
47 UN document E/CN.17/IFF/1999/12, 'Programme element II.d (vi), matters left pending and other issues arising from the programme elements of the Intergovernmental Panel on Forests process, issues that need further clarification: economic instruments, tax policies and land tenure,' 18 February 1999, para. 9(a), p.4.
48 UN document E/CN.17/IFF/1999/13, n.46 above, p.1.
49 UN document E/CN.17/IFF/1998/8, n.45 above, para. 17, p.6.
50 For an interesting analysis of Foucault's thought and its application to environmental politics see Toke, Dave (2000) *Green Politics and Neo-liberalism*, Basingstoke: MacMillan.
51 *Earth Negotiations Bulletin*, Vol. 13, No. 55, p.6; UN document E/CN.17/2000/14, n.4 above, para. 119.
52 *Earth Negotiations Bulletin*, Vol. 13, No. 55, p.10; UN document E/CN.17/2000/14, n.4 above, para. 117.
53 Similar price trends can be observed for other non-renewable resources, such as metals. See Clapp, Jennifer and Dauvergne, Peter (2005) *Paths to a Greener World: The Political Economy of the Global Environment*, Cambridge MA: MIT Press, p.103.
54 Taghi Shamekhi, former chair of LFCC process, interview, University of Toronto, 15 March 2004. Professor Shamekhi, a professional forester, chaired the LFCC process in 1998.
55 Anon., 'LFCCs: Low Forest Cover Countries' (undated leaflet).

56 List of forest cover in Low Forest Cover Countries (1995 figures). Available online at: www.lfccs.net/information_2.htm (accessed 15 December 2004).
57 Tehran Declaration. Available online at: www.lfccs.net/dec.htm (accessed 13 March 2004).
58 Taghi Shamekhi, former chair of LFCC process, interview, University of Toronto, 15 March 2004.
59 EU document COM(96) 569 final, 'Communication from the Commission, A Common Platform: Guidelines for European Union preparation for the United Nations General Assembly Special Session to be held in New York in June 1997, to review Agenda 21 and related outcomes of the United Nations Conference on Environment and Development held in Rio de Janeiro in June 1992,' Brussels, 12 November 1996, p.6; EU document 12713/96, ENV 407, DEVGEN 119, 'Council Conclusions, guidelines for European Union preparation for the United Nations General Assembly Special Session to be held in New York in June 1997,' Brussels, 10 December 1996, para. 9.
60 Interview with EU delegate, fourth session of the Intergovernmental Forum on Forests, New York, 9 February 2000.
61 Interview with EU delegate to the fourth session of the IFF, Stockholm, 17 September 2004.
62 I am grateful to John Bazill of DG Environment for explaining how the EU operates during international environmental negotiations: interview, fourth session of the United Nations Forum on Forests, Geneva, 6 May 2004. Box 4.4 also draws from John Vogler (1999) 'The European Union as an actor in international environmental politics,' *Environmental Politics*, Vol. 8, No. 3, pp.24–48.
63 *Earth Negotiations Bulletin*, Vol. 13, No. 66, p.11.
64 The ECOSOC has a rotating membership, with 54 member states each serving a three year term. It was unprecedented for an ECOSOC subsidiary body to have a membership larger than the ECOSOC itself. For example, the CSD, which reports to the ECOSOC, has 53 member states.
65 Jag Maini, former head and coordinator of the IFF secretariat, interview, New York, 24 May 2005; Jag Maini, email, 18 July 2005.
66 The proposal was signalled in a US paper circulated at the IFF: 'President Clinton and Vice President Gore: Protecting forests and biodiversity around the world,' 4 February 2000.
67 A spending proposal from the US president is no guarantee that funds will be approved by Congress.
68 UN document E/CN.17/2000/14, n.4 above, Annex, para. 3(c).
69 UN document E/CN.17/2000/14, n.4 above, Annex, para. 3(d).
70 UN press release ECOSOC 5934, 'Economic and Social Council establishes new UN Forum on Forests,' 18 October 2000. (I am grateful to Peter Willetts for passing this document to me.)
71 UN document E/2000/L/32, 'Draft resolution submitted by the President of the Council on the basis of informal consultations held on his behalf by Bagher Asadi (Islamic Republic of Iran), Report of the fourth session of the Intergovernmental Forum on Forests,' 18 October 2000, para. 18.

5 United Nations Forum on Forests

1 However, not too much should be made of this distinction: the IPF and IFF were 'open ended' fora, and all UN members could participate in their work, either as one of the 53 members of the Commission on Sustainable Development or as observers. During the IPF and IFF negotiations no distinction was made between member states and observer states.

2 This workshop was known as the Eight Country Initiative. The co-sponsors were Australia, Brazil, Canada, France, Germany, Iran, Malaysia and Nigeria.

3 World Rainforest Movement (2000) 'Report on the 8-Country Initiative: An international expert meeting to discuss "Shaping the programme of work for the United Nations Forum on Forests (UNFF)", Bonn, 27 November – 1 December 2000.' Available online at: www.wrm.org.uy/actors/IFF/8country.html (accessed 7 January 2003).

4 UN document E/CN.18/2001/2, 'Eight-Country Initiative: Shaping the programme of work for the United Nations Forum on Forests (UNFF),' 7 February 2001.

5 UN document E/2001/42(Part II)–E/CN.18/2001/3(Part II), 'Report of the United Nations Forum on Forests on its first session, New York, 11–22 June 2001,' Decision 1/2, pp.5–13.

6 People's Forest Forum (2004) 'Open letter for country delegations in [sic] fourth session of the UN Forum on Forests,' 14 May.

7 On the claimed advantages and potential problems with GM trees see *World Rainforest Movement Bulletin*, Issue 88, November 2004. Available online at: www.wrm.org.uy/bulletin/88/viewpoint.html (accessed 4 January 2005); Owusu, Rachel Asante (1999) *GM Technology in the Forest Sector: A Scoping Study for WWF*, Godalming: WWF-UK; Sedjo, Roger A. (2004) *Genetically Engineered Trees: Promise and Concerns*, Washington DC: Resources for the Future; Lang, Chris (2004) *Genetically Modified Trees: The Ultimate Threat to Forests*, Montevideo: World Rainforest Movement/Friends of the Earth; Humphreys, David, Gosens, Jorrit, Jackson, Michael J., Plasmeijer, Anouska, van Betuw, Wouter and Mohren, Frits (2005) 'Biotechnology in the forest?: Policy options on research on GM Trees,' *European Forest Institute Discussion Paper* 12, Joensuu: European Forest Institute.

8 The eight members of the Interagency Task Force on Forests joined the CPF, namely Food and Agriculture Organization (Chair), Center for International Forestry Research, Convention on Biological Diversity, International Tropical Timber Organization, UN Department of Economic and Social Affairs/UNFF secretariat, United Nations Development Programme, United Nations Environment Programme and the World Bank. In addition there were six new members: Convention to Combat Desertification, Framework Convention on Climate Change, Global Environment Facility, International Union of Forest Research Organizations, World Conservation Union (IUCN) and World Agroforestry Centre (formerly the International Centre for Research in Agroforestry).

9 UN document E/CN.18/2005/3/Add.6, 'Multi-stakeholder dialogue, discussion paper contributed by the indigenous peoples major group,' 7 March 2005, p.15.

10 NGO newsletter circulated at the fifth session of the UNFF: *UN Forest Frustrations*, Issue 2, 17 May 2005, p.2.

11 UN document E/CN.18/2004/INF/1, 'Collaborative Partnership on Forests framework 2004,' 4 March 2004, pp.4–7.

12 World Rainforest Movement (2001) 'UNFF: Little hope for forest peoples and forest biodiversity in this Forum.' Available online at: www.wrm.org.uy/bulletin/56UNFF.html (accessed 7 January 2003).

13 UN document A/CONF.199/PC/8, 'Ministerial declaration and message from the United Nations Forum on Forests to the World Summit on Sustainable Development,' 19 March 2002, para. 15.i.

14 This paragraph draws in part from work published earlier as Humphreys, David (2003) 'The United Nations Forum on Forests: Anatomy of a stalled international process,' *Global Environmental Change*, Vol. 13, No. 4, pp.319–323.

15 Hemmati, Minu (2002) *Multi-stakeholder Processes for Governance and Sustainability: Beyond Deadlock and Conflict*, London: Earthscan, p.33–34. I am indebted to Felix Dodds of the Stakeholder Forum for explaining the origins of the multi-stakeholder dialogue concept: interview, Stockholm, 29 August 2003.

16 UN document E/CN.17/2000/14, 'Report of the Intergovernmental Forum on Forests on its fourth session (New York, 31 January–11 February 2000),' Annex, para. 5.

17 Multi-stakeholder dialogue is increasingly being adopted outside the UN system. For example, the Ministerial Conference on the Protection of Forests in Europe held in Vienna in 2003 included a multi-stakeholder dialogue segment.

18 Words in quotation marks were noted by the author when Okrah delivered his opening statement on behalf of the NGOs at the UNFF's fourth session (Geneva, 6 May 2004). Okrah has written about his views on privatization in Lovera, Miguel, Avendaño, Tatiana Rosa and Torres, Irene Vélez (eds) (2005) *The New Merchants: Life as Commodity*, Bogotá: Global Forest Coalition/Censat Agua Viva.

19 Mary Coulombe, opening statement on behalf of business and industry, fourth session of the UNFF, Geneva, 6 May 2004. This last point was reiterated in the business and industry discussion paper, which stated that business and industry is 'concerned that more efforts need to be focused at the international level on creating secure, predictable, legally structured institutional frameworks within which businesses can operate': UN document E/CN.18/2004/4/Add.4, 'Discussion paper contributed by the business and industry major group,' 25 February 2004, pp.1, 8.

20 Marcus Colchester, Forest Peoples Programme, email, 11 March 2006.

21 UN document E/CN.18/2005/3/Add.4, 'Multi-stakeholder dialogue, discussion paper submitted by the non-governmental organizations major group,' 24 March 2005, para. 8, p.4.

22 Ole Henrik Magga, chair of the Permanent Forum on Indigenous Issues, verbal, statement to the fourth session of the UNFF, Geneva, 3 May 2004.

23 *Earth Negotiations Bulletin*, Vol. 13, No. 83, p.8. The language agreed was that targets are to be 'set by individual countries within the framework of national forest processes, as appropriate': UN document E/2001/42(Part II)-E/CN.18/2001/3(Part II), n.5 above, p.18.

24 Mia Söderlund, UNFF Secretariat, email, 20 January 2003.

25 Of these 16 were developed countries, 11 developing countries and 7 from countries with economies in transition.

26 UN document E/2002/42-E/CN.18/2002/14, 'United Nations Forum on Forests: Report of the second session (22 June 2001 and 4–15 March 2002),' *ECOSOC Official Records*, 2002, Supplement No. 22, p.30.

27 UN document E/2002/42-E/CN.18/2002/14, n.26 above, p.32.

28 Prior to this two country-led intersessional initiatives on this subject had taken place, in Japan (November 2001) and Italy (March 2003): UN documents E/CN.18/2002/12, 'International expert meeting on monitoring, assessment and reporting on the progress towards sustainable forest management, 5–8 November 2001, Yokohama, Japan,' 9 January 2002; and E/CN.18/2003/9, 'Lessons learned in monitoring, assessment and reporting on implementation of IPF/IFF proposals for Action, The Viterbo Report, 17–20 March 2003, Viterbo, Italy,' 4 April 2003.

29 UN document E/CN.18/2004/2, 'Report of the ad hoc expert group on approaches and mechanisms for monitoring, assessment and reporting,' 29 January 2004, pp.8, 10, 12–15.

30 UNFF Resolution 4/3, 'Forest-related monitoring, assessment and reporting: criteria and indicators for sustainable forest management,' para. 4, in UN document E/2004/42–E/CN.18/2004/17, 'United Nations Forum on Forests, report on the fourth session (6 June 2003 and 3–14 May 2004),' p.6.

31 There were also two intersessional initiatives on environmentally sound technologies hosted by the Republic of Congo and Nicaragua. However these initiatives appear to have operated in isolation both from the ad hoc expert group and from each other. The group reported as the 'ad hoc expert group on the [sic] finance and transfer of environmentally sound technologies.' This might suggest that the group was tasked not with producing options on finance per se, but only finance as it relates to environmentally sound technologies. However the group considers a broad range of financial matters, not all of which relate to technology. Note that the definite article is missing in the mandate of the group as established at the UNFF's first session: UN document E/2001/42/Rev.1– E/CN.18/2001/3/Rev.1, 'United Nations Forum on Forests, report of the organizational and first sessions (12 and 16 February and 1122 June 2001),' p.8.

32 UN document E/CN.18/2004/5, 'Report of the ad hoc expert group on the finance and transfer of environmentally sound technologies (15–19 December 2003, Geneva),' 11 February 2004, pp.21–22.

33 UN document E/CN.18/2004/5, n.32 above, pp.22–23.

34 'Statement on behalf of the Group of 77 and China delivered by Mr Faisal Abdullah al-Athba, Delegation of the Qatar Mission to the United Nations, New York, at the United Nations Forum on Forests, Geneva, 3 May 2004,' pp.3–4.

35 'Statement on behalf of the Group of 77 and China...,' n.34 above.

36 Peru for the G77, verbal intervention, fourth session of the UNFF, Geneva, 12 May 2004.

37 US, verbal intervention, fourth session of the UNFF, Geneva, 12 May 2004.

38 *Earth Negotiations Bulletin*, Vol. 13, No. 116, p.9.

39 UNFF Decision 4/2, 'Report of the ad hoc expert group on finance and transfer of environmentally sound technologies,' para.2, in UN document E/2004/42–E/CN.18/2004/17, n.30 above, p.11.

40 Anne Castle, Canadian delegate, interview, fourth session of the UNFF, Geneva, 12 May 2004. Section 35 of the Constitution Act of 1982 is available online at: http://laws.justice.gc.ca/en/const/annex_e.html (accessed 12 November 2005).

41 See, for example, Keal, Paul (2003) *European Conquest and the Rights of Indigenous Peoples: The Moral Backwardness of International Society*, Cambridge: Cambridge University Press, pp.217–223.

42 UN document E/CN.17/1997/12, 'Report of the ad hoc Intergovernmental Panel on Forests on its fourth session (New York, 11–21 February 1997),' 20 March 1997; UN document E/CN.18/2000/14, n.16 above.

43 Johannesburg Declaration on Sustainable Development, Article 25. Available online at: www.un.org/esa/sustdev/documents/WSSD_POI_PD/English/POI_PD.htm (accessed 15 March 2006).
44 UNFF Resolution 3/3, 'Maintaining forest cover to meet present and future needs,' in UN document E/2003/42-E/CN.18/2003/13, 'United Nations Forum on Forests, Report on the third session (15 March 2002 and 26 May to 6 June 2003),' p.17
45 UN document A/RES/60/1, 'Resolution adopted by the General Assembly [without reference to a Main Committee (A/60/L.1)],' 24 October 2005, para. 127, p.28. See also para. 56(d), p.13.
46 'Statement on behalf of the Group of 77 and China...,' n.34 above, p.5. Mitzi Gurgel Valente da Costa, Brazilian delegate and G77 spokesperson, interview, fourth session of the UNFF, Geneva, 13 May 2004.
47 US, verbal intervention, fourth session of the UNFF, Geneva, 12 May 2004.
48 Brazil for the G77, verbal intervention, fourth session of the UNFF, Geneva, 13 May 2004.
49 During a recess in the negotiations I asked the G77 spokesperson if the shift in negotiation strategy had taken place because the developing countries felt that their interests on this subject could be better realized at the CBD. The reply was 'yes.' Mitzi Gurgel Valente da Costa, Brazilian delegate and G77 spokesperson, pers comm, fourth session of the UNFF, Geneva, 13 May 2004.
50 As president of the EU the Irish delegation to the fourth session of the UNFF delivered the first EU statement in the UN system after the historic expansion of the EU on 1 May 2004. The addition of ten countries took the EU's membership to 25. The EU now has a forest cover that is approximately one fifth of the total global land area.
51 Those delegates that played the most active role in the evening session were Canada, the EU (the Irish presidency and the European Commission), the G77, New Zealand and the US. Some other delegations also attended but did not participate in the negotiations.
52 This was confirmed to the author after the negotiations had broken down by a member of the US delegation, pers comm, Geneva, 14 May 2004.
53 Brazil for the G77, verbal intervention, fourth session of the UNFF, Geneva, 13 May 2004.
54 US, verbal intervention, fourth session of the UNFF, Geneva, 13 May 2004.
55 Ireland for the EU, verbal intervention, fourth session of the UNFF, Geneva, 13 May 2004.
56 US, verbal intervention, fourth session of the UNFF, Geneva, 13 May 2004.
57 After the fourth session of the UNFF the International Alliance of Indigenous and Tribal Peoples of the Tropical Forests held an intersessional meeting in Costa Rica in December 2004 to evaluate progress on the implementation of the IPF/IFF proposals for action on TFRK, as well as on commitments made at the CBD. The findings, based on extensive case studies, are contained in Newing, Helen (ed) (2005) *Our Knowledge for Our Survival: Traditional Forest Related Knowledge and the Implementation of Related International Commitments*, Chiang Mai: International Alliance of Indigenous and Tribal Peoples of the Tropical Forests (2 volumes).
58 The third session of the UNFF agreed a resolution on enhanced cooperation that *inter alia* invited clarification of the concepts of the ecosystem approach and sustainable forest management. Resolution 3/4, 'Enhanced cooperation and policy and programme coordination,' in UN document E/2003/42-E/CN.18/2003/13, n.44 above, pp.19–21.

59 *Earth Negotiations Bulletin*, Vol. 13, No. 116, p.8.
60 UN document E/2001/42/Rev.1– E/CN..18/2001/3/Rev.1, 'United Nations Forum on Forests, Report of the organizational and first sessions (12 and 16 February and 11–22 June 2001),' p.8.
61 Libby Jones of the UK delegation kindly allowed me to make a verbal contribution to this meeting from the UK desk on 9 September 2004.
62 UN document E/CN.18/2005/2, 'Report of the ad hoc expert group on consideration with a view to recommending the parameters of a mandate for developing a legal framework on all types of forests (New York, 7–10 September 2004),' 29 September 2004, para. 55–65, pp.16–18; UN document E/CN.18/2005/11, 'Co-chairs' report on the country-led initiative in support of the United Nations Forum on Forests on the future of the international arrangement on forests (the Guadalajara Report),' pp.15–17.
63 UN document E/CN.18/2005/2, n.62 above, p.11; UN document E/CN.18/2005/11, n.62 above, para. 20, p.9.
64 The report of the ad hoc expert group meeting contains just two brief mentions of non-legally binding instruments: UN document E/CN.18/2005/2, n.62 above, para. 41, p.13, and para. 55, p.16. The report of the Guadalajara intersessional has a more extensive treatment of this subject: UN document E/CN.18/2005/11, n.62 above, pp. 14–15. See also *Sustainable Developments*, Vol. 101, No. 1, pp.4, 8.
65 Codex Alimentarius website: www.codexalimentarius.net/web/index_en.jsp (accessed 30 November 2005).
66 Kirton, J. J. and Trebilcock, M. J. (eds) (2004) *Hard Choices, Soft Law: Voluntary Standards in Global Trade, Environment and Social Governance*, Hampshire: Ashgate, p.22, cited in Tarasofsky, Richard and Flejzor, Lauren (2005) 'The potential role of a voluntary instrument on all types of forests,' background paper for UNFF country-led initiative, Berlin, 16–18 November, pp.3–4; Wood, Peter (2004) 'Soft law, hard law and the development of an international forest convention,' unpublished manuscript, dated 28 April.
67 UN document E/CN.18/2005/3/Add.1, 'Discussion paper contributed by the business and industry major group,' 24 February 2005, p.1.
68 The American Forest and Paper Association was prepared to support a code 'that would raise forest management standards worldwide': Mary Coulombe, pers comm, fifth session of the UNFF, New York, 25 May 2005.
69 US, verbal intervention, fifth session of the UNFF, New York, 20 May 2005.
70 Jamaica for the G77, verbal intervention, fifth session of the UNFF, New York, 20 May 2005. It is not clear whether the G77 was aware of the US proposal when it agreed this position.
71 Interview with G77 delegate, fifth session of the UNFF, New York, 26 May 2005.
72 The EU was bound to support a convention by Council Conclusions, although not all EU states were fully committed to this: 'Council Conclusions for UNNF-5,' 8305/05, Brussels, 21 April 2005, Annex, paras. 8, 14.
73 Framework Convention on Climate Change, Article 3.1; Kaminga, Menno T. (1995) 'Principles of international environmental law,' in Glasbergen, Pieter and Blowers, Andrew (eds) *Environmental Policy in an International Context: Perspectives on Environmental Problems*, London: Arnold, p.127.
74 Humphreys, David (1996) *Forest Politics: The Evolution of International Cooperation*, London: Earthscan, pp.96–98.
75 UN document E/CN.17/1997/12, n.42 above, para. 143.

76 Ecuador, verbal intervention, fifth session of the UNFF, New York, 24 May 2005; Syria, verbal intervention, fifth session of the UNFF, New York, 24 May 2005; 'Statement by Hon'ble Mr Namo Narain Meena, Minister of State for Environment and Forests [India] at the High Level Segment, fifth session of United Nations Forum on Forests,' 26 May 2005; 'Statement by H. E. Mr Mahmoud Hojjati, Minister of Agriculture of the Islamic Republic of Iran at the High Level Segment of the fifth session of the United Nations Forum on Forests,' 26 May 2005.

77 Quantifiable and time bound targets were proposed in two non-papers: 'Informal non-paper, potential elements of a Codex Sylvus' (undated); 'Informal non-paper – Towards the future international arrangement (IAF) on forests: A way forward,' 26 April 2005.

78 *Earth Negotiations Bulletin*, Vol. 13, No. 133, p.14.

79 *Earth Negotiations Bulletin*, Vol. 13, No. 133, p.14.

80 UN document E/2006/42-E/CN.18/2006/18, 'United Nations Forum on Forests, Report of the sixth session (27 May 2005 and 13–24 February 2006),' Chapter I, para. 3. (As this book went to press the resolution had yet to be passed by the UN Economic and Social Council.)

81 UN document E/2006/42-E/CN.18/2006/18, n.80 above, para. 32.

82 I have compiled this list from discussions at the sixth session of the UNFF with three delegates who attended this invitation only event.

83 UN document E/2006/42-E/CN.18/2006/18, n.80 above, para. 5(e).

84 UN document E/2006/42-E/CN.18/2006/18, n.80 above, para. 5(c).

85 There are five regional commissions of the UN Economic and Social Council: Economic Commission for Africa; Economic Commission for Europe; Economic Commission for Latin America and the Caribbean; Economic and Social Commission for Asia and the Pacific; and the Economic and Social Commission for Western Asia. There are six FAO regional forestry commissions: Africa; Asia and the Pacific; Europe; Latin America and the Caribbean; North America; and Near East.

86 Humphreys, n.74 above, pp.42–54.

87 One analyst sees the UNFF as a 'decoy institution' that 'does not fail to provide governance – it succeeds in denying it.' On this view, the UNFF's failures are not accidental, as the institution was intended to be ineffective: Dimitrov, Radoslav (2006) 'Hostage to norms: States, institutions and global forest politics,' *Global Environmental Politics*, Vol. 5, No. 1, pp.1–24.

6 The Certification Wars

1 Gulbrandsen, Lars (2004) 'Overlapping public and private governance: Can forest certification fill the gaps in the global forest regime?,' *Global Environmental Politics*, Vol. 4, No. 2, p.76.

2 Forests Forever (1996) 'Paper 3: Progress in timber certification schemes world-wide,' London, October 1996, p.4; and Ozinga, Saskia (2001) *Behind the Logo: An Environmental and Social Assessment of Forest Certification Schemes*, Moreton-in-Marsh, UK: FERN, pp.910.

3 World Wide Fund for Nature International (1988) *ITTO: Tropical Forest Conservation and the International Tropical Timber Organisation, Position Paper 1*, Gland, Switzerland: WWF International, p.10.

4 ITTO document PCM, PCF, PCI(V)/1, 'Pre-project proposal, labelling systems for the promotion of sustainably-produced tropical timber,' 15 August 1989. The proposal was prepared by Friends of the Earth, London, with input from the Oxford Forestry Institute.
5 ITTO document PCM(V)/D.1, 'Report to the International Tropical Timber Council, fifth session of the Permanent Committee on Economic Information and Market Intelligence,' 3 November 1989, p.6.
6 Elliott, Chris and Sullivan, Francis (1991) *Incentives and Sustainability: Where is ITTO Going?, WWF International Position Paper*, Gland, Switzerland: WWF International, pp.5–6.
7 Rainforest Alliance (1993) 'Description of the SmartWood Program,' 11 March.
8 Anon. 'Certification Working Group, San Francisco, 20–21 April 1991;' and 'Certification Working Group meeting notes, 20–21 April 1991, Notes by Pamela Wellner.'
9 This paragraph has benefited from the observations of Marcus Colchester, Forest Peoples Programme, email, 11 March 2006.
10 For an assessment of the arguments for and against plantations see Cossalter, Christian and Pye-Smith, Charlie (2003) *Fast-Wood Forestry: Myths and Realities*, Jakarta: CIFOR.
11 Roda, J-M (2002) 'Écocertification tropicale: Pour en finir avec les idées reçues!,' *Bois Mag*, No. 15, Février, cited in Eba'a Atyi, Richard and Simula, Markku (2002) 'Forest certification: Pending challenges for tropical timber,' *ITTO Technical Series No. 19*, Yokohama: International Tropical Timber Organization, p.12.
12 Rametsteiner, Ewald and Simula, Markku (2003) 'Forest certification: An instrument to promote sustainable forest management?,' *Journal of Environmental Management*, Vol. 67, pp.87–98.
13 World Rainforest Movement (2003) *Certifying the Uncertifiable: FSC Certification of Tree Plantations in Thailand and Brazil*, Montevideo: World Rainforest Movement.
14 'Certifying plantations: The FSC review process.' Available online at: www.certificationwatch.org/article.php3?id_article=2208 (accessed 27 October 2004).
15 Gulbrandsen, n.1 above, p.84.
16 Atyi and Simula, n.11 above, p.12.
17 Thornber, K. (1999) 'An overview of global trends in FSC certificates' (available online at: www.iied.org/psf) cited in Bass, Stephen, Thornber, Kirsti, Markopoulos, Matthew, Roberts, Sarah and Grieg-Grah, Maryanne (2001) *Certification's Impacts on Forests, Stakeholders and Supply Chains*, London: IIED, p.87.
18 Bill Mankin, formerly of the Global Forests Policy Project, interview, fourth session of the United Nations Forum on Forests, Geneva, 11 May 2004.
19 Poore, Duncan et al (1989) *No Timber Without Trees: Sustainability in the Tropical Forest*, London: Earthscan.
20 WWF (2004) 'Global Forest and Trade Network: Partnerships for responsible forestry' (information pamphlet), Gland, Switzerland: WWF.
21 Vilhunen, L., Hansen, E., Juslin, H. and Forsyth, K. (2001) 'Forest certification update for the ECE Region, summer 2001,' *Geneva Timber and Forest Discussion Papers*, ECE/TIM/DP/23, cited in Atyi and Simula, n.11 above, p.17.
22 'What is ISO 14000?.' Available online at: www.awm.net/iso/what.html (accessed 8 June 2000).
23 Gleckman, Harris and Krut, Riva (1996) 'Neither international nor standard: The limits of ISO 14001 as an instrument of global corporate environmental

management,' *Greener Management International*, No. 14 (April), p.113 (emphasis in original).

24 Gleckman and Krut, n.23 above, p.116. See also Krut, Riva and Gleckman, Harris (1998) *ISO 14001: A Missed Opportunity for Sustainable Global Industrial Development*, London: Earthscan.

25 Clapp, Jennifer (1998) 'The privatization of global environmental governance and the developing world,' *Global Governance*, Vol. 4, p.308.

26 Downes, David R. (1999) 'Global forest policy and selected instrument instruments: A preliminary review,' in Tarasofsky, Richard (ed) *Assessing the International Forests Regime*, Cambridge UK: IUCN, p.81.

27 Wenban-Smith, Matthew and Elliott, Chris (1996) 'The Forest Stewardship Council and the International Organization for Standardization,' *FSC Notes*, Vol. 1, No. 3, (June), p.7.

28 Gale, Fred (2002) '*Caveat Certificatum*: The case of forest certification,' in Princen, Thomas, Maniates, Michael and Conca, Ken (eds) *Confronting Consumption*, Cambridge MA: MIT Press, p.288.

29 Ozinga, Saskia (2001) *Behind the Logo: An Environmental and Social Assessment of Forest Certification Schemes*, Moreton-in-Marsh UK: FERN, p.19; Ozinga, Saskia with Krul, Leontien (2004) *Footprints in the Forest: Current Practices and Future Challenges in Forest Certification*, Moreton in Marsh UK: FERN, pp.20, 74.

30 'Questions and answers about ISO14000.' Available online at: www.scc.ca/iso14000/ (accessed 8 June 2000).

31 ISO 14001, cited in Wenban-Smith and Elliott, n.27 above, p.7

32 Brown Weiss, Edith (1999), 'The emerging structure of international environmental law,' in Vig, Norman J. and Axelrod, Regina S. (eds) *The Global Environment: Institutions, Law and Policy*, London: Earthscan, p.102.

33 The WTO's Technical Barriers to Trade (TBT) Agreement does not designate the ISO or any other body as competent to set international standards. However the TBT makes several references to the ISO, thus implicitly endorsing the ISO as an international standard setting body recognized by the WTO. See Downes, n.26 above, p.78.

34 Clapp, n.25 above, pp.295–316. See also Falkner, Robert (2003) 'Private environmental governance and international relations: Exploring the links,' *Global Environmental Politics*, Vol. 3, No. 2, p.77.

35 Gleckman and Krut, n.23 above, pp.114–115; Clapp, n.25 above, pp.295–316.

36 SFI website: www.aboutsfi.org/about.asp (accessed 7 February 2006). I am grateful to Ben Cashore for drawing this to my attention at the International Studies Association conference of 2005, during which he provided many useful comments on an early draft of this chapter (5 March 2005, Honolulu).

37 Mary Coulombe, American Forest and Paper Association, interview, fourth session of the United Nations Forum on Forests, Geneva, 11 May 2004.

38 Ozinga with Krul, n.29 above, p.23.

39 The letter is available online at: www.certificationwatch.org/print.php3?id_article=3098 (accessed 2 April 2005).

40 Ozinga with Krul, n.29 above, p.23.

41 Cashore, Benjamin, Auld, Graeme and Newsom, Deanna (2004) *Governing Through Markets: Forest Certification and the Emergence of Non-State Authority*, New Haven: Yale University Press, p.90.

42 Cashore, Auld and Newsom, n.41 above, Chapters 6 and 7; Gulbrandsen, Lars (2003) 'The evolving forest regime and domestic actors: Strategic or normative adaptation,' *Environmental Politics*, Vol. 12, No. 2, p.105; Gulbrandsen, Lars

(2005) 'Explaining different approaches to voluntary standards: A study of forest certification choices in Norway and Sweden,' *Journal of Environmental Policy and Planning*, Vol. 7, No. 1, pp.45–61.

43 Discussions took place within the Commission in four directorate generals; DG Industry, DG Agriculture, DG Development and DG Environment.

44 UN document ECE/TIM/DP/14, 'Geneva timber and forest discussion papers, the status of forest certification in the ECE region,' Hansen, Eric and Juslin, Heikki, pp.10–12. Available online at: www.unece.org/trade/timber/tc-publ.htm (accessed 20 February 2005).

45 PEFC website: www.pefc.org/lisbon/html (accessed 22 June 2001).

46 Annex 3, 'Pan-European Operational Level Guidelines for sustainable forest management,' in *Follow-Up Reports on the Ministerial Conferences for the Protection of Forests in Europe, Volume II, Third Ministerial Conference on the Protection of Forests in Europe, Lisbon, June 1998*, Lisbon: Ministry of Agriculture, Rural Development and Fisheries of Portugal, p.258.

47 Peter Mayer, former head of the MCPFE Liaison Unit, interview, fourth session of the United Nations Forum on Forests, Geneva, 13 May 2004.

48 Individual PEFC national schemes may prohibit or restrict the use of GM trees.

49 Vallejo, Nancy; Hauselmann, Pierre and Pi Environmental Consulting (2001) *PEFC: An Analysis, WWF Discussion Paper,* Zurich: WWF European Forest Team, p.19.

50 The rights of the Sami have only recently been recognized in Swedish law, although the interpretation of the legislation is unclear: Cashore, Auld and Newsom, n.41 above, p.202; Vallejo, Hauselmann and Pi Environmental Consulting, n.49 above, p.16.

51 Vallejo, Hauselmann and Pi Environmental Consulting, n.49 above, p.33.

52 By February 2005 a total of 51,320,494 hectares in 62 countries had been certified by the FSC: www.fsc.org/keepout/en/content_areas/77/55/files/ABU_REP_70_2005_02_03_FSC_Certified_Forests.pdf (accessed 10 February 2005). By December 2004 a total of 54,959,038 hectares in 19 countries had been certified by the PEFC: www.pefc.org/internet/resources/5_1184_1117_file.1137.pdf (accessed 10 February 2005).

53 The full figure is 186,798,580 hectares: http://register.pefc.cz/statistics.asp (accessed 7 February 2006).

54 The full figure is 68,125,087 hectares: www.fsc.org/keepout/en/content_areas/92/1/files/ABU_REP_70_2006_01_09_FSC_certified_forests.pdf (accessed 7 February 2006).

55 FSC (2006) 'The Random House Group gains FSC-CoC certification,' 18 January. Available online at: www.fsc.org/en/whats_new/news/news/56 (accessed 2 February 2006).

56 Jeanrenaud, Jean-Paul and Sullivan, Francis (1994) *Forest Certification and the Forest Stewardship Council: A WWF Perspective,* Godalming UK: WWF-UK; and Knight, Peter (1993) 'Timber watchdog ready to bark,' *Financial Times*, 6 October, p.18.

57 Colchester, Marcus, Sirait, Martua and Wijardo, Boedhi (2003) 'The application of FSC principles 2 & 3 in Indonesia: Obstacles and possibilities.' Available online at: www.forestpeoples.org/Briefings/UNFF/fsc_princip_2+3_indonesia_may03_eng.htm (accessed 14 February 2005).

58 Colchester, Sirait and Wijardo, n.57 above. See also Down to Earth (2003) 'Suspend FSC certification, says major new study.' Available online at: http://dte.gn.apc.org/57FSC.htm (accessed 15 June 2004).

59 Contreras-Hermosilla, Arnoldo and Fay, Chip (2005) *Strengthening Forest Management in Indonesia through Land Tenure Reform: Issues and Framework for Action*, Washington DC/Nairobi: Forest Trends/World Agroforestry Centre. Available online at: www.forest-trends.org/documents/publications/IndonesiaReport_final_11-4.pdf (accessed 17 March 2006).

60 Counsell, Simon and Terje Kloraas, Kim (2002) *Trading in Credibility: The myth and reality of the Forest Stewardship Council*, London and Oslo: Rainforest Foundation, p.5.

61 FSC (2003) Public Statement, 'Response to the Rainforest Foundation Report "Trading in Credibility",' 27 February.

62 FSC (2002) 'FSC response to the Rainforest Foundation's press release claiming major deficiencies in FSC's systems,' 20 November 2002; and FSC, n.61 above.

63 One exception, which argues that FSC has an insignificant impact and that it is supporting industrial-scale logging of primary forests, is Laschefski, Klemens and Feris, Nicole (2001) 'Saving the wood from the trees,' *The Ecologist*, Vol. 31 No. 6, July/August, pp.40–43.

64 Greenpeace Finland, Finnish Association for Nature and Finnish Nature League (2004) *Certifying Extinction?: An Assessment of the Revised Standards of the Finnish Forest Certification Scheme*, Helsinki: Greenpeace, September, p.21.

65 'Finnish Forest Certification Council responds to ENGO report on FFCS.' Available online at: www.certificationwatch.org/print.php3?id_article+2310 (accessed 28 October 2004).

66 International Forest Industry Roundtable (2001) 'Proposing an international mutual recognition framework: Report of the working group on mutual recognition between credible sustainable forest management systems and standards,' February 2001.

67 AFPA website: www.afandpa.org/Content/NavigationMenu/Environment_and_Recycling/SFI/Mutual_Recognition/Mutual_Recognition.htm (accessed 11 January 2005).

68 Synott, Timothy (2000) 'Forest Stewardship Council – Position on mutual recognition,' paper presented to the Second International Seminar on the Mutual Recognition of Credible Forest Certification Systems, Brussels, 28–29 November, p.3.

69 Canadian Sustainable Forestry Certification Coalition (2004) 'Many standards mean widespread application and a need for mutual recognition.' Available online at: www.sfms.com/recognition.htm (accessed 11 January 2005).

70 WWF (2001) *Certification: Mutual Recognition of Schemes: Position Paper*, Gland, Switzerland: WWF International.

71 Certification Watch website, 'Forest certification and mutual recognition: The fundamentals,' 19 February 2001. Available online at: http://sfcw.org/mutualrecognition/forest_certification_and_mutual_.htm (accessed 28 May 2004).

72 Certification Watch website, 'Environmental NGOs call for credible forest certification and reject IFIR mutual recognition proposal,' 19 February 2001. Available online at: http://sfcw.org/mutualrecognition/environment_ngos_call_for_cred.htm (accessed 28 May 2004).

73 Latvian Wood Industry Portal. Available online at: www.latvianwood.lv/default.aspx?tabid=2&id=2363&lang=2 (accessed 10 February 2006).

74 Fenner, Rudolf and Kill, Jutta (2001) *Behind the Logo: The Development, Standards and Procedures of the Pan-European Forest Certification Scheme and the Forest Stewardship Council in Germany*, Moreton-in-Marsh UK: FERN.

75 Brack, Duncan and Saunders, Jade (2004) *Public Procurement of Timber: EU Member State Initiatives For Sourcing Legal and Sustainable Timber*, London: Chatham House, p.5.
76 'The UK government's position in timber procurement.' Available online at: www.trade.co.uk/topics/environmental/government.html (accessed 9 February 2006). Three other schemes were accepted by the UK government as providing evidence of legality but not of sustainability, namely the SFI, PEFC and Malaysian Timber Certification Council.
77 International Accreditation Forum website, 'What is the IAF.' Available online at: www.compad.com.au/cms/iaf/public/13 (accessed 1 June 2004).
78 International Accreditation Forum website, 'Role of IAF.' Available online at: www.compad.com.au/cms/iaf/articles/14 (accessed 1 June 2004).
79 The final public draft of this code (dated January 2004) is available online at: http://inni.pacinst.org/inni/General/ISEALCodeFinal.pdf (accessed 9 February 2006).
80 International Organic Accreditation Service (IOAS) (2000) *Annual Report 1999 of the IOAS, International Organic Accreditation Service, Implementing the IFOAM Accreditation Programme*, Jamestown, ND: IOAS, p.15.
81 ISEAL Alliance (2003) 'International accreditation, E016 fact sheet #1,' July, para.10, p.2.
82 *PEFC News*, No. 19, March 2004, p.1; PEFC press release, 'PEFC Council joins International Accreditation Forum,' Luxembourg, 12 March 2004. The IAF members, including associate members, can be found online at: www.compad.com.au/cms/iaf/articles /145 (accessed 15 February 2005).
83 However the significance of this should not be overstated. It is not clear what benefits IAF associate membership actually entails, other than a recognition that the associate member has contacts with the IAF.
84 Sasha Courville, executive director of the ISEAL Alliance, interview, Oxford, 8 February 2006.
85 Gale, n.28 above.
86 World Business Council for Sustainable Development (2003) 'Discussion paper: Forest certification systems and the "Legitimacy" thresholds model (LTM).'
87 World Business Council for Sustainable Development, n.86 above, p.1.
88 Nussbaum, Ruth and Simula, Markku (2004) *Forest Certification: A Review of Impacts and Assessment Frameworks, A TFD Publication,* No. 1, Yale: The Forests Dialogue/Yale University, p.1.
89 WWF International and WBCSD Sustainable Forest Products Industry Working Group (2004) 'Joint statement on forest certification for The Forests Dialogue,' October.
90 UN document TD/B/WG.6/5, 'Trade, environment and development aspects of establishing and operating eco-labelling programmes: Report by the UNCTAD secretariat,' 28 March 1995, p.7.
91 *Earth Negotiations Bulletin*, Vol. 13, No. 25.
92 This formulation appears in the third report of the Intergovernmental Panel on Forests: UN document E/CN.17/IPF/1997/2, 'Report of the ad hoc Intergovernmental Panel on Forests on its third session (Geneva, 9–20 September 1996)' para.159bis. However it did not survive the final round of textual changes made during the IPF negotiations.
93 *Earth Negotiations Bulletin*, Vol. 13, No. 45, p.5.
94 UN document E/CN.17/2000/14, 'Report of the Intergovernmental Forum on Forests on its fourth session, New York, 31 January–11 February 2000,' para. 34.

95 WTO document WT/CTE/8, 11 July 2003 cited in Ozinga with Krul, n.29 above, p.17.
96 Gale, n.28 above, p.298.
97 Cashore, Auld and Newsom, n.41 above.
98 See, for example, Hurrell, Andrew (1993) 'International society and the study of regimes: A reflective approach,' in Rittberger, Volker (ed), *Regime Theory and International Relations*, Oxford: Clarendon Press, pp.51–52.

7 New Policies to Counter Illegal Logging

1 World Bank press release, 'Governments commit to action on forest law enforcement and governance in Europe and North Asia,' 25 November 2005.
2 On the range of illegal logging practices see Contreras-Hermosilla, Arnoldo (2002) *Law Compliance in the Forestry Sector: An Overview*, Washington DC: World Bank Institute, pp.6–8.
3 Taylor, Mark (2005) 'The Green Peace Prize,' *Adbusters: Journal of the Mental Environment*, Vol. 13, No. 3. (This article consists of comments made during a seminar with Wangari Maathai at the Nobel Institute in December 2004.)
4 Wesberry, J. (2001) 'Combating fraud in procurement and contracting,' in Kaufmann, D., Gonzalez de Asis, M. and Dininio, P. (eds) *Improving Governance and Controlling Corruption: Towards a Participatory and Action-oriented Approach Grounded on Empirical Rigour*, Washington: World Bank Institute, cited in Contreras-Hermosilla, Arnoldo (2002) 'Illegal forest production and trade: An overview' (unpublished manuscript). This paper draws in part from Contreras-Hermosilla, n.2 above.
5 *FIN Newsletter*, Issue 5, June 2003, p.5. (FIN stands for Forest Integrity Network.)
6 Rainforest Action Network press release, 'New report exposes illegal logging in Canada's boreal forests by Weyerhaeuser operations supplying Xerox,' 15 July 2005. Available online at: www.ems.org/nws/pf.php?p=1584 (accessed 21 July 2005) and www.ran.org/news/newsitem.php?id=1541&area=newsroom (accessed 7 February 2006).
7 Brianna Cayo Cotter, Rainforest Action Network, email, 8 February 2006; David Sone, Rainforest Action Network, email, 9 February 2006. Weyerhaeuser has responded that the operations highlighted by RAN are legal. Information on RAN's Weyerhaeuser campaign can be found at: www.ran.org/ (accessed 6 March 2006).
8 TRAFFIC stands for Trade Records Analysis of Fauna and Flora In Commerce. TRAFFIC International is financially supported by the WWF and IUCN.
9 Callister, Debra J. (1992) *Illegal Tropical Timber Trade: Asia Pacific, A TRAFFIC Network Report*, Cambridge UK: TRAFFIC International, p.iv.
10 Callister, n.9 above, p.6.
11 On Cambodia see, for example, Schweithelm, James and Chanthy, Srey (2004) *Cambodia: An Assessment of Forest Conflict at the Community Level*, June. Submitted to USAID/Cambodia and USAID/ANA/SPOTS, Washington.
12 Barnett Commission of Inquiry, Interim Report, cited in Marshall, George (1990) 'The political economy of logging: The Barnett Inquiry into corruption in the Papua New Guinea timber industry,' *The Ecologist*, Vol. 20, No. 5, September/October, p.174.

13 Marshall, George (1990) *The Barnett Report: A Summary of the Report of the Commission of Inquiry into Aspects of the Timber Industry in Papua New Guinea*, Hobart: Asia Pacific Action Group.
14 Greenpeace (2004) *The Untouchables: Rimbunan Hijau's World of Forest Crime and Political Patronage*, Amsterdam: Greenpeace International. See also Greenpeace (2005) *Partners in Crime: The UK Timber Trade, Chinese Sweatshops and Malaysian Robber Barons in Papua New Guinea's Rainforests*, London: Greenpeace.
15 Jasper Teulings, senior legal counsel of Greenpeace International, cited in Certification Watch (2004) 'World's largest forest destroyer takes on environmentalists,' 15 September. Available online at: www.certificationwatch.org/print.php3?id_article=2154 (accessed 27 October 2004).
16 Callister, n.9 above, p.65. The figures in question cover the years 1978–1987.
17 Asia Pulse (2001) 'Philippines steps up campaign against illegal logging,' 4 December. Available online at: http://forests.org/archive/asia/phstupca.htm (accessed 16 April 2004).
18 Hills, Jonathan (2005) 'Cause and effect: Illegal logging in the Asia Pacific,' *CSR Asia*, Vol. 1, Week 12, p.12. On the continuing illegal logging problem in the Philippines see Dauvergne, Peter (2001) *Loggers and Degradation in the Asia Pacific: Corporations and Environmental Management*, Cambridge: Cambridge University Press, pp.76–80.
19 Chua, Amy (2004) *World on Fire: How Exporting Free Market Democracy Breeds Ethnic Hatred and Global Instability*, New York: Anchor Books, p.43.
20 Vatikiotis, Michael R. J. (1993) *Indonesian Politics Under Suharto* (3rd edition), London and New York: Routledge, p.14, cited in Chua, n.19 above, p.44.
21 Borsuk, Richard (2003) 'Suharto crony stays busy behind bars: "Bob" Hasan starts business, pulls strings at Olympics.' Available online at: www.mongabay.com/external/bob_hasan_indonesia.htm (accessed 12 April 2004).
22 Smith, J., Obidzinski, K., Subarudi and Suramenggala, I. (2003) 'Illegal logging, collusive corruption and fragmented governments in Kalimantan, Indonesia,' *International Forestry Review*, Vol. 5, No. 3, pp.293–302.
23 WWF (2004) 'Most recent articles, 16 March 2004, Sumatran tiger on the brink of extinction.' Available online at: www.panda.org/about_wwf/what_we_do/forests/news/ (accessed 24 March 2004).
24 Environmental Investigation Agency and Telapak (2005) *The Last Frontier: Illegal Logging in Papua and China's Massive Timber Theft*, London: Environmental Investigation Agency; Aglionby, John (2005) 'Timber smuggling ring exposed,' *Guardian*, 18 February, p.19; Watts, Jonathan (2005) 'China consumes forests of smuggled timber,' *Guardian*, 22 April, p.17. See also Environmental Investigation Agency (1998) *The Politics of Extinction: The Orangutan Crisis – The Destruction of Indonesia's Forests*, London: EIA; Environmental Investigation Agency and Telapak (1999) *The Final Cut: Illegal Logging in Indonesia's Orangutan Parks*, London/Washington/Bogor: EIA/Telapak; Environmental Investigation Agency and Telapak (2002) *Above the Law: Corruption, Collusion, Nepotism and the Fate of Indonesia's Forests*, London/Washington/Bogor: EIA/Telapak.
25 Environmental Investigation Agency and Telapak (2004) *Profiting from Plunder: How Malaysia Smuggles Endangered Wood*, London: EIA.
26 Vidal, John (2002) 'UK plays key role in illegal logging,' *Guardian*, 19 April, p.11.
27 'Armed CAR to stop illegal logging,' 3 February 2004. Available online at: www.illegal-logging.info/news.php?newsId=158 (accessed 19 February 2004).

28 UN document S/2002/1146, 'Letter dated 15 October 2002 from the secretary-general addressed to the president of the Security Council,' Annex III, 'Business enterprises considered by the Panel to be in violation of the OECD Guidelines for Multinational Enterprises,' pp.7–8. See also Carroll, Rory (2002) 'Multinationals in scramble for Congo's wealth,' *Guardian*, 22 October, p.15.
29 UN document S/2002/1146, n.28 above, para. 101, p.20.
30 UN press release SC/7925, 'Security Council condemns continuing exploitation of natural resources in Democratic Republic of Congo,' 19 November 2003.
31 Rainforest Foundation (2004) 'New threats to the forests and forest peoples of the Democratic Republic of Congo,' Briefing Paper, February (with thanks to Simon Counsell).
32 UN document S/RES/1579(2004), 'Resolution 1579(2004), adopted by the Security Council at its 5105th meeting, on 21 December 2004,' p.2.
33 Anon. (2003)'Illegal logging in Ghana's Asubima Forest,' 16 December. Available online at: www.illegal-logging.info/news.php?newsId=112 (accessed 19 February 2004).
34 Bureau for Regional Oriental Campaigns, Friends of the Earth-Japan and Pacific Environment and Resources Centre (2000) *Plundering Russia's Far Eastern Taiga: Illegal Logging, Corruption and Trade*, Vladivostock: Bureau for Regional Oriental Campaigns.
35 Diamond, Jared (2005) *Collapse: How Societies Choose to Fail or Survive*, London: Allen Lane, pp.365–366, 372.
36 Global Witness (2003) *A Conflict of Interest: The Uncertain Future of Burma's Forests*, London: Global Witness; Vidal, John (2005) 'China "strips forests in Burma",' *Guardian*, 19 October, p.18.
37 Lopina, Olga; Ptichnikov, Andrei and Voropayev, Alexander (2003) *Illegal Logging in Northwestern Russia and Export of Russian Forest Products to Sweden*, Moscow: WWF Russia.
38 Sometimes called Siberian cedar or Russian cedar.
39 Elena Kulikova, WWF Russia, interview, Portugal, 11 February 2004.
40 European Forest Institute (2005) Press release, 'Illegal logging in Russia distorts roundwood markets in Russia and the EU,' 3 November.
41 WWF Latvia (2003) 'The features of illegal logging and related trade in the Baltic Sea region,' Discussion Paper; WWF Latvia (2003) 'Logging and trade of acquired timber: Legal regulation, procedures and ways to evade them,' Discussion Paper.
42 ITTO document ITTC(XXXV)/15, 'Achieving the ITTO objective 2000 and sustainable forest management in Peru – Report of the diagnostic mission,' 2 October 2003, pp.3, 6.
43 A mission of the International Tropical Timber Organization reported that in 2001 IBAMA seized 26,000 cubic meters of illegally felled mahogany: ITTO document ITTC(VVVII)/17, 'Achieving the ITTO objective 2000 and sustainable forest management in Brazil,' 25 September 2002, p.34.
44 Environmental News Service (2005) 'Brazil fells massive Amazon timber fraud ring,' 6 June. Available online at: www.ens-newswire.com/ens/jun2005/2005-06-06-03.asp (accessed 6 March 2006); Mongabay (2005) 'Timber traffickers arrested in Brazil,' 31 October. Available online at: http://news.mongabay.com/2005/1031-ap.html (accessed 6 March 2006). Greenpeace has found evidence in Brazil of widespread laundering of illegally logged mahogany through the fraudulent use of official documents: Greenpeace (2001) *Partners in Mahogany Crime: Amazon at the Mercy of 'Gentleman's Agreements,'* Amsterdam: Greenpeace International. See

also Greenpeace (1999) *Facing Destruction: A Greenpeace Briefing on the Timber Industry in the Brazilian Amazon*, Amsterdam: Greenpeace International.

45 'Illegal logging in Central America: Tackling its implications on governance and poverty.' Available online at: www.talailegal-centroamerica.org/eng_index.htm (accessed 20 December 2005).

46 Bill Mankin, Global Forest Policy Project, interview, fourth session of the Intergovernmental Forum on Forests, New York, 10 February 2000.

47 International Tropical Timber Agreement, 1994, Article 27.1(c). The previous agreement of 1983 contained no such mention.

48 McAlpine, J. L. (2003) 'Conservation diplomacy – one government's commitment and strategy to eliminate illegal logging,' *International Forestry Review*, Vol. 5, No. 3, pp.230–235; Jan McAlpine, US State Department, interview, fourth session of the United Nations Forum on Forests, Geneva, 4 May 2004.

49 UN document E/CN.17/1996/24, 'Review of sectoral clusters: Report of the ad hoc Intergovernmental Panel on Forests on its second session (Geneva, 11–22 March 1996),' para. 49.

50 UN document E/CN.17/1996/24, n.49 above, para. 112(b).

51 UN document E/CN.17/1997/12, 'Report of the ad hoc Intergovernmental Panel on Forests on its fourth session (New York, 11–21 February 1997),' para. 135(b). The Panel also noted the role as underlying causes of deforestation of 'illegal logging; illegal land occupation and illegal cultivation' (para. 20).

52 UN document E/CN.17/2000/14, 'Report of the Intergovernmental Forum on Forests at its fourth session New York, 31 January–11 February 2000,' 20 March 2000. para. 41(f). Other references to the illegal trade in the IFF's final report can be found in paras. 37 and 58. However these references are in those parts of the report that present conclusions and they do not form part of the proposals for action.

53 This section of the chapter benefited from some incisive observations from Carole Saint-Laurent of IUCN; email, 4 January 2006.

54 Palo, Matti (2001) 'World forests and the G8 economic powers: From imperialism to the Action Programme on Forests,' in Palo, Matti, Uusivuori, Jussi and Mery, Gerardo (eds) *World Forests, Markets and Policies*, Volume III, Dordrecht: Kluwer, p.184.

55 G7 Communique, Houston, July 1990, cited in Humphreys, David (1996) *Forest Politics: The Evolution of International Cooperation*, London: Earthscan, p.84. This is the only occasion that the US government has formally supported a forests convention.

56 G8 Communique, Denver, 22 June 1997, para. 19. Available online at: www.g8.utoronto.ca/summit/1997denver/g8final.htm (accessed 19 April 2004). The adoption of action programmes by the G8 is relatively recent but is becoming increasingly common. For example, at the 2002 summit in Kananaskis, Canada the G8 adopted the G8 Africa Action Plan, the G8 Global Partnership Against the Spread of Weapons and Materials of Mass Destruction, and the Cooperative G8 Action on Transport Security.

57 McAlpine, n.48 above, p.231.

58 'G8 Action Programme on Forests, backgrounders 2002,' and 'G8 Action Programme on Forests, final report 2002.' Both documents are available online at: www.g8.gc.ca//menu-en.asp (accessed 22 April 2004).

59 Horst, Alexander (2001) 'G8 Action Programme on Forests: Mere rhetoric?,' in Palo, Uusivuori and Mery (eds), n.54 above, p.204.

60 'G8 Action Programme on Forests – 9 May 1998,' paras 8–9. Available online at: http://birmingham.g8summit.gov.uk/forfin/forests.shtml (accessed 19 November 2002).
61 'Report on the implementation of the G8 Action Programme on Forests, Okinawa, 21 July, 2000, III. Implementation Highlights.' Available online at: www.g7.utoronto.ca/g7summit/2000okinawa/forest1.htm (accessed 19 November 2002).
62 The long standing concern of the American Forest and Paper Association on the effects of illegal logging on AFPA member organizations is noted in: '"Illegal" logging and global wood markets: The competitive impacts on the U.S. wood products industry,' prepared for the AFPA by Seneca Creek Associates and Wood Resources International, November 2004.
63 Jan McAlpine, US State Department, interview, fourth session of the United Nations Forum on Forests, Geneva, 4 May 2004.
64 David Cassells, World Bank, interview, Washington, 16 March 2004.
65 Jan McAlpine, US State Department, interview, fourth session of the United Nations Forum on Forests, Geneva, 4 May 2004; Jan McAlpine, email, 20 December 2005.
66 ITTO press release, 'Indonesia wants ITTO to address illegal logging,' 30 October 2000.
67 ITTO document ITTC(XXXI)/10, 'Achieving sustainable forest management in Indonesia: Report submitted to the International Tropical Timber Council by the Mission established pursuant to Decision 12 (XXIX) "Strengthening sustainable forest management in Indonesia",' 26 September 2001, pp.xix, xxiv–xxv.
68 ITTO document ITTC(XXXI)/10, n.67 above, p.xxv.
69 Freezailah, B. Che Yeom and Cherukat, Chandrasekharan (2002) 'Achieving sustainable forest management in Indonesia,' *ITTO Tropical Forest Update*, Vol. 12, No. 1, p.10. This article provides a condensed and amended version of the abstract to the mission's report.
70 Seven were from the Asia-Pacific region: Cambodia, China, Indonesia, Laos, Philippines, Thailand and Vietnam. In addition there were two from Africa, namely Congo-Brazzaville and Ghana: *Sustainable Developments*, Vol. 60, No. 1, p.1. Japan, the UK and the US also attended.
71 The EIA and its Indonesian partner Telapak released the following report for the FLEG meeting: EIA and Telapak (2001) *Timber Trafficking: Illegal Logging in Indonesia, South East Asia and International Consumption of Illegally-sourced Timber*, London/Washington/Bogor: EIA/Telapak.
72 For example, Global Witness (1999) 'The untouchables: Forest crimes and the concessionaires – can Cambodia afford to keep them?,' *Briefing Document*, December; Global Witness (2004) *Taking a Cut: Institutionalised Corruption and Illegal Logging in Cambodia's Aural Wildlife Sanctuary*, London: Global Witness.
73 *The Economist*, 9 August 2003, reprinted in *Commonwealth Forestry News*, No. 22, September 2003, p.12; Jon Buckrell, Global Witness, interview, Chatham House, London, 27 July 2005; Jon Buckrell, Global Witness, email, 11 January 2006.
74 'Forest law enforcement and governance, East Asia ministerial conference, Bali, Indonesia, 11–13 September 2001, Ministerial Declaration,' Preamble.
75 *Sustainable Developments*, Vol. 60, No. 1, p.10.
76 Marcus Colchester, Forest Peoples Programme, email, 11 March 2006.
77 UNFF Resolution 2/1 'Ministerial declaration and message from the United Nations Forum on Forests to the World Summit on Sustainable Development,'

para.15(d), in UN document E/2002/42-E/CN.18/2002/14, 'United Nations Forum on Forests, report on the second session (22 June 2001 and 4 to 15 March 2002),' p.6.

78 This committed states to '[t]ake immediate action on domestic forest law enforcement and illegal international trade in forest products....' See 'World Summit on Sustainable Development, Plan of Implementation,' Johannesburg, September 2002, para.43(c). Available online at: www.johannesburgsummit.org/html/documents/summit_docs/2309_planfinal.htm (accessed 6 March 2005).

79 Governments that were founding members of the Asia Forest Partnership are Australia, Cambodia, China, Finland, France, Indonesia, Japan, South Korea, Malaysia, Philippines, Switzerland, Thailand, UK, US and Vietnam.

80 'About Asia Forest Partnership.' Available online at: www.asiaforests.org/home/home.htm (accessed 22 April 2004).

81 *ITTO Tropical Forest Update*, Vol. 11, No. 4, p.19.

82 The four agreements are reproduced in full in *International Forestry Review*, Vol. 5, No. 3, 2003, pp.223–229. In 2003 Indonesia and South Korea signed a statement on 'The call for combating international trade in illegally harvested forest products.' Available online at: www.illegal-logging.info/papers/Indonesia_Korea_MoU.doc (accessed 12 December 2005).

83 Although Norway did not attend the Bali ministerial meeting, the Norwegian agreement with Indonesia drew its inspiration from FLEG, as well as from an earlier memorandum of understanding on the environment between Indonesia and Norway agreed in May 1990.

84 Speechly, H. (2003) 'Bilateral agreements to address illegal logging,' *International Forestry Review*, Vol. 5, No. 3, p.220.

85 *Sustainable Developments*, Vol. 60, No. 7, p.3.

86 'About the Congo Basin Forest Partnership.' Available online at: www.cbfp.org/en/about/aspx (accessed 26 April 2004).

87 'Congo Basin summit produces Africa's first ever region-wide conservation treaty.' Available online at: http://certificationwatch.org/print.php3?id_article=2913 (accessed 16 February 2005).

88 'Kinshasha Declaration on Great Apes.' Available online at: www.unep.org/ (accessed 13 December 2005).

89 'Report of the First Intergovernmental Meeting on Great Apes and the Great Apes Survival Project (GRASP) and the first meeting of the GRASP Council, Kinshasha, Democratic Republic of Congo, 5–9 September 2005,' pp.49, 54, 56, 76. Available online at: www.unep.org/grasp/Meetings/IGM-kinshasha/Outcomes/docs/Report.pdf (accessed 13 December 2005).

90 EU document COM (2003) 251 final, 'Communication from the Commission to the Council and the European Parliament: Forest Law Enforcement, Governance and Trade (FLEGT), Proposal for an EU Action Plan,' Brussels, 21 May. See also European Commission (2002) 'Communication on a global partnership for sustainable development.' Available online at: http://europa.eu.int/comm/external_relations/flegt/workshop/issue.htm (accessed 21 November 2002).

91 *Official Journal of the European Union*, 'Council Conclusions: Forest Law Enforcement, Governance and Trade (FLEGT) (2003/C 268/01).'

92 Recommended in Brack, Duncan, Marijnissen, Chantal and Ozinga, Saskia (2002) *Controlling Imports of Illegal Timber: Options for Europe*, London/Brussels: RIIA/FERN, pp.50–51. The recommendation appears in EU document COM (2003) 251 final, n.90 above, pp.11–12, 23.

93 Brack, Marijnissen and Ozinga, n.92 above, p.33; EU document COM (2003) 251 final, n.90 above, pp.16.

94 Brack, Marijnissen and Ozinga, n.92 above, p.34; EU document COM (2003) 251 final, n.90 above, pp.18.

95 Brack, Marijnissen and Ozinga, n.92 above, pp.35–36; EU document COM (2003) 251 final, n.90 above, pp.18.

96 This drew criticism from the WWF which, in addition to previous reports on illegal logging in the three Baltic states, noted widespread logging in Slovakia and the accession states of Bulgaria, where illegal logging accounts for 45 per cent of the timber harvest, and Romania. WWF's comments are available online at: www.certificationwatch.org/print.php3?id_article=3093 (accessed 2 April 2005).

97 EU document COM (2004) 515 final, 'Proposal for a Council regulation concerning the establishment of a voluntary FLEGT licensing scheme for imports of timber into the European Community,' Brussels, 20 July, Article 2(e), p.6. The Council regulation for the licensing scheme was adopted in 2005: *Official Journal of the European Union*, 'Council Regulation (EC) No 2173/2005 of 20 December 2005 on the establishment of a FLEGT licensing scheme for imports of timber into the European Community,' L 347/1, Article 2.5.

98 EU document COM (2004) 515 final, n.97 above, pp.2–3; and EU document SEC(2004) 977, 'Proposal for a Council regulation concerning the establishment of a voluntary FLEGT licensing scheme for imports of timber into the European Community,' 20 July, p.3.

99 Katharine Thoday kindly demystified this process for me: email, 27 January 2006.

100 The European Parliament has noted that 'it seems probable that even if an initial GATT violation was found, the licensing scheme would be found to be GATT-compatible under Article XX(d) or XX(g)': Letter from Luis Bergenuer Fuster, Chairman, Committee on Industry, External Trade, Research and Energy of the European Parliament to Patrick Cox, president of the European Parliament, 3 March 2004. A legal opinion provided to three environmental NGOs from a member of Matrix Chambers doubts that the scheme is GATT/WTO illegal: FERN/Greenpeace/WWF (2004) *Facing Reality: How to Halt the Import of Illegal Timber in the EU*, Brussels: FERN/Greenpeace/WWF, Annex II, p.33.

101 These NGOs include FERN, Friends of the Earth, Greenpeace and WWF: www.fern.org/pubs/articles/flegt.htm (accessed 19 April 2004); www.wwf.dk/4772774 (accessed 19 April 2004); Greenpeace (2004) 'Illegal timber trade and the EU: The stakes and the showdown,' June; FERN/Greenpeace/WWF 'Principles for FLEGT Partnership Agreements,' January 2005, p.2.

102 Duncan Brack argues that an 'outright ban' would be 'likely to run into WTO problems, depending on the degree of proof of legality required from importers': Brack, Duncan (2005) 'Controlling illegal logging and the trade in illegally harvested timber: The EU's Forest Law Enforcement, Governance and Trade initiative,' *Review of European Community and International Environmental Law (RECIEL)*, Vol. 14, No. 1, p.35.

103 On how trade restriction measures in existing multilateral environmental agreements may apply in the case of the trade in illegally sourced forest products see Brack, Duncan (2003) 'Lessons from international agreements,' *International Forestry Review*, Vol. 5, No. 3, pp.240–246.

104 Reported at a FLEG/FLEGT seminar conducted under the Chatham House rule, London, 27 July 2005.

105 *FLEGT Briefing Notes*, No. 7, 'Voluntary Partnership Agreements,' p.2.

106 Greenpeace (undated) press briefing, 'Greenpeace comments on the G8 foreign ministers' conclusions.'

107 'Shaping our common future: An action plan for EU-Japan cooperation,' European Union-Japan Summit, Brussels, 2001, p.16. Available online at: http://jpn.cec.eu.int/frame.asp?frame=/english/eu-relations/actionplan.pdf (accessed 27 April 2004).

108 Toyne, Paul, O'Brien, Cliona and Nelson, Rod (2002) *The Timber Footprint of the G8 and China: Making the Case For Green Procurement by Government*, Gland, Switzerland: WWF International.

109 EU document COM (2003) 251 final, n.90 above, p.15.

110 Brack, Duncan and Saunders, Jade (2004) *Public Procurement of Timber: EU Member State Initiatives for Sourcing Legal and Sustainable Timber*, London: Chatham House.

111 EU document COM (2003) 251 final, n.90 above, p.19. This provision was recommended by the RIIA/FERN report: Brack, Marijnissen and Ozinga, n.92 above, p.34.

112 Brack, n.102 above, p.37.

113 'TTF's responsible purchasing policy.' Available online at: www.illegal-logging.info/papers/TTF_Responsible_Purchasing_Policy.doc (accessed 12 December 2005).

114 Confederation of European Paper Industries (2005) 'Legal logging code of conduct for the paper industry.' Available online at: www.cepi.org/files/illegal%20logging-152955A.pdf (accessed 13 December 2005). CEPI launched this code of conduct at the Europe and North Asia FLEG meeting held in St Petersburg, 22–25 November 2005.

115 Brack, Marijnissen and Ozinga, n.92 above, p.35; and http://europa.eu.int/comm/external_relations/flegt/workshop/synthesis1.htm (accessed 25 November 2002).

116 Saunders, Jade (2005) *Improving Due Diligence in Forestry Investments: Restricting Legitimate Finance for Illegal Activities*, London: Chatham House.

117 Brack, Marijnissen and Ozinga, n.92 above, p.35.

118 EU document COM (2003) 251 final, n.90 above, p.18.

119 Brack, n.102 above, p.36.

120 EU document COM (2003) 251 final, n.90 above, pp.6–9.

121 Internal US State Department memorandum: 'Input to strategy paper for G-8 environment and development ministerial, March 2005, guidance for G-8 experts meeting on illegal logging, February 10–11, 2003 (sic), Draft January 31.' This document is marked 'confidential' and was authored by a State Department official who has played a key role in US delegations to the IPF, IFF and UNFF. The source of the leak is unknown.

122 'USAID: President's Initiative Against Illegal Logging.' Available online at: www.usaid.gov/about_usaid/presidential_initiative/logging.html (accessed 13 April 2004).

123 Brown, Paul and Harrabin, Roger (2005) 'US tries to sink forests plan,' *Guardian*, 16 March. Available online at: www.guardian.co.uk/international/story/0,3604,1438394,00.html (accessed 21 March 2005).

124 Smith, R. J. cited in Harrabin, Roger (2005) 'US blocks forest protection plan,' 15 March. Available online at: http://news.bbc.co.uk/1/hi/programmes/newsnight/4351863.stm (accessed 21 March 2005) and at: http://forests.org/articles/reader.asp?linkid=40065 (accessed 2 June 2005).

125 'G8 environment and development ministerial,' 18 March 2005, para. 10.
126 'G8 environment and development ministerial,' 18 March 2005, para. 12.
127 'G8: Gleneagles plan of action: Climate change, clean energy and sustainable development,' para. 38. Available online at: www.noticias.info/asp/PrintingVersionNot.asp?NOT=81992 (16 December 2005).
128 Verbal presentation, Japanese Ministry of the Environment, side event, sixth session of the United Nations Forum on Forests, New York, 22 February 2006.
129 The preparatory conference had been held in June 2005 in Moscow.
130 Reported at a FLEG/FLEGT seminar conducted under the Chatham House rule, London, 27 July 2005.
131 Carole Saint-Laurent, IUCN, email, 4 January 2006.
132 *Europe and North Asia FLEG Bulletin*, Vol. 110, No. 2, p.2. (This bulletin was produced by the International Institute for Sustainable Development, which also publishes the *Earth Negotiations Bulletin*.)
133 *Europe and North Asia FLEG Bulletin*, Vol. 110, No. 5, p.3.
134 *Europe and North Asia FLEG Bulletin*, Vol. 110, No. 5, p.3. The Ilim Pulp Enterprise spoke at the preparatory meeting in Moscow, June 2005: *Europe and North Asia FLEG Bulletin*, Vol. 110, No. 1, p.4.
135 *Europe and North Asia FLEG Bulletin*, Vol. 110, No. 3, p.2.
136 Carole Saint-Laurent, IUCN, email, 4 January 2006.
137 *Europe and North Asia FLEG Bulletin*, Vol. 110, No. 3, p.2.
138 'St Petersburg Declaration,' 25 November 2005, para. 17.
139 Secretariat of the Convention on Biological Diversity (2004) *Expanded Programme of Work on Forest Biological Diversity*, Montreal: Secretariat of the Convention on Biological Diversity, programme element 2, goal 1, objective 4, p.17.
140 International Tropical Timber Agreement, 2006, Article 1(n). Article 28.3(e) of the Agreement requires ITTO members to supply information on illegal harvesting and the illegal trade.
141 *Earth Negotiations Bulletin*, Vol. 13, No. 144, p.12.
142 UN document E/2006/42-E/CN.18/2006/18, 'United Nations Forum on Forests, Report of the sixth session (27 May 2005 and 13–24 February 2006),' Chapter I, para.6(g). (As this book went to press the resolution had yet to be passed by the UN Economic and Social Council.)
143 Environmental Investigation Agency (2005) 'CAFTA will flood Florida and U.S. markets with illegal timber products,' 6 June. Available online at: www.eia-international.org/cgi/news/news.cgi?t=template&a=246&source= (accessed 23 July 2005); Friends of the Earth (undated) 'Free Trade Area of the Americas, trading away our environment: Market "liberalization" without responsibility.' Available online at: www.foe.org/camps/intl/greentrade/ftaa.html (accessed 22 December 2005); Forest Action Network (undated) 'Trading away our forests: About the FTAA.' Available online at: www.fanweb.org/archives/ftaa/Files/about.html (accessed 22 December 2005).
144 The American Forest and Paper Association, which supports the FLEG processes because illegal logging undercuts the US forest products industry, has declared its support for CAFTA, as '[o]ur ability to sell our products in these markets will be improved as a result of this agreement': Letter (undated, probably 2004) by Maureen Smith, a consultant for the AFPA, to Christopher Padilla, Assistant US Trade Representative. Available online at: www.ustr.gov/assets/Trade_Agreements/Bilateral/CAFTA/CAFTA_Reports/asset_upload_file113_5958.pdf (accessed 18 December 2005).

8 The World Bank's Forests Strategy

1 David Cassells, formerly of the World Bank's forestry department, clarified the complexities of Bank forest policy and provided an invaluable insider's perspective on the preparation of the 2002 forests strategy in two extensive interviews (Washington DC, 16 March 2004; Geneva, 11 May 2004). He bears no responsibility for any errors in this chapter.

2 *Note on Nomenclature*: It is now common within the World Bank to refer to a *strategy* as a paper that outlines overall goals, objectives and options for a particular issue or sector, while *policy* refers to the internal operational procedures that Bank personnel are obliged to follow in pursuit of a strategy. (Confusingly, internal policies are sometimes referred to as *safeguards*.) The 1978 document was labelled a 'policy,' which is how it is referred to here. In 1991 the Bank issued *The Forest Sector: A World Bank Policy Paper*. Recent Bank literature refers to this as the 1991 strategy, which is how it is referred to here. It was supported by a set of operational policies issued in 1993 (which are not examined in this chapter). In 2002 the Bank issued its 'revised forest strategy,' in effect a completely new strategy. It is supported by a revised set of safeguard policies. The shift from *forestry* in 1978 to *forests* in subsequent strategies reflects a broadening of thinking within the Bank from forestry as a separate, relatively enclosed, sector towards a more intersectoral approach.

3 Mallaby, Sebastian (2004) *The World's Banker: A Story of Failed States, Financial Crises and the Wealth and Poverty of Nations*, New Haven: Yale University Press, p.44.

4 World Bank (2001) 'Forestry development: A review of Bank experience.' Available online at: http://lnweb18.worldbank.org/oed/oeddoclib.nsf/DocUNIDViewForJavaSearch/6314B428033F6A54852567F5005D81ED (accessed 12 May 2005).

5 World Bank, n.4 above. See also World Bank (undated) 'Forestry development: The World Bank experience.' Available online at: http://lnweb18.worldbank.org/oed/oeddoclib.nsf/DocUNIDViewForJavaSearch/5EF3153BDD265CE28525681C00697107 (accessed 11 May 2005). This summarizes a review of the Operations Evaluation Department of the 1978 policy and was probably written in the late 1980s.

6 World Bank (1978) *Forestry*, sector policy paper, p.33, cited in Payer, Cheryl (1982) *The World Bank: A Critical Analysis*, New York: Monthly Review Press, p.295.

7 World Bank (undated), n.5 above.

8 Payer, n.6 above, p.296.

9 Payer, n.6 above, p.289.

10 Poore, Duncan et al (1989) *No Timber Without Trees: Sustainability in the Tropical Forest*, London: Earthscan.

11 Price, David (1989) *Before the Bulldozer: The Nambiquara Indians and the World Bank*, Washington DC: Seven Locks Press.

12 Rich, Bruce (1994) *Mortgaging the Earth: The World Bank, Environmental Impoverishment and the Crisis of Development*, Boston: Beacon Press, pp.29–33. See also Hall, Anthony L. (1989) *Developing Amazonia: Deforestation and Social Conflict in Brazil's Carajás Programme*, Manchester: Manchester University Press.

13 Rich, n.12 above, p.94.

14 Colchester, Marcus and Lohmann, Larry (1990) *The Tropical Forestry Action Plan: What Progress?*, Penang/Sturminster Newton UK: World Rainforest Movement/ The Ecologist; Halpin, Elizabeth A. (1990) *Indigenous Peoples and the Tropical Forestry Action Plan*, Washington DC: World Resources Institute; Lynch, Owen J. (1990) *Whither the People?: Demographic, Tenurial and Agricultural Aspects of the Tropical Forestry Action Plan*, Washington DC: World Resources Institute; Winterbottom, Robert (1990) *Taking Stock: The Tropical Forestry Action Plan After Five Years*, Washington DC: World Resources Institute.
15 World Bank (1991) *The Forest Sector: A World Bank Policy Paper*, Washington DC: The World Bank, p.65.
16 World Bank, n.15 above, p.66.
17 World Bank, n.15 above, pp.62–63.
18 World Bank, n.15 above, p.65.
19 World Bank, n.15 above, p.12.
20 World Bank, n.15 above, p.47.
21 World Bank, n.15 above, p.37.
22 Kolk, Ans (1996) *Forests in International Environmental Politics: International Organizations, NGOs and the Brazilian Amazon*, Utrecht: International Books, p.227.
23 World Bank Operations Evaluation Department (2000) *Financing the Global Benefits of Forests: The Bank's GEF Portfolio and the 1991 Forest Strategy*, Washington DC: World Bank, p.xiv
24 Global Environment Facility (2004) *Forests Matter: GEF's Contribution to Conserving and Sustaining Forest Ecosystems*, Washington DC: GEF, p.32.
25 Rich, n.12 above, pp.164–165.
26 World Bank/WWF Alliance (2002) *Progress Through Partnership: Catalyzing Change in Forest Policy and Practice*, Washington DC, p.3. 'About the Congo Basin Forest Partnership.' Available online at: www.cbfp.org/en/about/aspx (accessed 26 April 2004).
27 World Bank/WWF Alliance, n.26 above, p.3. World Bank/WWF Alliance (2003) *Annual Report 03*, Washington DC, p.3.
28 World Bank, n.15 above, p.17.
29 David Cassells, World Bank, interview, Washington, 16 March 2004.
30 World Bank/WWF Alliance (2002), n.26 above, p.18.
31 World Bank/WWF Alliance (2003), n.27 above, p.13.
32 World Bank/WWF Alliance (2003), n.27 above, p.27.
33 Marcus Colchester, Forest Peoples Programme, telephone interview, 15 February 2006.
34 World Bank Operations Evaluation Department (2000) *The World Bank Forest Strategy: Striking the Right Balance*, Washington DC: World Bank, p.xxi.
35 The phrase 'chilling effect' is sometimes cited out of context in Bank literature. The original OED report noted that if the logging ban were to be extended to the forests of Europe and Central Asia this 'could have a chilling effect on the Bank's ability to mobilize much-needed funding for continued responsible forest management in this region': World Bank Operations Evaluation Department, n.34 above, p.47. However a 2002 Bank publication claimed a broader interpretation: '... the [OED] review found that a "chilling effect" permeated forest-related operations in the World Bank Group, including the International Finance Corporation (IFC) and the Multilateral Investment Guarantee Agency (MIGA)': World Bank (2002) *Sustaining Forests: A World Bank Strategy*, Washington DC: World Bank, p.6. While

this is not precisely what the OED review said, it is certainly a view that found favour within the Bank's forestry department.

36 World Bank (2002) *A Revised Forest Strategy for the World Bank Group*, Washington DC: World Bank, p.2.

37 World Bank Operations Evaluation Department, n.34 above, p.46.

38 Flejzor, Lauren (2006) 'Explaining strategic change in international organizations,' PhD thesis, London School of Economics and Political Science, p.238.

39 Marcus Colchester, Forest Peoples Programme, telephone interview, 15 February 2006.

40 David Cassells, World Bank, interview, Washington, 16 March 2004.

41 World Bank Operations Evaluation Department, n.34 above, p.44.

42 David Cassells, World Bank, interview, Washington, 16 March 2004.

43 World Rainforest Movement (undated) 'The World Bank in the forest.' Available online at: www.wrm.org.uy/publications/briefings/worldbank.html (accessed 19 February 2004).

44 Letter to Jim Douglas and David Cassells, World Bank, from Stewart Maginnis, IUCN, 'Suggested textual amendments to the proposed revision of OP/BP 4.36,' 2 August 2002.

45 World Bank, n.36 above, p.9.

46 World Bank (undated) 'Revising the Bank's forest policy: Key questions and answers,' fact sheet.

47 World Bank, n.36 above, p.47.

48 World Bank Operations Evaluation Department, n.34 above, p.46

49 World Bank, n.35 above, pp.12–13.

50 World Bank, n.35 above, p.8.

51 Rich, Bruce (2002) 'The World Bank under James Wolfensohn,' in Pincus, Jonathan R. and Winters, Jeffrey A. (eds) *Reinventing the World Bank*, Ithaca: Cornell University Press, pp.26–38.

52 World Bank website (undated) 'The World Bank Operational Manual.' Available online at: http://wbln0018.worldbank.org/institutional/manuals/opmanual.nsf/TextDefinition1?OpenNavigator (accessed 1 April 2004).

53 World Bank website (undated) 'Operational and Safeguard Policies.' Available online at: http://web.worldbank.org/WBSITE/EXTERNAL/TOPICS/ENVIRONMENT/0,,contentMDK:20124313~menuPK:549278~pagePK:148956~piPK:216618~theSitePK:244381,00.html (accessed 7 June 2004).

54 The ten World Bank safeguard policies are forests, environmental assessment, natural habitats, pest management, involuntary resettlement, indigenous peoples, safety of dams, cultural property, international waterways and disputed areas.

55 The IUCN also proposed several other amendments to early Bank drafts. Letter to Jim Douglas and David Cassells, World Bank, from Stewart Maginnis, n.44 above.

56 The World Bank's environmental categories are available online at: http:/web.worldbank.org/WBSITE/EXTERNAL/PROJECTS/0,,contentMDK:20061220~menuPK:51563~pagePK:41367~piPK:51533~theSitePK:40941,00.html (accessed 19 March 2006).

57 World Bank Group Intranet, Home > Operational Manual > BP 4.36 Forests, para. 3, accessed 16 March 2004.

58 World Bank Operational Manual, section OP 4.36 'Forests,' para.5. Available online at: http://wbln0018.worldbank.org/Institutional/Manuals/OpManual.nsf/

tocall/C972D5438F4D1FB78525672C007D077A?OpenDocument (accessed 1 February 2005).

59 David Cassells, World Bank, interview, Washington, 16 March 2004.

60 World Bank Operational Manual, section OP 4.36 – Forests, n.58 above, para. 7.

61 The proposals that NGO members of the technical advisory group advocated are contained in *inter alia*: Colchester, Marcus (2000) 'Some thoughts about the World Bank's new forest policy: Input to the safeguards and definitions focus group of the technical advisory group,' 4 August; and Colchester, Marcus (2000) 'Towards a socially appropriate notion of "High Conservation Value Forests": 2nd input to safeguards and definitions focus group of the technical advisory group,' 24 October. These documents and other inputs to the World Bank's 2002 forests strategy from the Forests Peoples Programme are available online at: www.wrm.org.uy > International Processes/Actors > International Financial Institutions > World Bank (accessed 16 February 2006).

62 Victor Teplyakov, IUCN Russia, interview, Vienna, 13 February 2006.

63 Marcus Colchester, Forest Peoples Programme, telephone interview, 15 February 2006. The final draft of OP 4.36 states that 'The Bank does not finance projects that contravene applicable international environmental agreements': World Bank Operational Manual, section OP 4.36 – Forests, n.58 above, para. 6.

64 Marcus Colchester, Forest Peoples Programme, telephone interview, 15 February 2006.

65 David Cassells, World Bank, interview, Washington, 16 March 2004. The Bank's safeguards on indigenous peoples are outlined in OP 4.20 and BP 4.20. Other Bank policies that relate to forests include OP 4.01/BP 4.01 Environmental Assessment, OP 4.04/BP 4.04 Natural Habitats, OP 4.11/BP 4.11 Cultural Property and OP 4.12/BP 4.12 Involuntary Resettlement.

66 World Bank Group Intranet, Home > Operational Manual > BP 4.36 Forests, paras.6(b) and 6(c) (accessed 16 March 2004).

67 World Bank Operational Manual, section OP 4.36 – Forests, n.58 above, para. 11.

68 David Cassells, World Bank, interview, Washington, 16 March 2004. This position can also be found in World Bank, n.36 above, p.40.

69 World Bank Operational Manual, section OP 4.36 – Forests, n.58 above, para. 9(b).

70 Colchester, Marcus (2002) 'Does the World Bank's new draft forest policy protect "critical forests"?,' Forest Peoples Programme paper, 25 June 2002; Forest Peoples Programme (2005) 'The invisible sourcebook: a "critical" omission,' in Rainforest Foundation et al (2005) *Broken Promises: How World Bank Group Policies Fail to Protect Forests and Forest Peoples' Rights*, Rainforest Foundation, April, p.46.

71 Stiglitz, Joseph (2002) *Globalization and its Discontents*, London: Penguin Books, especially p.13.

72 Mallaby, n.3 above, pp.208–209; Stiglitz, n.71 above, pp.89–132.

73 See, for example, Wunder, Sven (2003) *Oil Wealth and the Fate of the Forest*, London: CIFOR/Routledge.

74 OP 8.60 and BP 8.60 on 'Development Policy Lending' replaced the Bank's Operational Directive on 'Adjustment Lending Policy' (OD 8.60).

75 Letter from Marcus Colchester, Forest Peoples Programme, Moreton-in-March, UK, to Stephen Lintner, World Bank, Washington DC, 'Draft OP 8.60 on Development Policy Lending,' 18 May 2004.

76 OP 8.60 'Development Policy Lending,' August 2004, para.11. Available online at: http://wbln0018.worldbank.org/Institutional/Manuals/OpManual.nsf/tocall/AD55139DFE937EE585256EEF00504282?OpenDocument (accessed 9 May 2005).
77 OP 8.60 'Development Policy Lending,' n.76 above.
78 BP 8.60 'Development Policy Lending,' August 2004. Available online at: http://wbln0018.worldbank.org/Institutional/Manuals/OpManual.nsf/whatnewvirt/3371A4146FAB65F985256EEF0066C54F?OpenDocument (accessed 15 July 2005).
79 Horta, Korinna (2005) 'Forests and structural adjustment: the World Bank's steamrolling of stakeholders and its own board,' in Rainforest Foundation et al, n.70 above, pp.44–45.
80 David Cassells, World Bank, interview, Washington, 16 March 2004.
81 World Rainforest Movement, 'The World Bank in the forest,' n.43 above.
82 World Rainforest Movement, 'The World Bank in the forest,' n.43 above.
83 Rich, n.51 above, p.29.
84 World Bank website: http://wbln0018.worldbank.org/dgf/dgf.hsf/0/a088b04a45cbe4e885256d9b005d57af?OpenDocument (accessed 9 May 2005).
85 World Business Council for Sustainable Development (2005) *Executive Brief: Sustainable Forest Products Industry*, p.1. Available online at: www.wbcsd.org/DocRoot/oY8ixp7MjSJvIqZ0dmG/Forest.pdf (accessed 7 June 2006).
86 International Finance Corporation Operational Policies, OP 4.36, November 1998, para. 1(a).
87 International Finance Corporation, n.86 above, paras. 1(b) and 1(c).
88 International Finance Corporation, n.86 above, paras. 1(c) (iii) and (iv).
89 David Cassells, World Bank, interview, fifth session of the United Nations Forum on Forests, New York, 18 May 2005.
90 Lazarus, Suelen (2004) 'The Equator Principles: a milestone or just good PR?,' 26 January. Available online at: www.equator-principles.com/ga1.shtml (accessed 27 May 2004).
91 Lazarus, n.90 above.
92 The ten areas covered by IFC safeguard policies are environmental assessment, natural habitats, pest management, forestry, safety of dams, indigenous peoples, involuntary resettlement, cultural property, child and forced labour, and international waterways. The areas covered by the IFC safeguards thus differ slightly from those covered by other World Bank Group institutions (as shown in n.54 above).
93 David Cassells, World Bank, interview, fourth session of the United Nations Forum on Forests, Geneva, 11 May 2004.
94 Letter from Ben Gunneberg, PEFC Council Secretary, to David Cassells and Bruce Cabarle, co-chairs of World Bank/WWF Alliance, dated 7 October 2004. Gunneberg wrote that PEFC would 'welcome new forms of cooperation with the World Bank and WWF e.g. in a broader Alliance and with other stakeholders.'
95 World Bank/WWF Alliance for Forest Conservation and Sustainable Use (2005) 'New alliance targets,' paper disseminated at the formal launch of the renewed Alliance, fifth session of the United Nations Forum on Forests, New York, 25 May 2005.

9 The International Forests Regime

1 Krasner, Stephen D. (1983) 'Structural causes and regime consequences: Regimes as intervening variables,' in Krasner, Stephen D. (ed) *International Regimes*, Ithaca NY: Cornell University Press, p.2.

2 The view that a regime is broadly equivalent to hard international legal instruments has been the dominant approach in the extensive regime theory literature. See, for example, the collection of essays in Krasner (ed), n.1 above. See also De Fontaubert, Charlotte (2003) *Achieving Sustainable Fisheries: Implementing the New International Legal Regime*, Gland, Switzerland: IUCN; Friedheim, Robert L. (1992) *Negotiating the New Ocean Regime*, Columbia: University of South Carolina Press; Joyner, Christopher C. (1998) *Governing the Frozen Commons: The Antarctic Regime and Environmental Protection*, Columbia: University of South Carolina Press; Rittberger, Volker (ed) (1993) *Regime Theory and International Relations*, Oxford: Clarendon Press; Yamin, Farhana and Depledge, Joanna (2004) *The International Climate Change Regime: A Guide to Rules, Institutions and Procedures*, Cambridge: Cambridge University Press; and Young, Oran R. and Osherenko, Gail (eds) (1993) *Polar Politics: Creating International Environmental Regimes*, Ithaca NY: Cornell University Press.

3 For example, Glück, Peter; Tarasofsky, Richard; Byron, Neil and Tikkanen, Ilpo (1997) *Options for Strengthening the International Legal Regime for Forests*, Joensuu: European Forest Institute; Tarasofsky, Richard (1995) *The International Forests Regime: Legal and Policy Issues*, Gland, Switzerland: WWF/IUCN; and Tarasofsky, Richard (ed) (1999) *Assessing the International Forests Regime. IUCN Environmental Law and Policy Paper No. 37*, Cambridge: IUCN Publications.

4 Skala-Kuhmann, Astrid (1996) 'Legal instruments to enhance the conservation and sustainable management of forest resources at the international level,' report commissioned by the German Federal Ministry for Economic Co-operation and Development (MMZ) and Deutsche Gesellschaft für Technische Zusammenarbeit (GTZ), July.

5 CBD Decision IV/7, 'Forest biological diversity,' in Secretariat of the Convention on Biological Diversity (2003) *Handbook of the Convention on Biological Diversity* (2nd edition), Montreal: Secretariat of the Convention on Biological Diversity, pp.479–487. The CBD has agreed five thematic programmes of work: agricultural biodiversity; dry and sub-humid lands; inland water biological diversity; marine and coastal biodiversity; and forest biodiversity. See also UN document UNEP/CBD/SBSTTA/3/Inf.5, 'Report of the Meeting of the Liaison Group on Forest Biological Diversity,' 14 July 1997.

6 CBD Decision VI/26, 'Strategic plan for the convention on biological diversity,' Annex, in Secretariat of the Convention on Biological Diversity, n.5 above, pp.883–887.

7 Humphreys, David (1999) 'The evolving forests regime,' *Global Environmental Change*, Vol. 9, No. 3, pp.251–254.

8 'WCPA members meeting, Agenda and background documentation, Bangkok, Thailand, 16–17 November 2004,' p.12.

9 Colchester, Marcus (2003) *Salvaging Nature: Indigenous Peoples, Protected Areas and Biodiversity Conservation*, Montevideo/Moreton-in-Marsh: World Rainforest Movement.

10 MacKay, Fergus and Caruso, Emily (2004) 'Indigenous lands or national parks,' *Cultural Survival Quarterly*, Issue 28.1, 15 March. Available online at: http://209.200.101.189/publications/csq/csq-article.cfm?id=1737 (accessed 26 January 2006).

11 In particular from the Theme on Indigenous and Local Communities, Equity and Protected Areas (TILCEPA), a joint theme of the IUCN Commission on Environmental, Economic and Social Policy (CEESP) and the World Commission on Protected Areas (WCPA). See: www.iucn.org/themes/ceesp/Wkg_grp/TILCEPA/community.htm (accessed 14 April 2005). The reports of the TILCEPA for the 2003 World Parks Congress held in Durban are available online at: www.tilcepa.org (accessed 26 January 2006). TILCEPA is composed entirely of volunteers. In articulating the concept of community conserved areas TILCEPA was assisted by professionals in the IUCN secretariat (in particular Maria Fernanda Espinosa), the World Alliance of Mobile Indigenous Peoples (WAMIP), indigenous peoples representatives at the Convention on Bioligical Diversity, and volunteer groups associated with the IUCN, including the Theme on Governance, Equity and Rights (TGER): Grazia Borrini-Feyerabend, email, 9 November 2005.

12 Borrini-Feyerabend, Grazia (2003) 'Governance of protected areas ... innovations in the air,' *Policy Matters*, Issue 12, pp.92–101. Available online at: www.iucn.org/themes/ceesp/Publications/newsletter/PM12.pdf (accessed 26 January 2006). See also Pathak, Neema; Bhatt, Seema; B., Tasneem (sic); Kothari, Ashish and Borrini-Feyerabend, Grazia (2004) 'Community conserved areas: A bold frontier for conservation,' *Briefing Note 5*, November 2004.

13 Pathak et al, n.12 above. See also Borrini-Feyerabend, G. (2003) 'Community conserved areas (CCAs) and co-managed protected areas (CMPAs) – towards equitable and effective conservation in the context of global change,' Draft report of the IUCN joint CCESP/WCPA Theme on Indigenous and Local Community, Equity and Protected Areas (TILCEPA) for the Ecosystems, Protected Areas and People (EPP) Project, Gland, Switzerland: IUCN. Available online at: www.iucn.org/themes/ceesp/Wkg_grp/TILCEPA/community.htm (accessed 17 June 2005).

14 UN document UNEP/CBD/SBSTTA/9/INF/3, 'Report of the ad hoc technical expert group on protected areas,' 22 September 2003.

15 'The Durban Accord: Our global commitment for people and earth's protected areas.' Available online at: www.iucn.org/themes/wcpa/wpc2003/english/outputs/durban.htm (accessed 10 June 2004); 'The Durban Action Plan, Revised Version, March 2004.' Available online at: www.iucn.org/themes/wcpa/wpc2003/pdfs/outputs/wpc/durbanactionplan.pdf (accessed 10 June 2004).

16 SBSTTA recommendation IX/4, 'Protected areas,' para. 2.1.3. Available online at: www.biodiv.org/recommendations/default.aspx?m+SBSTTA-09&id=7460&lg+0 (accessed 12 April 2005).

17 CBD Decision VII/28, 'Protected areas.' Available online at: www.biodiv.org/decisions/?dec+Vii/28 (accessed 26 January 2006). Community conserved areas feature in the second element – 'Governance, participation, equity and benefit sharing' – of the CBD's programme of work on protected areas. Available online at: www.biodiv.org/programmes/cross-cutting/protected/wopo.asp?prog=p2 (accessed 26 January 2006).

18 It was more usual for a technical expert group to be created when analysis or evaluation was required. Such groups tend to have a small, closed membership, whereas all CBD members can participate in an open-ended working group, which is a more substantive follow-up mechanism.

19 Gijs Van Tol, Secretariat of the Convention on Biological Diversity, interview, Montreal, 19 March 2004.
20 CBD Decision VII/30 established the working group on review of implementation, while Decision VII/28 established the working group on protected areas.
21 In 2004 the WCPA issued guidelines policy and practice for community conserved areas that emphasized the rights of indigenous and local communities: Borrini-Feyerabend, Grazia, Kothari, Ashish, Oviedo, Gonzalo et al (2004) 'Indigenous and local communities and protected areas: Towards equity and enhanced conservation – Guidance on policy and practice for co-managed protected areas and community conserved areas,' *WCPA Best Practice Protected Area Guidelines Series* No. 11, Gland, Switzerland and Cambridge, UK: IUCN.
22 I am grateful to Grazia Borrini-Feyerabend for this last point: email, 9 November 2005.
23 See, for example, Secretariat of the Convention on Biological Diversity (2004) 'Biodiversity issues for consideration in the planning, establishment and management of protected area sites and networks,' *CBD Technical Series* No. 15, Montreal: Secretariat of the Convention on Biological Diversity.
24 The UNEP World Conservation Monitoring Centre is also an important actor. See Mulongoy, Kalemani Jo and Chape, Stuart (eds) (2004) *Protected Areas and Biodiversity*, Montreal/Cambridge: CBD/WCMC.
25 Sayer, Jeffrey, Isharan, Natarajan, Thorsell, James and Sigaty, Todd (2000) 'Tropical forest biodiversity and the World Heritage Convention,' *Ambio: A Journal of the Human Environment*, Vol. 29, Issue 6, pp.302–309.
26 Child, Brian (2004) 'Parks in transition: Biodiversity development and the bottom line,' in Child, Brian (ed) *Parks in Transition: Biodiversity, Rural Development and the Bottom Line*, London: Earthscan, pp.245–246.
27 Okrah, Lambert and Kofie, Richard (2005) 'Multi-stakeholder involvement in national parks management: The case of private sector in Mole national park,' in Lovera, Miguel, Avendaño, Tatiana Rosa and Torres, Irene Vélez (eds) *The New Merchants: Life as commodity*, Bogotá: Global Forest Coalition/Censat Agua Viva, pp.55–61.
28 Friends of the Earth International (2005) *Privatization: Nature for Sale – The Impacts of Privatizing Water and Biodiversity*, Amsterdam: Friends of the Earth International, pp.13, 33.
29 Convention on Biological Diversity 1992, Article 2.
30 CBD Decision V/6, 'Ecosystem approach,' Annex, A.1, in Secretariat of the Convention on Biological Diversity, n.5 above, p.560.
31 Steiner, Achim (2003) 'Trouble in paradise,' *New Scientist*, 18 October, p.21.
32 In 2001 the CBD signalled its interest in sustainable forest management with publication of a technical paper: Secretariat of the Convention on Biological Diversity (2001) 'Sustainable management of non-timber forest resources,' *CBD Technical Series* No. 6, Montreal: Secretariat of the Convention on Biological Diversity.
33 Text in quotation marks is from CBD Decision VII/11 'Ecosystem approach,' Annex II 'Consideration of the relationship between sustainable forest management and ecosystem approach, and review of, and development of strategies for, the integration of the ecosystem approach into the programmes of work of the convention.' Available online at: www.biodiv.org/decisions/default.aspx?dec=VII/11 (accessed 7 June 2004).

34 Republic of Congo, verbal intervention, fourth session of the United Nations Forum on Forests, Geneva, 4 May 2004.
35 Convention on Biological Diversity 1992, Article 10(c). The CBD moved towards operationalizing the concept of sustainable use by adopting in 2004 the Addis Ababa Principles and Guidelines for the Sustainable Use of Biodiversity, which are intended to provide biodiversity managers with practical advice on how to achieve sustainable use. CBD Decision VII/12 'Sustainable Use (Article 10),' Annex II 'Addis Ababa Principles and Guidelines for the Sustainable Use of Biodiversity.' Available online at: www.biodiv.org/decisions/default.aspx?m=COP-7&id=774981g+0 (accessed 12 April 2005). See also Secretariat of the Convention on Biological Diversity (2003) 'Facilitating Conservation and Sustainable Use of Biological Diversity,' *CBD Technical Series* No. 9, Montreal: Secretariat of the Convention on Biological Diversity.
36 UN document UNEP/CBD/TKBD/1/2, 'Traditional knowledge and biological diversity,' 18 October 1997, para. 76, p.18.
37 UN document UNEP/CBD/TKBD/1/2, n.36 above, para. 102, p.24.
38 Colchester, Marcus, Silva Monterrey, Nalúa and Tomedes, Ramón (2004) *Protecting and Encouraging Customary Use of Biological Resources: The Upper Caura, Venezuela*, Moreton-in-Marsh UK: Forest Peoples Programme, p.53. These recommendations apply specifically to Venezuela, although they also have a more generic applicability.
39 UN International Covenant on Economic, Social and Cultural Rights, Article 1.1. Available online at: www.hrweb.org/legal/escr.html (accessed 31 March 2006); and UN International Covenant on Civil and Political Rights, Article 1.1. Available online at: www.hrweb.org/legal/cpr.html (accessed 31 March 2006).
40 UN document E/CN.4/2004/81, 'Report of the working group established in accordance with Commission on Human Rights Resolution 1995/322,' 7 January 2004, Article 3, p.21.
41 UN document A/RES/60/1, 'Resolution adopted by the General Assembly [without reference to a Main Committee (A/60/L.1)],' 24 October 2005, para. 127, p.28.
42 UN document E/CN.4/Sub.2/1993/29/Annex 1, 'Report of the working group on indigenous populations on its eleventh session,' 23 August 1993, Article 29. Available online at: www.cwis.org/drft9329.html (accessed 7 June 2004).
43 ILO Convention 169, Convention Concerning Indigenous and Tribal Peoples in Independent Countries, Article 7.1. See also Richardson, Benjamin J. (2001) 'Indigenous peoples, international law and sustainability,' *Review of European Community and International Environmental Law (RECIEL)*, Vol. 10, No. 1, pp.1–12.
44 Colchester, Marcus and MacKay, Fergus (2004) *In Search of Middle Ground: Indigenous Peoples, Collective Representation and the Right to Free, Prior and Informed Consent*, Moreton-in-March UK: Forest Peoples Programme, p.9.
45 Colchester and MacKay, n.44 above, pp.10–11. The UN Centre for Transnational Corporations was closed in 1993 and its activities transferred to the UNCTAD, a move that represented a downgrading of oversight of corporations within the UN system.
46 Forest Stewardship Council, 'Policy & standards: FSC principles & criteria of forest stewardship.' Available online at: www.fsc.org/en/about/policy_standards/princ_criteria/5 (accessed 3 April 2006).
47 Colchester and MacKay, n.44 above, pp.12; World Commission on Dams (2000) *Dams and Development: A New Framework for Decision-Making*, London: Earthscan, pp.xxxiv, 281–282.

48 Colchester and MacKay, n.44 above.
49 UN document E/CN.4/Sub.2/1993/29/Annex 1, n.42 above, Article 30.
50 UN document E/CN.4/2004/81, n.40 above, Article 30, p.26. The word 'obtain' also appears in one alternative for draft Article 20, where it is not bracketed. However this does not signify agreement, as in the second alternative, which merges draft Articles 19 and 20, the word 'obtain' does not appear at all (p.24).
51 The Forest Principles state that '[a]ppropriate indigenous capacity and local knowledge regarding the conservation and sustainable development of forests should ... be recognized, respected, recorded, developed and, as appropriate, introduced in the implementation of programmes.' See 'Non-legally binding authoritative statement of principles for a global consensus on the management, conservation and sustainable development of all types of forests,' para. 12(d).
52 Convention on Biological Diversity 1992, Article 8(j).
53 'Strategic Plan 1997–2002' (adopted by the 6th Conference of Parties to the Ramsar convention), paras. 2.7.3 and 2.7.4, cited by Downes, David R. (1999) 'Global forest policy and selected international instruments: A preliminary review,' in Tarasofsky (ed), n.3 above, p.68.
54 Argentina, Bolivia, Colombia, Costa Rica, Denmark, Ecuador, Fiji, Guatemala, Honduras, Mexico, the Netherlands, Norway, Paraguay and Peru. Noted in Richardson, n.43 above.
55 Meyer, Anja (2001) 'International environmental law and human rights: Towards the explicit recognition of traditional knowledge,' *Review of European Community and International Environmental Law (RECIEL)*, Vol. 10, No. 1, pp.37–46.
56 Tvedt, Morten Walløe (2005) 'How will a substantive patent law treaty affect the public domain for genetic resources and biological material?,' *Journal of World Intellectual Property*, Vol. 8, No. 3, pp.311–344.
57 Sell, Susan K. (2003) *Private Power, Public Law: The Globalization of Intellectual Property Rights*, Cambridge: Cambridge University Press, p.175.
58 Anuradha, R.V. (2001) 'IPRs: Implications for biodiversity and local and indigenous communities,' *Review of European Community and International Environmental Law (RECIEL)*, Vol. 10, No. 1, pp.27–36.
59 CBD Decision IV/9 'Implementation of Article 8(j) and related provisions,' in Secretariat of the Convention on Biological Diversity, n.5 above, pp.487–489.
60 CBD Decision V/26 'Access to genetic resources,' in Secretariat of the Convention on Biological Diversity, n.5 above, pp.645–647.
61 CBD Decision VI/24 'Access and benefit sharing as related to genetic resources,' Annex, I.a, para.11(j), and section 5, sub-section c, para.3(f), in Secretariat of the Convention on Biological Diversity, n.5 above, pp.834, 850.
62 World Summit on Sustainable Development, Johannesburg, September 2002, 'Plan of Implementation,' para. 44(o). Available online at: www.johannesburgsummit.org/html/documents/summit_docs/2309_planfinal.htm (accessed 27 April 2005).
63 *Earth Negotiations Bulletin*, Vol. 9, No. 311, p.4.
64 Richardson, n.43 above, p.10.
65 Press Statement of the International Indigenous Forum on Biodiversity, 'ABS ignores indigenous human rights,' 17 February 2005. See also Friends of the Earth International, n.28 above, p.14.
66 Convention to Combat Desertification, Article 2.
67 Convention to Combat Desertification, Annex I, 'Regional implementation annex for Africa,' Article 8.3(b)(i).

68 CCD Decision 12/COP.6, in CCD document ICCD/COP(6)/11/Add.1, 'Report of the Conference of the Parties on its sixth session, held in Havana from 25 August to 5 September 2003,' p.29.
69 This account has been obtained from two interviews on 4 May 2004 at the fourth session of the UNFF in Geneva: Kanta Kumari, Biodiversity Program Manager of the Global Environment Facility; and Jan McAlpine, US delegation.
70 The GEF Assembly meets every three to four years and comprises all GEF members, currently about 170 governments. The GEF Council meets twice a year. It has a rotating membership of 32 governments, namely 16 developing countries, 14 from developed countries and 2 from countries with economies in transition. The 2002 GEF Assembly adopted as a focal area 'Land degradation, primarily desertification and deforestation.' See 'Beijing Declaration of the Second GEF Assembly,' 18 October 2002, p.6.
71 Global Environment Facility, 'Operational program on sustainable land management (OP15),' revised on 18 December 2003.
72 UN document GEF C.23 Inf.13, 'Progress report on implementation of the GEF operational program on sustainable land management,' 28 April 2004.
73 US verbal intervention by Jan McAlpine, fourth session of the UNFF, Geneva, 4 May 2004.
74 GEF verbal intervention by Kanta Kumari, fourth session of the UNFF, Geneva, 4 May 2004.
75 Tickell, Oliver (2000) 'Stalemate on key carbon forestry principles,' *Independent* (special report on Global Warming), 15 December, p.5.
76 See, for example, Carrere, Ricardo (1999) *Ten Replies to Ten Lies: Briefing Paper, Plantations Campaign*, Moreton-in-March/Montevideo: World Rainforest Movement; Lohmann, Larry (1999) *The Carbon Shop: Planting New Problems, Briefing Paper, Plantations Campaign*, Moreton-in-March/Montevideo: World Rainforest Movement.
77 Carrere, n.76 above, p.15.
78 *Earth Negotiations Bulletin*, Vol. 12, No. 231, p.15; Greenpeace (2003) 'Sinks in the CDM: After the climate, biodiversity goes down the drain – An analysis of the CDM sinks agreement at CoP-9' (unpublished paper).
79 Lohmann, n.76 above, p.21.
80 Lohmann, n.76 above, p.24.
81 Kill, Jutta (2001) *Sinks in the Kyoto Protocol: A Dirty Deal for Forests, Forest Peoples and the Climate*, Brussels: FERN, p.9.
82 UN document FCCC/CP/2003/6/Add.2, 'Report of the Conference of the Parties on its ninth session, held at Milan from 1 to 12 December 2003,' 30 March 2004, Decision 19/CP.9, Annex, p.16, para. 1(g)–(h).
83 CBD Decision V/4, 'Progress report on the implementation of the programme of work for forest biological diversity,' in Secretariat of the Convention on Biological Diversity, n.5 above, para. 16, p.543. On the CBD's efforts to ensure that international climate law takes into account the provisions of the CBD see Secretariat of the Convention on Biological Diversity (2003) 'Interlinkages between biological diversity and climate change: Advice on the integration of biodiversity considerations into the implementation of the United Nations Framework Convention on Climate Change and its Kyoto Protocol,' *CBD Technical Series* No. 10, Montreal: Secretariat of the Convention on Biological Diversity.
84 Greenpeace, n.78 above; Kill, n.81 above.

85 *CIFOR Infobrief*, No. 2 October 2002, 4pp; CIFOR (2001) 'Capturing the value of forest carbon for local livelihoods,' Jakarta: Indonesia, 16pp.
86 Cited in McCarthy, Michael (2003) 'A year of extremes provides evidence of global warming,' *Independent*, 3 December, p.3.
87 Convention on International Trade in Endangered Species of Wild Fauna and Flora, 1973, Article IV.2(a).
88 IUCN and TRAFFIC (2002) 'Timber and the twelfth meeting of the Conference of the Parties to CITES, Chile 2002,' briefing document.
89 IUCN and TRAFFIC, n.88 above. (*Perscopsis elata* is sometimes referred to as African teak.)
90 Forestry Research Programme, *Research Summary* 003, *Saving the Fate of Big-leaved Mahogany*, Aylesford: Department for International Development and Oxford Forestry Institute, March 2003, p.1. (Bigleaf mahogany is sometimes referred to as Big-leaved mahogany.)
91 US Fish and Wildlife Service (undated) 'International affairs: Mahogany Working Group.' Available online at: www.fws.gov/citestimber/mahogany/mahoganyworkinggroup.html (accessed 18 April 2004).
92 'Recommendations made by the Mahogany Working Group.' Available online at: www.cites.org/eng/decis/valid12/annex3.shtml (accessed 31 August 2004).
93 Chen, H. K. and Zain, S. (2005) 'CITES, trade and sustainable forest management,' *Unasylva* 219, Vol. 55, pp.44–45.
94 Website of UNEP's Great Apes Survival Project (GRASP): www.unep.org/grasp (accessed 18 December 2003).
95 Wilkie, D. S. and Carpenter, J. F. (1999) 'Bushmeat hunting in the Congo Basin: An assessment of impacts and options for migration,' *Biodiversity and Conservation*, Vol. 8, No. 7, pp.927–955.
96 Secretariat of the Convention on Biological Diversity (2004) *Expanded Programme of Work on Forest Biological Diversity*, Montreal: Secretariat of the Convention on Biological Diversity, p.11.
97 CITES document CoP 13 Inf.12, 'Memorandum of cooperation between the secretariat of CITES and the secretariat of the CBD,' Article 4.b.
98 CITES document PC13 WG5 Doc.1, 'Global strategy for plant conservation (CBD): Analysis and links with CBD [Decisions 12.12 and 10.86].'
99 See for example, World Rainforest Movement (2000) 'The World Trade Organization and forests.' Available online at: www.wrm.org.uy/publications/briefings/wto.html (accessed 19 February 2004).
100 As noted in Brack, Duncan and Gray, Kevin (2003) *Multilateral Environmental Agreements and the WTO*, London: Royal Institute of International Affairs, p.22.
101 A counterargument is that the original General Agreement on Tariffs and Trade of 1947 predates most multilateral agreements: Brack and Gray, n.100 above, p.22.
102 The long standing support of the AFPA for tariff elimination is made clear in a letter (undated, probably 2004) by Maureen Smith, a consultant for the AFPA, to Christopher Padilla, the Assistant US Trade Representative. Available online at: www.ustr.gov/assets/Trade_Agreements/Bilateral/CAFTA/CAFTA_Reports/asset_upload_file113_5958.pdf (accessed 18 December 2005).
103 'Federal Register Notice on WTO Forest Products Agreement.' Available online at: http://forests.org/archive/general/federreg.htm (accessed 29 July 2003).
104 Humphreys, David (2004) 'Redefining the issues: NGO influence on international forest negotiations,' *Global Environmental Politics*, Vol. 4, No. 2, pp.51–74.

105 American Lands Alliance (2002) 'Impact of the Free Trade Area of the Americas on forests.' Available online at: http://americanlands.org/IMF/free_trade.htm (accessed 29 July 2003).
106 Weber, Martin (2001) 'Competing political visions: WTO governance and green politics,' *Global Environmental Politics*, Vol. 1, No. 3, p.105.
107 Saunders, J. Owen (1992) 'Trade and environment: The fine line between environmental protection and environmental protectionism,' *International Journal*, Vol. XLVII, pp.723–750. A certification system such as the FSC could play a role here, although under current WTO rules no mandatory certification system is permissible, and even the voluntarism of the FSC could be ruled WTO illegal (Chapter 6 this volume).
108 Jenkins, Leesteffy (1996) 'Trade sanctions: Effective enforcement tools,' in Cameroon, James, Werksman, Jacob and Roderick, Peter (eds) *Improving Compliance with International Environmental Law*, London: Earthscan, pp.221–228.
109 Gill, Stephen (2002) *Power and Resistance in the New World Order*, London: Palgrave Macmillan.
110 Eckersley, Robin (2004) 'The big chill: The WTO and multilateral environmental agreements,' *Global Environmental Politics*, Vol. 4, No. 2, pp.25–26. Eckersley refers here to Gill, Stephen (1995) 'Globalization, market civilization and disciplinary neoliberalism,' *Millennium: Journal of International Studies*, Vol. 24, No. 3, pp.399–423.
111 Eckersley, n.110 above, p.32

10 The Crisis of Global Governance

1 I make this statement based on numerous informal conversations with government delegates between 2000 and 2006 at four UN negotiation sessions on forests and one UN expert level meeting.
2 Ruggie, John Gerard (2004) 'Reconstituting the global public domain – Issues, actors, and practices,' *European Journal of International Relations*, Vol. 10, No. 4, p.504.
3 Wapner, Paul (1995) 'Politics beyond the state: Environmental activism and world civic politics,' *World Politics*, Vol. 47, No. 3, pp.311–341; Wapner, Paul (1996) *Environmental Activism and World Civic Politics*, Albany: State University of New York Press.
4 Bass, Stephen (2002) 'Global forest governance: Emerging impacts of the Forest Stewardship Council,' paper presented to the SUSTRA workshop on Architecture of the Global System of Governance of Trade and Development,' Berlin, 9–10 December 2002.
5 See, for example, Bernstein, Steven (2000) 'Ideas, social structure and the compromise of liberal environmentalism,' *European Journal of International Relations*, Vol. 6, No. 4, pp.464–512, especially p.474; Bernstein, Steven (2001) *The Compromise of Liberal Environmentalism*, New York: Columbia University Press; Bernstein, Steven (2002), 'Liberal environmentalism and global environmental governance,' *Global Environmental Politics*, Vol. 2, No. 3, pp.1–16, especially p.11.
6 Bernstein (2001), n.5 above, pp.111–114, 236.

7 On expanding the trade in forest products see UN document E/CN.17/1997/12, 'Report of the Intergovernmental Panel on Forests on its fourth session (New York, 11–21 February 1997),' para. 108; and UN document E/CN.17/2000/14, 'Report of the Intergovernmental Forum on Forests at its fourth session New York, 31 January–11 February 2000,' para. 33. On promoting private sector forest investment see UN document E/CN.17/1997/12, paras. 13, 57, 61-63, 67(f), 69; and UN document E/CN.17/2000/14, paras. 21, 22, 28, 30(c), 50, 56(b), 111.
8 World Bank (2003) 'Forest Investment Forum.' Available online at: http://lnweb18.worldbank.org/ESSD/ardext.nsf/14ByDocName/EventsForestInvestmentForum Oct2003 (accessed 18 December 2003).
9 International Tropical Timber Organization (2006) 'International tropical forest investment forum: Issues and opportunities for investment in natural tropical forests.' Available online at: www.itto.or.jp/live/PageDisplayHandler?pageId+223 &id=1022 (accessed 22 March 2006).
10 Saurin, Julian (1994) 'Global environmental degradation, modernity and environmental knowledge,' *Environmental Politics*, Vol. 2, No. 4, pp.46–64, also published in Thomas, Caroline (ed) (1994) *Rio: Unravelling the Consequences*, London: Frank Cass, pp.46–64. See also Paterson, Matthew (2001) *Understanding Global Environmental Politics: Domination, Accumulation, Resistance*, London: Palgrave, pp.35–65.
11 On dynamic disequilibrium and economic crisis see Soros, George (1998) *The Crisis of Global Capitalism: Open Society Endangered*, London: Little, Brown and Company.
12 On the IMF's role in contributing to the 1997 Asian and 1998 Russian economic crises through prescribing inappropriate economic policies see Stiglitz, Joseph (2002) *Globalization and its Discontents*, London: Allen Lane, pp.89–165.
13 The argument that global financial stability is a global public good is made in Underhill, Geoffrey (2001) 'The public good versus private interests and the global financial and monetary system,' in Drache, Daniel (ed) *The Market or the Public Domain: Global Governance and the Asymmetry of Power*, London: Routledge, pp.274–295; Wyplosz, Charles (1999) 'International financial instability,' in Kaul, Inge, Grunberg, Isabelle and Stern, Marc A. (eds) (1999) *Global Public Goods: International Cooperation in the 21st Century*, Oxford: Oxford University Press/ United Nations Development Programme, pp.152–189.
14 WWF/CIFOR report cited in Knight, Danielle (2001) 'World Bank, IMF chided on Indonesian forests,' *Asia Times*, 27 October. Available online at: www.atimes.com/se-asia/CJ27Ae02.html (accessed 15 April 2004).
15 Cited in Environmental Defense Fund press release, 'Groups warn IMF, World Bank that proposed loans will worsen Indonesian crisis,' 16 April 1998. Available online at: www.edf.org/pressrelease.cfm?ContentID=1521 (accessed 15 April 2004).
16 Environmental Investigation Agency (1998) *The Politics of Extinction: The Orangutan Crisis – The Destruction of Indonesia's Forests*, London: Environmental Investigation Agency. See also Young, O. R. (ed) 'Science plan for the project on the Institutional Dimensions of Global Environmental Change,' Box C, p.20. Available online at: http://fiesta.bren.ucsb.edu/~idgec/html/publications/html/scienceplan/txt/Plan.txt (accessed 15 April 2004).
17 Welch, Carol (undated) 'Indonesia: Crisis and the IMF: Talking points.' Available online at: www.inannareturns.com/indonesi.htm (accessed 15 April 2004). The argument that the IMF operates principally in the interests of the international

financial sector is also made in Stiglitz, n.12 above. See also Brazier, Chris (2004) 'The power and the folly,' *New Internationalist*, No. 365, March, pp.9–11.

18 Menotti, Victor (1998a) 'Globalization and the acceleration of forest destruction since Rio,' *The Ecologist*, Vol. 28, No. 6, November/December, pp.354–362. See also Sachs, Wolfgang (1999) *Planet Dialectics: Explorations in Environment and Development*, London: Zed Books, p.144.

19 See for example Faulkner, J. Hugh (1995) 'Public–private partnerships in sustainable development,' *Review of European Community and International Environmental Law (RECIEL)*, Vol. 4, No. 2, pp.133–136.

20 Joel Bakan makes the point that 'If corporations and governments are indeed partners, we should be worried about the state of our democracy, for it means that government has effectively abdicated its sovereignty over the corporation': Bakan, Joel (2004) *The Corporation: The Pathological Pursuit of Profit and Power*, London: Constable, p.108.

21 Van der Pijl, Kees (1998) *Transnational Classes and International Relations*, London: Routledge, especially pp.46–47.

22 These are just some of the mergers and acquisitions that International Paper has been involved in. See International Paper website: www.international paper.com/Our%20Company/About%20Us/ (accessed 13 October 2005).

23 Lee, Kelley, Humphreys, David and Pugh, Michael (1997), '"Privatization" in the United Nations system: Patterns of influence in three intergovernmental organizations,' *Global Society: Journal of Interdisciplinary International Relations*, Vol. 11, No. 3, pp.339–357.

24 Korten, David (1995) *When Corporations Rule the World*, London: Earthscan, p.59; Drutman, Lee and Cray, Charlie (2004) *The People's Business: Controlling Corporations and Restoring Democracy*, San Francisco: Berrett-Koehler, p.63.

25 Bollier, David (2003) *Silent Theft: The Private Plunder of Our Common Wealth*, London: Routledge.

26 Harvey, David (2003) *The New Imperialism*, Oxford: Oxford University Press, p.158.

27 Prashad, Vijay (2002) *Fat Cats and Running Dogs: The Enron Stage of Capitalism*, London: Zed, p.73.

28 Drutman and Cray, n.24 above, pp.213–215. Freedland, Jonathan (2006) 'Big business, not religion, is the real power in the White House,' *Guardian*, 7 June, p.27.

29 Prashad, n.27 above, pp.113–114. For an account of Enron's collapse from an insider see Curver, Brian (2003) *Enron: Anatomy of Greed*, London: Arrow.

30 An observation made in Drutman and Cray, n.24 above, p.171.

31 US Forest Service (undated) 'Healthy Forests Initiative.' Available online at: www.fs.fed.us/projects/hfi (accessed 18 October 2005). See also 'Healthy forests: An initiative for wildfire prevention and stronger communities.' Available online at: www.whitehouse.gov/infocus/healthyforests/Healthy_Forests_v2.pdf (accessed 18 October 2005).

32 Sierra Club (2003) 'Debunking the "Healthy Forests Initiative".' Available online at: www.sierraclub.org/forests/fires/healthyforests_initiative.asp (accessed 18 October 2005).

33 Behan, Richard W. (2001) *Plundered Promise: Capitalism, Politics and the Fate of the Federal Lands*, Washington DC: Island Press.

34 US General Accounting Office figures, cited in Pye-Smith, Charlie (2002) *The Subsidy Scandal: How Your Government Wastes Your Money to Wreck Your Environment*, London: Earthscan, p.19.
35 Shutt, Harry (1998) *The Trouble With Capitalism: An Enquiry into the Causes of Global Economic Failure*, London: Zed Books. See also Elliott, Larry and Atkinson, Dan (1998) *The Age of Insecurity*, London: Verso.
36 Derber, Charles (2002) *People Before Profit: The New Globalization in an Age of Terror, Big Money, and Economic Crisis*, London: Souvenir Press, p.162. See also the United States Council for International Business website: www.uscib.org/index.asp?documentID=697 (accessed 27 September 2005).
37 This quote is cited in: Chomsky, Noam (1999) *People Over Profit: Neoliberalism and the Global Order*, New York: Seven Stories Press, p.141; and Derber, n.36 above, p.161.
38 The UK Labour government backed the MAI, and supports the promotion of international investment rules in the WTO: Cromwell, David (2001) *Private Planet: Corporate Plunder and the Fight Back*, Charlbury UK: Jon Carpenter, p.30; Curtis, Mark (2003) *Web of Deceit: Britain's Real Role in the World*, London: Vintage, pp.225–226; Monbiot, George (2000) *Captive State: The Corporate Takeover of Britain*, London: Macmillan, Chapter 10, especially p.303.
39 Chapter 11 of the NAFTA is available online at: www.sice.oas.org/trade/nafta/chap-112.asp (accessed 25 September 2005).
40 Klein, Naomi (2002) *Fences and Windows: Dispatches from the Front Lines of the Globalization Debate*, London: Flamingo, p.56–57; Wallach, Lori and Woodall, Patrick/Public Citizen (2004) *Whose Trade Organization: A Comprehensive Guide to the WTO*, New York: The New Press, p.270; Chomsky, n.37 above, p.141; Public Citizen press release, 'NAFTA Chapter 11 investor-state cases: Lessons for the CAFTA,' 23 February 2005. Available online at: www.bilaterals.org/article.php3?id_article=1353 (accessed 26 September 2005); Greenfield, Gerard (2000) 'The NAFTA ruling on Metalclad *vs.* Mexico – The broader context,' 18 September. Available online at: www.nadir.org/nadir/initiativ/agp/free/nafta/000918metalclad.htm (accessed 26 September 2005).
41 Choudry, Aziz (2004) 'The bilateral bypass,' *New Internationalist*, No. 369, July, p.6. The term 'competitive liberalization' has been attributed to Robert Zoellick, cited in *New Internationalist*, No. 376, March 2005, p.29.
42 McMurtry, John (2002) *Value Wars: The Global Market Versus the Life Economy*, London: Pluto, p.165. See also McMurtry, John (1999) *The Cancer Stage of Capitalism*, London: Pluto.
43 Gill, Stephen (2002) *Power and Resistance in the New World Order*, London: Palgrave Macmillan; Gill, Stephen (1995) 'Globalization, market civilization and disciplinary neoliberalism,' *Millennium: Journal of International Studies*, Vol. 24, No. 3, pp.399–423.
44 Derber, n.36 above, pp.105–126
45 Derber, n.36 above, pp.113–114.
46 Brazil stated its support for a Forest Products Agreement at the fourth session of the Intergovernmental Panel on Forests in 1997: *Earth Negotiations Bulletin*, Vol. 13, No. 34, p.7.
47 On the relationship between US corporations and US environmental policies see, for example, Martin-Brown, Joan (1996) 'The U.S. private sector and the environment,' *International Negotiation*, Vol. 1, pp.247–253; Paarlberg, Robert L. (1999) 'Lapsed leadership: U.S. International environmental policy since Rio,' in Vig, Norman J. and Axelrod, Regina S. (eds) *The Global Environment:*

Institutions, Law and Policy, London: Earthscan, pp.236–255; Falkner, Robert (2001) 'Business conflict and U.S. international environmental policy: Ozone, climate, and biodiversity,' in Harris, Paul G. (ed) *The Environment, International Relations and U.S. Foreign Policy*, Washington DC: Georgetown University Press, pp.157–177.

48 For an account of how the US weakened the Rome Statute before voting against it see Robertson, Geoffrey (2002) *Crimes Against Humanity: The Struggle for Global Justice*, London: Penguin, Chapter 9, especially p.369.

49 Bush senior refused to sign the Convention on Biological Diversity as its provisions on access and benefit sharing were opposed by US industry. President Clinton subsequently signed the convention, which has yet to be ratified by Congress, and the Kyoto Protocol, which was subsequently disowned by Bush junior.

50 Other treaties that the US has not adopted include the 1989 UN Convention on the Rights of the Child and the 1997 Mine Ban Treaty. As at September 2005 the US had signed but not ratified the Convention on the Rights of the Child, and had not signed the Mine Ban Treaty.

51 For a persuasive argument that the invasion of Iraq was counter to international law see Sands, Philippe (2005) *Lawless World: America and the Making and Breaking of Global Rules*, London: Penguin, pp.174–203.

52 Krugman argues that the Bush administration is a 'revolutionary power' that no longer accepts the legitimacy of the international political system: Krugman, Paul (2003) *The Great Unravelling: From Boom to Bust in Three Scandalous Years*, London: Allen Lane/Penguin, pp.3–26. Defining rogues states as 'states that do not regard themselves as bound by international norms' Chomsky argues that the US was a rogue state well before Bush junior was elected: Chomsky, Noam (2000) *Rogue States: The Rule of Force in World Affairs*, London: Pluto.

53 For a discussion of the different property rights traditions in the US and Europe see Hutton, Will (2002) *The World We're In*, London: Little, Brown, pp.49–85.

54 Beder, Sharon (2002) *Global Spin: The Corporate Assault on Environmentalism* (revised edition), Totnes: Green Books, pp.48–60; Bollier, n.25 above, pp.73–4.

55 The full text of the fifth amendment is available online at: http://caselaw.lp.findlaw.com/data/constitution/amendment05/ (accessed 27 September 2005).

56 Drutman and Cray, n.24 above, pp.58–61.

57 Beder, n.54 above, p.61.

58 Hutton, n.53 above, pp.69–73.

59 Article 14 of the German constitution, cited in Hutton, n.53 above, p.50. (Hutton incorrectly refers to Article 13 of the German constitution, rather than Article 14.)

60 Welford, Richard (1996) 'Business and environmental policies,' in Blowers, Andrew and Glasbergen, Pieter (eds) *Environmental Policy in an International Context: Prospects for Environmental Change*, London: Arnold, pp.66–67.

61 The tenth principle was added after the negotiation of the UN Convention Against Corruption in 2003.

62 The EU considers that voluntary action is 'the frame of reference for further European initiatives in support of companies' efforts to act in a socially responsible way': EU document COM(2002) 347 final, 'Communication from the Commission concerning Corporate Social Responsibility: A business contribution to Sustainable Development,' 2 July 2002, p.4. At the time of writing the EU has yet to agree a common set of guidelines on CSR policy and practice.

63 Christian Aid (2004) *Behind the Mask: The Real Face of Corporate Social Responsibility*, London: Christian Aid, p.50.

64 Assadourian, Erik (2006) 'Transforming corporations,' in Worldwatch Institute, *State of the World 2006*, New York, WW Norton, p.186.

65 Ruggie, John Gerard (2003) 'Taking embedded liberalism global: The corporate connection,' in Held, David and Koenig-Archibugi, Mathias (eds) *Taming Globalization: Frontiers of Governance*, Cambridge: Polity, p.115. Ruggie is a Harvard professor and a member of the Global Compact Advisory Council.

66 UN Global Compact website: www.unglobalcompact.org/ > Search Participants (search carried out 22 March 2006).

67 Nadeau, Robert L. (2003) *The Wealth of Nature: How Mainstream Economics Has Failed the Environment*, New York: Columbia University Press, especially pp.2–11, 154–6, 167–72.

68 Jeff Gates makes the point that a democratized capitalism that is effective in conserving the environment requires 'immediate feedback to speed up systems learning': Gates, Jeff (2000) *Democracy at Risk: Rescuing Main Street from Wall Street*, Cambridge MA: Perseus, p.270. See also p.251.

69 This distinction is drawn in Monbiot, n.38 above, pp.51–52. See also Sklair, Lesley (2002) *Globalization: Capitalism and its Alternatives*, Oxford: Oxford University Press; and Gates, n.68 above.

70 See, for example, Hines, Colin (2000) *Localization: A Global Manifesto*, London: Earthscan; Mander, Edward and Goldsmith, Jerry (eds) (2001) *The Case Against the Global Economy and for a Turn Towards Localization*, London: Earthscan.

71 I proposed an International Convention on Transnational Corporations in Humphreys, David (2001) 'Environmental accountability and transnational corporations,' in Gleeson, Brendan and Low, Nicholas (eds) *Governing for the Environment: Global Problems, Ethics and Democracy*, London: Palgrave, pp.88–101. This chapter was based on a paper presented at the International Academic Conference on Environmental Ethics and Justice for the Twenty-first Century, Melbourne, 1–3 October 1997.

72 This requirement would most likely prove too difficult to enforce initially. It would probably be necessary to apply the Convention on Transnational Corporations in the first instance only to transnational corporations that exceeded certain thresholds, for example, those trading or investing in more than x countries with transnational trade or investment flows exceeding \$$y$ million.

73 This argument assumes that local community groups always act in the local public interest. In the vast majority of instances this is so; it is not in the interests of a community knowingly to engage in the degradation of its economy or resource base. However, a cautionary note sounded by Doreen Massey should be heard. Massey notes a tendency 'to glorify "resistance," to assume it is always ranged against "domination".' Massey makes this point with respect to urban politics: Massey, Doreen (1999) 'Living in Wythenshawe,' in Borden, I., Kerr, J., Pivaro, A. and Rendell, J. (eds) *The Unknown City*, Cambridge MA: MIT Press. Following on from this point; there is always the risk that local elites or minority groups can claim to act in the interests of a local community in order to promote narrow sectional interests.

74 On the possible constitution and functions of an international environmental court see Postiglione, Amedeo (2001) 'An international court on the environment,' in Gleeson and Low (eds), n.71 above, pp.211–220.

75 Foresters generally accept that after three rotations (i.e. when a forest has been selectively logged three times) that the ecological composition of the forest changes and the forest does not revert entirely to its original condition.

76 On this point see also Drutman and Cray, n.24 above, p.121.
77 Held, David (1995) *Democracy and the Global Order: From the Modern State to Cosmopolitan Governance*, Cambridge: Polity, p.171 [definition italicized in original].
78 Falk, Richard (2001) 'Human governance and the environment: Overcoming neo-liberalism,' in Gleeson and Low (eds), n.71 above, p.224 [emphasis in original].
79 Stewart, Keith (2001) 'Avoiding the tragedy of the commons: Greening governance through the market or the public domain?,' in Drache (ed), n.13 above, pp.202–228.
80 Underhill is specifically writing here about financial markets. His point also applies to corporate activity: Underhill, n.13 above, p.277.
81 Derber, n.36 above, p.60.
82 Carroll, Rory (2006) 'Shell told to pay Nigerians $1.5bn pollution damages,' *Guardian*, 25 February, p.18.
83 Morales, Evo (2006) 'Bolivia: From neoliberalism to a homeland for all,' *Green Left Weekly*, 13 March. Available online at: www.zmag.org/content/showarticle.cfm?ItemID=9897 (accessed 2 May 2006); Owen, Paul and agencies (2006) 'Bolivia orders troops to seize gas and oil supplies, *Guardian*, 2 May, p.14.
84 George, Susan (2004) *Another World is Possible If…*, London: Verso, p.110. See also p.107.
85 The nine states are Connecticut, Delaware, Maine, Massachusetts, New Hampshire, New Jersey, New York, Rhode Island and Vermont: DePalma, Anthony (2005) '9 states in plan to cut emissions by power plants,' *New York Times*, 24 August. Available online at: www.nytimes.com/2005/08/24/nyregion/24air.html (accessed 20 March 2006).
86 Hutton, n.53 above, p.367.
87 McGrew, Anthony (2002) 'Between two worlds: Europe in a globalizing era,' *Government and Opposition*, Vol. 37, No. 3, pp.343–358. Effective environmental policy involves a blend of regulation and market incentives. See, for example, Harrington, Winston, Morgenstern, Richard D. and Sterner, Thomas (2004) *Choosing Environmental Policy: Comparing Instruments and Outcomes in the United States and Europe*, Washington DC: Resources for the Future; Desai, Uday (ed) (2002) *Environmental Politics and Policy in Industrialized Countries*, Cambridge MA: MIT Press.
88 Biermann, Frank and Sohn, Hans-Dieter (2004) 'Europe and multipolar global governance: India and East Asia as new partners?,' *Global Governance Working Paper* No. 10, Amsterdam/Berlin/Oldenburg/Potsdam: The Global Governance Project.
89 Arguably the main reason why the G77's demand for a New International Economic Order failed is that the G77 insisted on negotiating on this subject within the UN. For an account of how the developed states blocked the demands of the G77 see Renninger, John P. (1989) 'The failure to launch global negotiations at the 11th special session of the General Assembly,' in Kaufmann, Johan (ed) *Effective Negotiation: Case Studies in Conference Diplomacy*, Dordrecht: Martinus Nijhoff/Kluwer Academic, pp.231–254.
90 Coleman, David (2003) 'The United Nations and transnational corporations: from an inter-nation to a "beyond-state" model of engagement,' *Global Society: Journal of Interdisciplinary International Relations*, Vol. 17, No. 4, pp.339–357, especially p.353.

Index

Printed in Great Britain
by Amazon